FIBER OPTIC COMPONENT DESIGN, FABRICATION, TESTING, OPERATION, RELIABILITY AND MAINTAINABILITY

FIBER OPTIC COMPONENT DESIGN, FABRICATION, TESTING, OPERATION, RELIABILITY AND MAINTAINABILITY

by

N.L. Christian and L.K. Passauer

Vitro Corporation
Silver Spring, Maryland

NOYES DATA CORPORATION
Park Ridge, New Jersey, U.S.A.

Library of Congress Catalog Card Number 89-30276
ISBN: 0-8155-1203-1
Printed in the United States

Published in the United States of America by
Noyes Data Corporation
Mill Road, Park Ridge, New Jersey 07656

10 9 8 7 6 5 4 3 2 1

Library of Congress Cataloging-in-Publication Data

Christian, N.L.
Fiber optic component design, fabrication, testing, operation, reliability, and maintainability / by N.L. Christian and L.K. Passauer.
p. cm.
Bibliography : p.
Includes index.
ISBN 0-8155-1203-1 :
1. Fiber optics. 2. Fiber optics--Reliability. I. Passauer, L. K. II. Title.
TA1800.C49 1989
621.36'92--dc19 89-30276
CIP

Foreword

This book provides a thorough analysis of the currently available information on fiber optic reliability, maintainability and supportability. It covers the full range of fiber optic technology, from component fabrication, screening and accelerated testing, to failure mechanisms and modes, to maintenance philosophy and logistics considerations. It focuses on major components of fiber optic systems which include fiber and cable (transmission medium); connectors and splices (physical interconnecting devices); emitters, transmitters, detectors, and receivers (electro-optic and opto-electronic conversion devices); couplers, multiplexers/demultiplexers, and switches (signal combining and routing devices); and enclosures, splice trays, and organizers (ancillary items). Specifically excluded from consideration are those items which are predominantly non-fiber optic, such as bus interface units and telephone or video signal multiplexer/demultiplexer units. In addition to the primary fiber optic components, military fiber optic systems are addressed. These include the Communication, Navigation and Interrogation (CNI) system on the AV-8B Harrier fighter aircraft, the Fiber Optic Transmission System-Long Haul (FOTS-LH) field deployable communications link, and the AN/FAC-2 and 3 fixed communications link.

The approach taken was three-fold. First, voluminous amounts of technical data were gathered from many sources using a variety of techniques. Second, as the information was obtained, analysis was performed to ascertain whether it was current and relevant to the study. At this point, the data was cataloged for easy retrieval. This provided a baseline from which recommendations and guidelines were developed for system designers and planners based on engineering analysis.

The qualitative information was used to develop considerations which address the many phases of fiber optic component design, fabrication, testing, and operation which are important in the component selection process. In addition, considerations were developed for system installation, check-out, environmental factors, maintenance and physical lay-out. The quantitative data received from the results of laboratory research and from deployed systems were used to calculate component lifetimes and failure rates.

The book also contains a significant amount of information on the fundamentals of fiber optic component design, construction and operation. This provides the background necessary to understand the detailed reliability concepts which are presented, thus making the book useful as a fiber optics training aid.

The availability of Reliability, Maintainability and Logistic (R,M&L) data on fiber optic components and systems was limited, if not non-existent, in military data collection activities. Also, many military systems were billed as being "fiber optic systems" when in reality they were still in the conceptual design phase, development phase, or were simply one-time demonstrations. However, in several cases the lack of information on military systems indicated that there simply were no operational failures to report. For some of the components, the populations on which data were available were not large and/or the actual number of elapsed operating hours was low. These conditions make it necessary to exercise caution when using the resulting calculations. That is, they should not be used in detailed design analysis, but they are appropriate for use in narrowing the wide range of component types to those that are suitable for a given application. Once the device type is selected, its reliability and failure characteristics presented in the book can be studied. This will tell the designer which features of the intended application are prone to impact the component the most.

The R,M&L recommendations and guidelines put forth in this study provide the reliability and maintainability engineer with the necessary direction to enhance system design. Also,those wishing to acquire an understanding of the basic fiber optic discipline can benefit by studying this document.

The information in the book is from *Impact of Fiber Optics on System Reliability and Maintainability,* prepared by N.L. Christian and L.K. Passauer of Vitro Corporation for the U.S. Department of Defense Rome Air Development Center, June 1988.

The table of contents is organized in such a way as to serve as a subject index and provides easy access to the information contained in the book.

Advanced composition and production methods developed by Noyes Data Corporation are employed to bring this durably bound book to you in a minimum of time. Special techniques are used to close the gap between "manuscript" and "completed book." In order to keep the price of the book to a reasonable level, it has been partially reproduced by photo-offset directly from the original report and the cost saving passed on to the reader. Due to this method of publishing, certain portions of the book may be less legible than desired.

ACKNOWLEDGMENTS

The authors would like to acknowledge the many contributors of technical data used within this report. The data was obtained through numerous phone contacts, survey respondees and through previously published technical papers.

Acknowledgment is also in order for the many people who assisted in the preparation of this report. Many thanks for her endless patience go to Catherine S. Baechtel in performing the word processing for this report. Thanks go to Carol A. Smith for the preparation of the figures. We would also like to thank Theodore B. Smith, Jr. and Alan L. Weiser for their suggestions and guidance.

NOTICE

Contents and Subject Index

SECTION II
NUMERICAL RELIABILITY SUMMARY

SECTION III
COMPONENTS

Introduction

Photonics is the technology that deals with the manipulation and transfer of photons, packets of electromagnetic radiation with frequencies in the optical range. Although the term "photonics" is relatively new, the field of study encompassed by photonics is not. It is a wide-ranging technology with applications in military, industrial, and consumer markets. Photonics is much more pervasive than we may realize. Compact disk players, the current rage in stereo equipment, use lasers to read the information stored on the disk. Wireless remote controls for TV and stereo equipment use infrared signals to transmit the control information. Infrared light is also used in security systems for motion and intrusion detection. Most modern grocery stores have laser scanners at the checkout line to read the bar code information printed on the packages. Infrared radar systems and laser guided missiles are just two military applications of photonics. Some portions of photonic technology are well developed and understood, as evidenced by those systems listed above. Other areas of the technology are the subject of intense research and development efforts. Work is underway now to develop photonic logic gates and switching devices and all-optical computers. These will lead to new products and applications of photonics unheard of even a few years ago.

One segment of photonics that has experienced tremendous growth and extensive change over the last twenty years is the field of fiber optics. Fiber optics involves the transmission of information, in the form of optical signals, over optical waveguides or fibers. The concept of using light to carry information is not new. In the 1870s Alexander Graham Bell proposed a "photophone" in which sunlight would be used as a carrier, modulated by audio or voice signals. A contemporary version of Bell's idea can be seen in free space infrared optical links used in many urban areas, where running cables may not be cost effective and the use of microwaves may not be permitted for safety reasons.

In the early 1970s, with the advent of relatively low-loss optical fibers, it became feasible to utilize the high bandwidth capacity of an optical carrier in a guided or confined transmission medium. Since that time, the use of fiber optics has been explosive. Today, networks of optical fibers crisscross the world. Optical fibers compete with satellites and terrestrial microwave links for the transmission of long distance telephone calls and they are used to tie together computers in office buildings and campuses.

The technical advantages of fiber optics, particularly those beneficial to the military, are numerous. First and foremost, fiber has an extremely high bandwidth. With the ever increasing reliance on computers in military systems, the need to move large volumes of data rapidly is increasing. Fiber offers a way to meet this need with a decrease in cabling volume and weight as compared to copper cables. Fiber is also non-conductive to electrical signals. This means that interconnected equipment can be electrically isolated and ground loop problems eliminated. In addition, electrical hazards, such as shock and arcing in explosive environments, are also eliminated by the use of fiber. Fiber is immune to electromagnetic interference and largely immune to permanent damage from the energy contained in electromagnetic pulses. Fiber cables need not be shielded when routed in electrically noisy areas nor be protected from electronic countermeasures. Fiber is very difficult to tap, especially in a non-invasive manner, which means that it can be used to enhance system security. Finally, fiber has a very low signal attenuation, enabling long, repeaterless links to be built with optical fibers.

Potential military applications for fiber optic technology abound. The military, however, has lagged behind the commercial world in using fiber optics. One reason for this has been a lack of definitive information regarding the reliability of fiber optic components and systems and the related issues of maintainability and logistics support. The information contained in this report will allow reliability engineers to address these technical issues in greater detail than has been conveniently possible before.

Considering the large amount of information contained in this report, it was important that the contents be formatted in a manner which makes the information easy to locate. This was accomplished in part by making the first two sections summaries.

Section I, RECOMMENDATIONS AND GUIDELINES, is a synopsis of the technical issues impacting R,M,&L that need to be considered by systems designers and planners. Most of the considerations are broken down into three items: 1) each issue is stated, 2) it is then followed by a statement of the possible cause(s) in the case of a problem, and 3) where known, a method of preventing the problem is given. These issues are organized in an easy-to-use fashion which gives the reader the ability to quickly find the information he needs on a specific design consideration. These guidelines are divided into the following areas of interest: environmental, physical lay-out and installation, operational/optical requirements, manufacturing, logistics, and maintenance. Page numbers to the associated chapters in the report are referenced so further information can be found.

Section II, NUMERICAL RELIABILITY SUMMARY, contains the reliability parameters calculated for each of the components and systems discussed in the individual chapters. Where applicable, these parameters are given for different component constructions. Again, references point the reader to more detailed information in the associated chapters.

Section III, COMPONENTS, contains seven chapters on the fundamental fiber optic components. These are FIBER AND CABLE; CONNECTORS AND SPLICES; EMITTERS AND TRANSMITTERS; DETECTORS AND RECEIVERS; COUPLERS, MULTIPLEXERS AND DEMULTIPLEXERS; SWITCHES; and ENCLOSURES/SPLICE TRAYS AND ORGANIZERS. Each chapter contains detailed information on the component's design and operation, and a thorough discussion on the failure modes and mechanisms of that component. Also, an explanation is given regarding the reliability data obtained and any calculations performed to develop reliability parameters from that data. Any assumptions that were made are clearly stated. Maintenance and logistics information pertinent to each component is also included.

Section IV, SYSTEMS, contains three chapters each of which discusses a military system which uses fiber optics. These systems are the AN/FAC-2 and 3 fixed communications set, the Communication, Navigation and Interrogation System on the AV-8B Harrier fighter aircraft, and the Fiber Optic Transmission System-Long Haul (FOTS-LH) field deployable communications link. The FOTS-LH is a system which is still in the development stage. These chapters describe the systems, their design, construction, operation and exhibited performance.

Section V, TESTING, contains chapters on the two prevalent methods used to test fiber optic systems: the optical source/power meter test set, and the Optical Time Domain Reflectometer (OTDRs). The operation of the test equipment and how that test equipment is used in system testing is addressed. Information received from manufacturers regarding repair, predominant failures, storage and operating conditions, and anticipated lifetimes is also included.

Section VI, SUMMARY, concludes the report and provides recommendations for future activity based on our experiences and the information obtained during the course of this study.

This report can be used in four ways. First, it can be used as a quick reference guide for determining those R,M&L areas which need to be considered when planning a fiber optic system. Section I provides the necessary information for this purpose. Second, it can be used to quickly determine the lifetime of the fiber optic components addressed. Section II would be used to find these values. Third, detailed information on the R,M&L issues associated with each component can be studied in the individual chapters. And fourth, the information found throughout the report on the basics of these fiber optic components can be used in teaching those who are unfamiliar with the technology.

These applications of this report provide a substantial start to understanding and improving fiber optic system reliability, maintainability and supportability.

Section I

Recommendations and Guidelines

RECOMMENDATIONS AND GUIDELINES

The following Recommendations and Guidelines are to be used much like a tickler file in that they will alert system designers and planners to those areas that need to be considered for their impact on fiber optic system reliability, maintenance and logistics (R,M&L). The user can look up in the index a particular parameter, condition or component and be referenced to areas in the document which provide summarized information on those considerations to which attention should be paid.

These recommendations are organized in the following major headings each of which are aspects of system design. Component screening is discussed in the section on manufacturing, whereas quality assurance testing is only addressed in the individual chapters.

Each of these headings is broken down into various sub-headings to make it easier to locate specific areas of concern. The components which are impacted by the individual sub-headings are then addressed. Each of these components is followed by the associated recommendations which should be considered for effective system design. Generally, each recommendation is broken down into three items: a statement of the problem or consideration, its impact, i.e., what condition is created, and finally, how the condition affects the specific system or component with respect to R,M&L. If known, preventive measures are given to correct a problem. Associated with each recommendation is a reference which directs the reader to the pages in the report where additional information can be found. The user of these Recommendations and Guidelines should reference its index to identify all of the areas that address the components or parameters under investigation. After locating these items, the system designer should review the information with respect to the R,M&L requirements of the application at hand. Components make up a system, therefore the designer will need to read the details about each system component with respect to every major heading listed above. For the sake of brevity, abbreviations for the elements and chemical compounds are used in these Recommendations and Guidelines. Their complete spelling is given in Table I-1 of the Appendix.

These Recommendations and Guidelines provide an easy to use method of identifying specific areas of concern which impact R,M&L. The system designer who is faced with the task of performing a trade-off of conventional hardwired systems versus fiber optic linked systems can use this information to that end. When fiber optics has been selected as the technology of choice, the Recommendations and Guidelines will assure that reliable, maintainable and supportable systems are designed for the United States military.

INDEX TO RECOMMENDATIONS AND GUIDELINES

Page

RECOMMENDATIONS AND GUIDELINES

I. ENVIRONMENTAL

1. TEMPERATURE

Ref.

a. Cable

78 (1) **Gel Filling Viscosity Changes**

- High temperature causes gel filling in cable interstices to become less viscous.
- Results in filling running into the back of the connector which can wick to the fiber endface increasing attenuation. This will degrade performance and result in unscheduled maintenance.

363 (2) **Increase In Fiber Attenuation At Low Temperatures**

- Increases in attenuation occur at cold temperatures if cabled improperly.
- The cable must be closely monitored during cabling process to insure that no undue mechanical stress is placed on the fiber. Excess mechanical stress can cause microbends in the fiber which may be increased when the cable is subjected to cold temperatures. Low temperatures can also cause a stiffening of the fluids used in the cable which can also cause microbends, and thus attenuation.

b. Connector

138 (1) **Epoxy Bonding Failure**

- High temperatures beyond the epoxy manufacturer's specified curing temperatures causes incomplete bonding of epoxy during assembly and causes the epoxy to soften after curing.
- This allows the movement of the fiber causing misalignment thereby increasing attenuation which can degrade the system.

I. ENVIRONMENTAL

1. TEMPERATURE (con't)

Ref. b. Connector (con't)

138- (2) **Proper Application of Epoxy**

140 - High temperatures cause variations in epoxy viscosity which makes it more plastic and a poor securing agent. The epoxy characteristics can vary from lot-to-lot and the uniformity of the mixing may vary from application to application. Uniformity is needed for proper adhesion. The epoxy must be applied in a thin layer and not used as a potting agent if proper adhesion is desired. The lap shear strength and adhesion are degraded by improper curing. The glass transition temperature for most epoxies is 100-110°C.

- These conditions result in attenuation increase due to discoloration, or an air gap induced by a separation of the fiber endfaces commonly called pistoning.

130 (3) **Index Matching Fluid Viscosity**

- High temperature causes variations in index matching fluid viscosity and transparency. The fluid must remain soft and pliable in all environments to prevent separation from the fiber endface.

- An increase in viscosity results in an attenuation increase.

363 (4) **Temperature Cycling Impact on Connector**

- Causes improper seating of fibers with a secondary coating in the ferrule.

- This results in broken fibers which requires connector replacement.

c. Splice

131 (1) **Splice Epoxy Bond Degradation**

- High temperatures beyond those specified by the epoxy manufacturer causes incomplete bonding of epoxy.

I. **ENVIRONMENTAL**

1. TEMPERATURE (con't)

Ref. c. Splice (con't)

(1) **Splice Epoxy Bond Degradation** (con't)

- Allows the movement of fiber endfaces in the splice thereby increasing attenuation which may degrade system performance.

131 (2) **Temperature Induced Movement of Coiled Spare Fiber**

- Temperature cycling causes expansion and contraction of coiled spare fiber and splice piece parts in splice case.
- Results in movement of spare fiber which induces attenuation due to microbends.

131 (3) **Splice Sensitivity to Temperature Cycling**

- Causes the piece parts of mechanical splices to go through thermal expansion and contraction.
- Results in relative movement of fiber endfaces which may degrade system performance due to increased attenuation.

d. Source

182, 183, 186 (1) **Laser Temperature Sensitivity**

- Laser threshold current increases with temperature which causes an increase in output power.
- Results in thermal runaway unless feedback circuitry or thermoelectric cooling is used.

181, 182 (2) **Temperature Effects On Laser Leakage Current**

- Leakage current increases with temperature if adequate burn-in has not been performed.
- An increase in leakage current causes a reduction in power output which is compensated for by increasing the laser threshold current in a constant output application; this increased threshold decreases device life.

187 (3) **Temperature Enhanced Whisker Formation In Lasers**

- Ambient temperature of 50°C or more enhances the formation of whisker growth in tin-rich solder.

I. **ENVIRONMENTAL**

1. TEMPERATURE (con't)

Ref.

d. Source (con't)

(3) **Temperature Enhanced Whisker Formation In Lasers (con't)**

- Whiskers form effective electrical shorts causing catastrophic damage. Adding lead to solder significantly reduces whisker formation.

152, 196 (4) **Laser Material Temperature Sensitivity Comparison**

- AlGaAs lasers are about 1/4 as sensitive to temperature fluctuations as are InGaAsP lasers.

161 (5) **Trade-Off Between Surface LED Emitting Area and Current Density.**

- A large emitting area reduces thermal dissipation constraints since the current density is decreased, however, coupling efficiency is also reduced.
- A small emitting area requires a low current density (reduced output power) to prevent thermal damage but provides high coupling efficiency.

163 (6) **Peak Wavelength Shift Reduction**

- Temperature fluctuations result in emitter peak wavelength shifts for devices not using temperature compensation circuitry.
- Very good thermal conduction is achieved by constructing the active layer very close to the heat sink which reduces device thermal resistance. This is done in the Etched-well LED design.

170, 181, 184 (7) **Extended Laser Lifetime**

- Operating lasers at room temperature, at the lowest possible threshold current allowed by the application and within the specified duty cycle for digital applications, will significantly extend the life of a quality laser.
- All lasers expected to exhibit long lifetimes must be fabricated using defect free and contamination free materials.

I. **ENVIRONMENTAL**

1. TEMPERATURE (con't)

Ref. d. Source (con't)

178 (8) **Causes of LED Rapid Degradation**

- **High current density, high temperature and high impurity concentration in the active region contribute to the formation of darkline defects and dark spot defects.**
- **These defects in the active region are non-radiative which necessitates a higher current density to achieve a given output power level. This increases junction temperature which reduces device lifetime.**

178, (9) **Causes of LED Slow Degradation (Wear Out)**

179
- **High temperature and impurities in the active region contribute to the onset of LED wear out even if dark line or dark spot defects don't exist.**
- These conditions contribute to long-term contamination of the active region which reduces device lifetime.

178- (10) **Emitter Temperature Sensitivity**

184
- As the emitter temperature increases, the output power decreases.
- This will result in an increased driving current to maintain the same output power level which decreases the device lifetime.

204 (11) **Reducing Transmitter Heat Build-Up**

- Thermoelectric coolers use the Peltier effect to reduce the emitter temperature as ambient temperatures rise.
- Maintaining a reduced emitter temperature prevents early device failure.

204 (12) **Transmitter Thermoelectric Heat Sink Size**

- The larger the heat sink, the cooler the emitter can be kept, thus extending emitter lifetime.
- Heat sink size can not be too bulky as to interfere with package installation or available real estate.

I. **ENVIRONMENTAL**

1. TEMPERATURE (con't)

Ref. d. Source (con't)

206 (13) **Transmitter Elements and The Use Of Epoxy**

- Epoxy is widely used to secure the individual optical elements of a transmitter.
- Epoxy can shrink over time and softens at high temperatures (typically 70°C). These characteristics cause relative movement between the elements thus decreasing optical coupling. Epoxies having these characteristics should not be used in harsh environments or critical applications.

e. Detector

218, 224-226 (1) **Required APD Reverse Voltage Is 100-400Vdc**

- At this level of bias, the APD gain is very temperature sensitive.
- Compensating circuitry is required for stability in the form of thermoelectric cooling, automatic bias control circuitry or ambient temperature control which increases circuit complexity and decreases reliability by increasing the number of components.

218 (2) **APD Temperature Sensitivity At High Reverse Voltages**

- Electron and hole ionization rates are temperature sensitive in the 100-400Vdc APD reverse voltage range.
- Temperature variations result in gain fluctuations unless bias compensating circuitry or thermal control is used. These measures increase circuit complexity and decrease reliability by increasing parts count.

231 (3) **Thermally Induced Permanent APD Degradation**

- AuZn contact deterioration occurs at elevated temperatures in the form of precipitates entering the depletion region.
- Resulting increase in dark current is permanent which degrades receiver sensitivity.

I. ENVIRONMENTAL

1. TEMPERATURE (con't)

Ref. e. Detector (con't)

232 (4) **Photodetector Failure During Temperature Cycling**

- -55°C to 125°C has induced open lead-bonds in plated contacts. This is catastrophic failure.
- Evaporated contacts eliminate this problem.

f. Coupler

273 (1) **FBT Coupler Polarization Variation With Temperature**

- Elevating temperature to 90°C has caused birefringence induced by thermal strain in FBTs fabricated using twisted fibers.
- This causes output instability and variations in coupling ratio which can degrade system performance.

(2) **Temperature Cycling On Bulk Optic Couplers**

- Causes optical materials used in multiplexers, demultiplexers and couplers to experience fogging and condensation if the temperature changes are too dramatic and rapid.
- Fogging and condensation degrade the performance of the device by increasing its loss which affects system operation.

272 (3) **Temperature Cycling Effects On FBT Couplers**

- Coupling ratio is relatively independent of temperature cycling.
- Resultant variations in coupling ratio have been reported to be ±1%.

273 (4) **Temperature Dependence of Wavelength In FBT Technology Versus Bulk Optics Technology**

- FBT devices are very temperature independent whereas bulk optic devices using interference filters exhibit wavelength shifts linearly tracking temperature.
- These shifts occur when operated outside temperature specifications.

I. ENVIRONMENTAL

1. TEMPERATURE (con't)

Ref.

g. Switches

291 (1) **Temperature Dependence Of Prism/Mirror Movement Switches**

- Optical elements will frost if exposed to too rapid a temperature change.
- This thermal characteristic limits the use of these switches and their transportation mode to prevent switch degradation.

2. HUMIDITY

a. Source

179, (1) **Laser Degradation Due To Humidity**

190 - Absence of a protective coating (passivation) will result in the formation of an oxide layer on the facets in a humid environment.
- Laser output power is degraded which may lead to system malfunction.

b. Photodetector

232 (1) **Humidity Induced Photodetector Degradation**

- Photodetectors have experienced electro-chemical oxidation when tested at relative humidities up to 85%.
- As humidity increases, failures increase proportionately.
- This results in effective electrical shorts which cause catastrophic failure unless the critical concentration of moisture is sealed out by packaging.

c. Coupler

275 (1) **Coupler Degradation due To Humidity**

- Dielectric thin-film filters (dichroic) are hygroscopic and therefore very moisture sensitive.
- Moisture penetration degrades filter performance with time unless properly sealed.

I. **ENVIRONMENTAL**

2. HUMIDITY (con't)

Ref.

d. Fiber/Cable

92, 93 (1) **Attenuation Degradation Due To Humidity and Moisture**

- Cables must be designed to prohibit moisture from contacting the fiber.
- If moisture enters the cable, the water may cause existing microcracks to propagate (enlarge) which increases attenuation. If water inside the cable freezes, sufficient pressure on the fiber will result in microbend induced attenuation.

3. ULTRAVIOLET RADIATION

a. Cable

94 (1) **Attenuation Degradation Due To Ultraviolet Rays**

- Unless designed to withstand ultraviolet rays, cracks will form in the cable jacket allowing water to enter.
- As the water freezes it can exert sufficient pressure on the fiber to cause microbend-induced attenuation. The water can also cause existing microcracks in the fiber to propagate which will increase attenuation over time.

4. RODENT POPULATION

a. Cable

94 (1) **Rodents Gnawing On Cable**

- Gnawing destroys the cable sheath and inner elements of the cable. This also allows water and contaminants to enter the cable.
- Gnawing degrades system performance either catastrophically or gradually over a period of time.

5. CHEMICALS

a. Cable

93 (1) **Sheath Deterioration Due To Chemical Reactions**

- Chemicals deteriorate the cable sheath composition which provides a means for both chemical and environmental contamination of the fiber.

I. **ENVIRONMENTAL**

5. CHEMICALS (con't)

Ref.

a. Cable (con't)

(1) **Sheath Deterioration Due To Chemical Reactions**

- Fiber exposure to these contaminants leads to easy breakage, increased attenuation and unpredictable performance.

b. Connectors

129 (1) **Connector Seal Deterioration Due To Chemical Reactions**

- Exposure to oils, fuels, cleaning compounds, and other chemicals may cause degradation of the connector seals.
- Contaminants can enter the connector through the bad seals causing an increase in attenuation which will require repair or replacement.

c. Fiber

92, 93 (1) **Microcrack Growth In Fiber**

- Chemical exposure can stress the fiber surface which may cause microcracks to propagate.
- This can result in a failure or reduced service life of the fiber. The probability of failure is dependent upon the residual stress in the fiber, the length of the fiber, duration of the stress and the amount of moisture present.

6. FLAME/FIRE

a. Cable

94 (1) **Cable Damage Due To Flame**

- Causes degradation of cable sheath leading to destruction of the entire cable and possibly the emission of toxic fumes.
- Damaged cable requires replacement to prevent future failures or to return the system to an operating status. Personnel may require treatment for inhalation of toxic fumes.

I. ENVIRONMENTAL

6. FLAME/FIRE (con't)

Ref. b. Connectors

129 (1) **Connector Damage Due To Flame**

- Causes degradation of the connector seals or total destruction of the connector unless a flame-proof design is employed.
- Seal damage allows contamination of optical interface which degrades system performance.

7. NUCLEAR RADIATION

a. Fiber

82-84 (1) **Susceptibility To Nuclear Radiation**

- Fiber not designed to be radiation hardened, may become opaque when exposed to nuclear radiation for a short duration (several microseconds) or the discoloration may be permanent. The duration depends upon the fiber composition and amount of radiation.
- The fiber/cable must receive prior nuclear irradiation in order to operate in such an environment.

8. LIGHTNING

a. Cable

79 (1) **Cable Susceptibility To Lightning Strikes**

- Cables containing metallic elements are prone to lightning strikes which will burn and destroy the cable jacket and EMI induced by lightning.
- This results in catastrophic failure of the system as a result of optical and mechanical cable failure.

9. EXPLOSIVE FUMES

a. Switches

286-288 (1) **Manual Switching In Hazardous Environments Safety**

- Manual optical shutter switches can be used in explosive environments without electrical sparks.
- Optical systems can be used in industrial environments where manual switching is necessary without the need for elaborate electrical explosion proof switches.

II. PHYSICAL LAYOUT & INSTALLATION

1. ANCILLARY EQUIPMENT

<u>Ref.</u> a. Environment

304 (1) **Installation Considerations For A Benign Environment**

- Complex installation problems do not arise unless as a result of special constraints.
- Therefore, equipment which does not require elaborate protection can be used.

304 (2) **Installation Considerations For A Harsh Environment**

- Requires complex installation considerations.
- The use of ruggedized components is required to meet the stringent demands imposed by the environment.

311 b. Closures

(1) **Closure Seal Degradation Due To Ultraviolet Exposure**

- May cause wear and cracking of moisture proof closure seals.
- Seal degradation results in moisture contaminating the splices and spare fiber which will cause microcracks to propagate thus increasing attenuation.

c. Splices

311 (1) **Fiber Movement Due To Temperature Cycling**

- Movement of fibers and splices occurs with temperature cycling which can affect the adhesion of the epoxy used in the splice or the protection sleeve. Strain relief must be used to limit movement.
- Movement causes an increase in attenuation due to misalignment of the fiber.

2. LINK LENGTH

a. Fiber

62- (1) **Long Link Lengths and Optical Fiber Characteristics**

65
- Characteristics of the fiber in terms of attenuation/km and dispersion will vary depending upon the link length. These characteristics impact the selection of the operating wavelength.
- Longer wavelengths (1100-1500nm) offer low attenuation and dispersion but require the use of circuit intensive and shorter lived laser sources.

II. PHYSICAL LAYOUT & INSTALLATION

2. LINK LENGTH (con't)

Ref. a. Fiber (con't)

64 (2) **Long Link Length and Fiber Concatenation**

- The longer the link length, the more concatenation of fiber will be required in the form of fusion or mechanical splicing.
- Concatenation alters the bandwidth characteristics of the fiber from those specified by the manufacturer due to splices performed in the field. This effect means additional testing must be performed where bandwidth limiting is not tolerable.

b. Source

(1) **Long Link Length Impact On Source Type**

- Long links operate with fewer repeaters when laser sources are used. This means temperature compensation circuitry is needed for laser stability.
- Additional laser circuitry increases overall complexity of system resulting in reduced circuit reliability. This places higher lifetime demands on individual components or greater derating.

161 (2) **Large Emitting Area Device Limitations**

- This type of device has high capacitance which does not permit high data rate use ($\leq$ 10Mb/s).
- Large emitting area reduces coupling efficiency, thereby preventing use in long-haul applications which use small core fibers.

3. INSTALLATION CONSTRAINTS

a. Cable

(1) **Long Link Length Repairability**

- Makes access difficult to entire cable run due to location, terrain and equipment portability.
- This impacts maintainability and logistics in terms of locating repair equipment/spare parts, vehicular equipment and the type of equipment needed to perform repairs.

II. PHYSICAL LAYOUT & INSTALLATION

Ref. 3. INSTALLATION CONSTRAINTS (con't)

a. Cable (con't)

94, (2) **Cable Location Marking**

95 - During installation, maintenance and parallel construction, it must be ensured that the location of the cable is well mar.ed. This will assure that the cable will not be inadvertently damaged during these activities.

- If the cable location is not marked, it may be severed accidentally causing a catastrophic system failure.

92 (3) **Cable Residual Tension Considerations**

- Cable must not be installed so that it is under residual tension. If the residual tension is less than 33% of the fiber proof test value, the service life of the fiber should not be affected. If the residual tension is too great, the fiber will be under stress.

- This results in cable jacket tension which can cause it to stretch and eventually crack or break. Residual tension also results in propagation of microcracks existing in the fiber which will increase fiber attenuation.

93 (4) **Cable Pulling Tensions**

- Cables should be able to withstand a maximum installation tension of 600 pounds or the level specified for the particular application.

- The minimum level of force will depend upon the application. Pulling around corners, pulling long runs, pulling in ducts with existing cable all increase the tensile load exerted on the cable.

II. PHYSICAL LAYOUT & INSTALLATION

Ref. 3. INSTALLATION CONSTRAINTS (con't)

b. Splices

131 (1) **Splice Handling**

- During installation, improper splice handling such as exceeding the minimum bend radius of the spare fiber or improper cable anchoring within the splice closure can cause excess residual tension or induced microbends or macrobends in the fiber.
- This can result in fiber failure within the splice case at a point other than the actual splice itself.

141 (2) **Splice Installation Environment**

- Cleanliness of the fiber endface, alignment of the fiber, and environmental conditions all impact the final splice loss.
- Paying attention to these factors will improve the final quality of the splice.

c. Ancillary Items

308-310 (1) **Ancillary Item Installation Considerations**

- Number of splices per organizer should be limited to a reasonable level for ease of maintenance.
- This allows for easier removal of splices for repair without entangling other splices or spare fiber.

d. Sources

191 (1) **Accurate Laser Lifetime Predictions**

- Wear out mode must be reached before laser degradation rate can be accurately assessed or before the laser is installed.
- If wear out mode is not reached, device characteristics will vary or the device may not be beyond the early failure stage.

179-182, 190, 191 (2) **Laser Saturable Current Mode Caused By Leakage Current**

- As more dark spot defects form, leakage current increases. This reduces the threshold current which causes output power to decrease.
- Dark spot defects stop forming when wear out mode is reached. This stabilized level of operation must be reached during stress testing before installation.

III. OPERATIONAL/OPTICAL REQUIREMENTS

Ref. 1. OPTICAL POWER BUDGET

a. Connector

142 (1) **Link Length And Connector Selection**

- Impacts connector style selection in terms of maximum allowable loss.
- The greater the link length, the lower the connector loss must be and the higher the emitter output power. This accommodates for increased fiber loss.

2. BANDWIDTH ANALYSIS

a. Fiber

(1) **Link Lengths Greater Than 2km And Effect On Bandwidth**

- For most fiber manufacturers, it becomes necessary to concatenate fibers to attain lengths of 2km or more.
- Concatenation impacts bandwidth of the fiber unpredictably. After concatenation, the bandwidth may not meet system requirements even if they were met prior to concatenation.

b. Detectors

217 (1) **APD Current Gain Versus Bandwidth Trade-Off**

- The APD gain is increased by raising the reverse bias voltage. This increases the occurrence of avalanche multiplication.
- Multiplication requires time to build up, which reduces the bandwidth of the device.

224, 227 (2) **APD Optimum Gain-Bandwidth Product**

- Selecting a high ionization rate will provide high gain and slow response time
- Selecting a low ionization rate will provide medium gain and fast response time which yields an optimum APD gain-bandwidth product.

c. Receivers

228, 229 (1) **Low Noise, Low Dynamic Range, High Sensitivity Receivers**

- High impedance bipolar transistor and FET receivers have low noise, low dynamic range and an abbreviated bandwidth.

III. OPERATIONAL/OPTICAL REQUIREMENTS

2. BANDWIDTH ANALYSIS (con't)

228- c. Receivers (con't)

229 (1) **Low Noise, Low Dynamic Range, High Sensitivity Receivers (con't)**

- Bandwidth can be extended for broadband applications by including an equalization filter which increases circuit complexity and reduces reliability.

IV. MANUFACTURING

1. SCREENING

Ref. a. Sources

176, (1) **Interpretation Of Emitter Lifetime Test Results**

177
- Activation energy is a value used in characterizing an emitter's lifetime as a function of temperature. This value typically ranges from 0.3 to 1.0 eV.
- The technique for assigning a value to the activation energy has not been standardized which has already resulted in significantly different lifetime predictions. Therefore, correlating test results must take this into consideration.
- Failure to do so can end up with meaningless comparisons.

190 (2) **Laser Prescreening**

- Lasers must initially be subjected to a high temperature burn-in at normal bias for approximately 100 hours to eliminate those devices prone to infant mortality.
- Prescreening only is not sufficient to reach a stabilized degradation rate.

191 (3) **Laser Active Stress Screening**

- Laser active testing is performed after prescreening while operating at high temperatures and high currents (100°C/250mA) (60mA normal operation). This results in stabilized wear out mode operation in 20 hours. Over 1000 hours are required at lasing currents and 60°C.
- Weak lasers are identified and removed from the lot.

IV. **MANUFACTURING**

1. SCREENING (con't)

Ref.

a. Sources (con't)

191 (4) **Wear Out Degradation Screening**

- After wear out is reached by prior active screening, elevated temperature (60°C) burn-in at normal bias is performed.
- After 1000 hours, further changes in degradation rate are insignificant.

b. Detectors

233 (1) **Photodetector Conventional Long Duration Screening**

- Photodetectors do not follow predictable dark current degradation rates; some take much longer to stabilize with no early indication of stability onset.
- Reliable screening must, therefore, be performed for 500-1000 hours under conventional conditions of elevated temperature and normal bias voltage.

233, (2) **Accurate Short Duration Photodetector Screening**

234

- High-temperature (200°C), high reverse bias voltage (2-3 times rated) positively screens weak and unstable photodetectors.
- Duration of screening ranges from 10-20 hours without causing damage to good devices.

c. Optical Fiber

88 (1) **Fiber Proof Testing**

- Fiber should be proof-tested to a minimum level of 50 kpsi to ensure that all existing surface flaws larger than a given size are forced to fail.
- If the fiber is not proof tested, flaws may exist in the fiber which may fail prematurely.

IV. MANUFACTURING

Ref. 2. DESIGN CONSIDERATIONS

a. Fiber

88 (1) **Fiber Coating Protection**

- The coating of the fiber must not be damaged because it minimizes the occurrence of microbends.
- The cabling process can bend the coated fiber. If the coating is damaged, the bare fiber is exposed which degrades the strength and allows moisture to penetrate into existing fiber microcracks.

63 (2) **Multimode Fiber Applications**

- Use multimode fiber for short to intermediate length systems (1-10km) where data rates of a low to medium level are used.

64 (3) **Single Mode Fiber Applications**

- Use single mode fiber for long haul systems (> 10km) where high data rates may be needed and where minimal repeater use would result in a substantial cost savings.

b. Cable

100 (1) **Fiber Overfill and Overlength Design Considerations**

- The trigonometric constraints of the cable/fiber interface, such as fiber overfill and overlength, must be carefully monitored.
- If the amount of overfill or overlength is too great, the fiber may fail due to microbends; yet if the overlength is too small, the fiber may fail due to unequal amounts of tensile elongation between the cable and the fiber.

78 (2) **Materials Used In Cable Design**

- Materials used in cable construction should not contain hydrogen nor outgas hydrogen when subjected to extreme pressures.
- If hydrogen is present under extreme pressures, the attenuation of the fiber will increase which will degrade system performance.

IV. MANUFACTURING

Ref. 2. DESIGN CONSIDERATIONS (con't)

c. Ancillary Equipment

308 (1) **Fiber Storage Design Considerations**

- Splice trays must limit the minimum bend radius of the spare coiled fiber.
- Excess residual tension will occur if the minimum bend radius is exceeded.

d. Source

182, (1) **Laser Facet Oxidation**

183

- Exposure to air prior to passivation causes facet/mirror photo-oxidation which is accelerated by moisture.
- This exposure results in a reduction of facet/mirror reflectivity. This raises the internal temperature of the laser thus increasing the threshold current. This reduces device lifetime.

152 (2) **Laser Output Power Comparison**

- Single mode 0.8 - 0.9 μm wavelength laser output power is an order of magnitude greater than operation at 1.2-1.7 μm.

172 (3) **Laser Double Heterojunction Stripe Geometry Attributes**

- Stripe geometry decreases threshold current density and decreases the number of lateral modes.
- This allows stable single mode operation for narrow stripe widths (up to ≅ 29 μm).

172 (4) **Narrow Double Heterojunction Laser Stripe Width Trade-Off**

- Narrow stripe widths provides stable single mode operation but drive current is reduced which decreases output power.
- Wide stripe widths allow greater drive current, therefore, increases power output but single mode operation becomes unstable.

IV. **MANUFACTURING**

Ref. 2. DESIGN CONSIDERATIONS (con't)

d. Source (con't)

179, 181 (5) **Increased Lifetime Through Passivation**

- The application of a passivation layer will minimize emitter surface contamination and in-migration of metal atoms from contact deterioration.
- Reduced surface contamination leads to decreased surface leakage currents. This decreases threshold current which will extend device life.

181, 182 (6) **Cause Of Saturable Current**

- Leakage current around the active region increases as non-radiative dark spot defects are formed which have a higher current density. This causes a reduction in threshold current and output power.
- Output power level is raised by increasing threshold current.

182-186 (7) **Wear Out Degradation Causes**

- Photo-oxidation of facets, contact degradation and crystal grown-in defects contribute to wear out.
- Photo-oxidation results from moisture or oxygen contacting facets before passivation. Contact degradation is caused by an increased thermal resistance between laser contact and heat sink. Gold contacts and indium solder form high thermal resistance compounds. Grown-in or dark line defects are the primary cause of wear out caused by material lattice defects introduced during fabrication.

185 (8) **Acceleration Of Laser Dark Line Defect Formation**

- These defects are non-radiative centers which increase in size over time.
- Once formed, the rate of formation and thus degradation is directly proportional to the current density.

IV. MANUFACTURING

2. DESIGN CONSIDERATIONS (con't)

Ref. d. Source (con't)

181- (9) **Laser Wear Out Degradation Prevention**

183, - Facet photo-oxidation is minimized by coating facets

185- with Al_2O_3, Si_3N_4 or SiO_2. Contact

186 degradation is minimized by reducing temperature and current density, by using p-side up construction and by reducing the gold content in contacts. Grown-in defects are minimized by using high quality materials and by avoiding high Zn diffusion concentrations in the active region.

- These precautions will extend a quality laser's lifetime.

186 (10) **Laser Facet Damage**

- As a result of excessive optical power density (several milliwatts per micrometer of facet width), the laser facet becomes physically damaged.
- The threshold current increases, quantum efficiency decreases and temperature abruptly increases to the point where the laser is destroyed.

186 (11) **Laser Facet Damage Prevention**

- Anti-reflective coating applied to facets reduces facet power absorption and the internal power needed to produce a given output power.
- This causes a corresponding increase in threshold current, therefore this technique is only good for pulsed mode operation.

200 (12) **LED Lifetimes**

- Room temperature InGaAsP family LED lifetimes increase exponentially with decreasing band gap.
- Therefore, long wavelength devices should outlast short wavelength devices since long wavelength devices have a small band gap.

202 (13) **Laser Instability Due To Reflections**

- Transmitters using lenses to optically couple output power from the laser experience reflections back into the laser.

IV. MANUFACTURING

2. DESIGN CONSIDERATIONS (con't)

d. Source (con't)

202 (13) **Laser Instability Due to Reflections** (con't)

- Reflections cause laser instability and decreases coupling efficiency.
- Antireflection coatings and slightly defocusing the graded index coupling element such that reflections are divergent significantly reduces reflections.

202 (14) **High Efficiency Transmitter Taper Fiber Coupling And Low Reflection Into Emitter**

- Using a taper fiber with a conical lens formed onto the fiber tip eliminates most output power from reflecting off of the fiber endface back into the transmitter source.
- Taper fiber designs have a 75% theoretical and a 25% typical coupling efficiency.

202 (15) **Transmitter Butt Couple Efficiency**

- Reflections off of fiber endfaces from emitters are a major noise source and cause of device instability.
- Coupling efficiency ranges from 10-12%.

204- (16) **Transmitter Monitor Photodetector Purpose**

205
- These photodetectors provide a means of monitoring the output level of transmitter laser sources.
- Feedback is provided which adjusts the driving current thereby maintaining a constant output power. This helps to keep the system operating within its power budget and is a means of providing feedback for temperature control.

204- (17) **Transmitter Laser Instability**

205
- Reflections off of the monitor photodetector into the laser backface will cause output power instability and unwanted wavelength shift.
- This unstable operation will degrade system performance unless antireflection coatings are used and/or reflections are prevented from reaching the laser backface.

IV. MANUFACTURING

2. DESIGN CONSIDERATIONS (con't)

Ref. d. Source (con't)

205- (18) **Transmitter Material Selection**

206 - The materials used to mount the emitter, monitor photodetector and fiber carrier must have compatible coefficients of thermal expansion in the proper directions.

- This allows optical alignment between these elements to be maintained during temperature variations which prevents signal degradation.

206- (19) **Transmitter Solder Type Trade-Offs**

207 - Low melting point solders, like Indium, often creep, resulting in misalignment between elements. High melting point solders make active alignment very difficult. All solder shrinks with age.

- Anticipated system lifetime and ambient temperature extremes must be considered and compared to solder tolerances when selecting a transmitter.

207 (20) **Securing Fiber Pigtail To Transmitter By Welding**

- Spot welds are performed at very high temperatures for short durations, these welds do not creep nor are they affected by high temperatures.

- This is a very expensive technique due to the specialized equipment but it performs extremely well in securing the fiber.

207- (21) **Transmitter Hermeticity At Fiber/Housing Interface**

208 - Epoxy is not typically used to form a hermetic seal.

- Either glass or low melting point (eutectic) solder is used at the fiber/housing interface to realize hermeticity.

e. Connectors

137- (1) **Repeated Mating**

138 - The connector chosen for a given system must be able to withstand the anticipated number of matings. The connector must be covered with a protective cap when unmated.

IV. **MANUFACTURING**

2. DESIGN CONSIDERATIONS (con't)

Ref. e. Connectors (con't)

(1) **Repeated Mating (con't)**

- If the connector is mated and demated frequently, it results in wear on the contacting mating surfaces of the connector which increases attenuation due to eventual misalignment. Repeated mating also allows dirt and dust to contaminate the fiber endface which is one of the primary failure mechanisms of connectors. The endfaces must be protected when in an unmated position to eliminate this contamination and scratching of the endface. Connectors should be cleaned periodically.

140 (2) **Cable/Connector Interface Weakness**

- During connector mating, the cable/connector interface flexes.
- After repeatedly flexing, this may result in broken fibers at the interface. The connector must be designed with cable strain reliefs which limit the degree of bend radius the cable can experience.

142, 144 (3) **Installation Environment Considerations During Connector Selection**

- The installation site defines environmental conditions such as the presence of gases, contaminating sand, high or low temperature, humidity level, etc.
- The presence or absence of these conditions impacts the choice of connector to be used. These factors must be considered to insure the connector can withstand the anticipated operating environment and installation stresses.

IV. **MANUFACTURING**

2. DESIGN CONSIDERATIONS (con't)

Ref. f. Photodetectors

214, 222-223 (1) **Photodetectors Responsivity Versus Response Time Trade-Off**

- Increasing responsivity is achieved by widening the depletion region or optimizing the interband absorption coefficient.
- Widening the photodetector depletion region results in a longer response time due to increased photocarrier transit time. This makes a slower device and, therefore, one with a smaller bandwidth.

215 (2) **Leakage Current In Mesa Depletion Layer p-n Photodetector**

- The mesa walls are not conducive to passivation. If passivation is not performed, the device is prone to contamination.
- Contaminants on the mesa walls and the in-migration of contaminants through the mesa walls, will increase leakage current resulting in an increased threshold current. This reduces laser lifetime.

216 (3) **APD Has High Responsivity**

- The current gain achieved by the avalanche effect produces high gain resulting in high responsivity.
- This enhancement can be used to increase repeater spacing.

218, 219 (4) **APD Gain Characteristics Increases Device Noise**

- Quantum noise, excess noise and noise due to bulk current increase with gain.
- The resultant performance degradation due to this noise decreases sensitivity, thereby decreasing device dynamic range.

220 (5) **PIN Has Low Reverse Bias Of 5-12Vdc**

- This bias level is insufficient to produce a current gain.

IV. **MANUFACTURING**

2. DESIGN CONSIDERATIONS (con't)

Ref.

f. Photodetectors (con't)

(5) **PIN Has Low Reverse Bias Of 5-12Vdc (con't)**

- The result is low noise which is good and significantly reduced responsivity which is undesirable. Low responsivity means closer repeater spacing.

221 (6) **PIN Responsivity Improved By Back-Illumination Construction**

- The depletion layer extends to the substrate in this design which allows photocarrier generation away from device surfaces where they may contribute to leakage current.
- Resulting increase in carriers reaching depletion region improves quantum efficiency. This reduces the threshold current for a given output power.

221 (7) **Photodetector Passivation Increases Lifetime**

- Passivation seals and protects device surface.
- Sealing decreases contamination and surface leakage current, thereby extending operating life.

222 (8) **PIN Dark Current Reduction**

- InGaAs or InP is often grown onto the active layer.
- The relative bandgap of these compounds is greater than that of the active layer which reduces surface dark currents. This reduces device noise.

224 (9) **Photodetector Characteristics Improvements**

- Device material ionization rate must be small to achieve lower noise levels, better sensitivity and higher gain-bandwidth product.
- Si, Ge and InGaAs are rated in increasing order of ionization rate.

229-230 (10) **Low Noise, Large Dynamic Range, Low Sensitivity Receiver**

- The transimpedance amplifier has high bandwidth due to a low input and feedback resistance combination.

IV. MANUFACTURING

2. DESIGN CONSIDERATIONS (con't)

Ref. f. Photodetectors (con't)

(10) **Low Noise, Large Dynamic Range, Low Sensitivity Receiver (con't)**

- These low resistances beneficially reduce the sensitivity 2dB - 10dB relative to the high impedance amplifier. This eliminates the need for equalization circuitry which decreases circuit complexity and increases circuit reliability.

231 (11) **PIN and APD Photodetector Dark Current**

- APD has a high dark current at 0.8 μm and low dark current at 1.3 μm and vice versa for the PIN.
- Selecting an APD for 0.8 μm and a PIN for 1.3 μm wavelength operation will adversely reduce receiver sensitivity due to increased noise.

218, 235 (12) **Accelerated APD Degradation**

- High reverse bias voltages impose stress on the p-n junction.
- The absence of a guard ring makes the perimeter of an illuminated p-n junction prone to breaking down which accelerates dark current degradation and reduces device life.

232 (13) **PIN Photodetector Dark Current Degradation**

- Contamination of passivation layer results in mobile ions.
- Accumulation of mobile ions contribute to dark current which increases noise and decreases receiver sensitivity.

g. Switches

284 (1) **Switching Into Operation Redundant Emitters Or Photodetectors**

- In remote applications (undersea, desert, mountainous) where immediate repair is not possible, automatically switching into operation one or more active elements can be achieved with silicon chip array switches which utilize fiber movement technology.

IV. MANUFACTURING

2. DESIGN CONSIDERATIONS (con't)

Ref.

g. Switches (con't)

(1) **Switching Into Operation Redundant Emitters Or Photodetectors (con't)**

- This partial redundancy prevents the occurrence of long downtimes prior to repair and enhances the reliability of the overall system.

292 (2) **Low Voltage, Low Crosstalk In Electro-Optic Switch**

- A low crosstalk can be achieved at a low voltage by tapering the interaction region of the electro-optic switch channels.
- This provides a more efficient switch and reduces the probability of system degradation due to signal crosstalk.

293-294 (3) **Trade-Off Between Switching Unpolarized Light And Polarized Light**

- Switching polarized light requires the use of polarization maintaining fiber which is not manufactured in large quantities. Switching unpolarized light requires no special fiber but does require three times the switching voltage to attain the same crosstalk level.
- Higher switching speeds will be attainable with polarized light than with unpolarized light due to the lower switching voltage.

299 (4) **Reducing Attenuation In Fiber Alignment Switches**

- Immersion of the switch fibers in reflection controlling fluid will reduce switch attenuation.
- To extend the range of applications the reflection controlling fluid must be compatible with the application temperatures.

IV. MANUFACTURING

2. DESIGN CONSIDERATIONS (con't)

Ref. g. Switches (con't)

299 (5) **False Switching**

- Shock has resulted in fiber alignment switches experiencing false switches.
- This is intolerable in critical applications and can be reduced by using a latching switch.

299 (6) **Wavelength Constraint In Prism/Mirror Movement Switches**

- Optical elements are wavelength dependent due to limitations imposed by materials.
- Usefulness of the switch is limited to a specific wavelength band which means a variety of switches will have to be stocked for use in the entire fiber optics communication wavelength spectrum.

300 (7) **Predominant Shutter Switch Failure**

- Activating spring faults have been identified by switch manufacturers as the most common failure.

301 (8) **Excessive Input Power For Electro-Optic Switches**

- Typical optical power levels of < 5mW have no detrimental impact on this switch's operation. Optical powers at the 50mW level cause increased crosstalk and degraded switching between channels.
- Having a short channel interaction length will reduce the susceptibility to crosstalk.

302 (9) **Electro-Optic Switch Crosstalk**

- An insufficient buffer layer between the switch electrodes and channels can result in crosstalk.
- A 200nm buffer layer of SiO_2 has been reported as being sufficiently thick.

IV. **MANUFACTURING**

2. DESIGN CONSIDERATIONS (con't)

Ref. h. Couplers

245 (1) **Prism And Grating Based Multiplexer And Demultiplexer Coupling Efficiency**

- This type of device performs better as a demultiplexer since the trunk transmission fiber core which is the demultiplexer input is generally smaller than the demultiplexer input pigtail core resulting in high coupling efficiency.
- Selecting this type of device for multiplexing requires that the multiplexer pigtail core be large to accommodate variations in source wavelengths. This results in poor coupling efficiency when multiplexing to a smaller core trunk transmission fiber.

245 (2) **Prism And Grating Based Multiplexer And Demultiplexer Wavelength Incompatibilities**

- The optical element materials in these devices exhibit different attenuations at 0.8 μm and 1.3 μm wavelengths. If the material has low attenuation at 0.8 μm, then it exhibits high attenuation at 1.3 μm and vice versa.
- These particular wavelengths are in the low attenuation bands of most optical fibers. This significantly reduces system flexibility by restricting the wavelengths that can be multiplexed or demultiplexed.

248 (3) **Slant Rod Grating Multiplexer And Demultiplexer Shortcomings**

- Chromatic aberration results from the construction of the device.
- This reduces the coupling efficiency of the device which lowers the available system optical power. Prism grating devices reduce this aberration.

IV. MANUFACTURING

2. DESIGN CONSIDERATIONS (con't)

Ref. h. Couplers (con't)

250 (4) **Interference Filter Dependence On Position**

- The angle of incidence of light if not normal to the filter will alter the wavelength of peak transmission.
- This characteristic increases attenuation and must be guarded against in high vibration and shock environments. Also, temperature changes may cause different material expansions or contractions resulting in increased attenuation at the operating wavelength.

252 (5) **Comparison Of Edge And Bandpass Interference Filters**

- Bandpass filters can be designed to have much narrower pass bands than edge filters.
- Narrow pass bands allow more signals to be multiplexed in a given bandwidth. This can reduce system cost and complexity.

273, 275 (6) **Environmental Sensitivity Of Bulk Optic Device And Fused Biconical Taper Device**

- Bulk optic devices require precise mechanical alignment for the various elements which are sensitive to material coefficients of thermal expansion differences. These devices are more prone to failure as a result of the greater number of parts inside the housing.
- Fused biconical taper devices are manufactured from fiber which is then protected with a housing. This device has fewer parts than bulk optic devices which increases its reliability.

263, 264 (7) **Reflective Star Coupler Noise**

- These devices direct input signals to all output ports as well as to all input ports.

IV. **MANUFACTURING**

2. DESIGN CONSIDERATIONS (con't)

Ref.

h. Couplers (con't)

(7) **Reflective Star Coupler Noise (con't)**

- The signal reflected into the input ports can increase noise especially if laser sources are used. Increased noise levels decrease the link dynamic range which reduces the maximum distance between transmitter and receiver. This complicates system design.

267 (8) **Impact Of Laser Instability On Multiplexer Performance**

- Temperature variations cause laser peak wavelength to shift unless artificial cooling is used. Bandpass interference filter multiplexers have finite pass bands designed to operate at specific wavelengths.
- Laser wavelength shifts outside of this band result in reduced dynamic range and increased crosstalk.

275 (9) **Interference Filter Type Multiplexer And Demultiplexer Materials**

- These devices are fabricated using hygroscopic materials which must be sealed for moisture protection.
- Materials which proved to be very unstable to moisture are chiolite and ZnS. Stable materials are SiO_2 and TiO_2.

275, 276 (10) **FBT Versus Bulk Optic Multiplexer/Demultiplexer And Coupler Vibration Tolerance**

- FBT devices are manufactured with fiber in a housing. Bulk optic devices have individual optical elements requiring precise alignment. These elements can move causing misalignment.
- FBT devices have fewer alignment requirements which makes them inherently more reliable than bulk optic devices.

IV. MANUFACTURING

2. DESIGN CONSIDERATIONS (con't)

Ref. h. Couplers (con't)

277, (11) **FBT Device Prevalent Failures**

278 - The fiber pigtails have been reported as failing most often in handling.

- These devices require care in handling to prevent exceeding the fiber minimum bend radius. Fiber strain relief is not typically provided.

V. LOGISTICS

1. STORAGE, TRANSPORTATION AND HANDLING

a. Sources

190 (1) **Laser Storage**

- Unbiased lasers can be stored at temperatures below the solder melting point with no failure.
- Consideration must be paid to the temperature at which solder creeps to avoid poor optical coupling.

b. Fiber

99 (1) **Fiber Storage**

- Fiber must be stored at -60°C to +55°C and less than 98% RH prior to cabling.

c. Cable

99 (1) **Cable Storage**

- Cable must be stored at -45°C to +70°C and the reels should be kept in a vertical position.
- This eliminates microbends and stress points being exerted on the cable.

d. Connector

146 (1) **Connector Storage**

- Connectors must be stored and operated at the manufacturer specified temperatures. An average of four manufacturer storage temperature specifications ranged from -51.25°C to +117.5°C. The average operating temperature specifications ranged from -49.25°C to +102.5°C.

V. LOGISTICS

Ref. 1. STORAGE, TRANSPORTATION AND HANDLING (con't)

e. Splices

146 (1) **Splice Storage And Operating Temperatures**

- Splices must be stored and operated at the manufacturers specified temperatures. An average of three manufacturer storage temperature specifications ranged from -30°C to +70°C and operating temperatures ranged from -30°C to +70°C.

308 (2) **Splice Protection During Operation**

- Splices must be stored in a splice organizer in such a way that they can be removed easily in an emergency restoration situation. The splices must be labeled to properly identify each fiber.
- If the splices are difficult to manipulate, in the attempt to repair one damaged splice, another one can be broken. Labeling minimizes downtime.

2. SUPPORTABILITY

a. Cable

101-103 (1) **Link Length And Terrain**

- The link length and its location can make accessibility to the entire cable run difficult.
- This impacts the maintainability and logistics of making repairs due to problems with locating equipment/spare parts and gaining access to the repair point.

102-104 (2) **Logistics**

- In order to repair a damaged fiber optic cable system quickly, an emergency repair kit needs to be on-site. The kit must contain all necessary tools, supplies and test equipment needed to perform both temporary and permanent repairs. Trained personnel must be available to eliminate lengthy downtime.

V. LOGISTICS

Ref. 2. SUPPORTABILITY (con't)

b. **Training**

102- (1) **Training Of Personnel**

104 - Personnel require training to install and maintain fiber optic systems.

- Training should include theory of operation, troubleshooting, and actual hands-on training in installing connectors and splices and using test equipment. If training is given to all personnel, they will have a common base from which to perform installation, troubleshooting, and repair work. This training will insure that technicians follow proper installation practices such as not exceeding the minimum bend radius and proper storage of the splices.

c. **Spares**

146 (1) **Availability Of Spares**

- To have a fully supportable system, there must be adequate spares available on-site to facilitate quick repair.

146 - The emergency repair kit must have spare piece parts for connectors and splices. The kit must also contain temporary elastomeric splices for emergency repairs. The splices must be compatible with the fiber size used in the system. A spare piece of cable with fiber of the same optical and mechanical properties must be available for replacement of a damaged section of cable.

VI. MAINTENANCE

1. ACCESSIBILITY AND REPAIRABILITY

a. System

101- (1) **Link Length And Terrain**

103 - The link length and its location can make accessibility to the entire cable run difficult.

103 - This impacts maintainability and logistics due to problems with locating equipment and spare parts and gaining access to the repair point.

VI. **MAINTENANCE (con't)**

Ref. 1. ACCESSIBILITY AND REPAIRABILITY (con't)

a. System (con't)

101- (2) **Maintenance Activity On Ring Network Nodes**

- Ensuring continuity of data transmission for active nodes while performing repairs on an inactive node is accomplished by designing the ring with a bypass switch between each node and the ring.
- The bypass switch makes the node/ring network interface fault tolerant and maintainable without compromising the performance of the entire network.

b. Cable

102, (1) **Cable System Repair**

104

- An emergency repair kit should be available to perform emergency repairs and routine maintenance.
- The kit should contain a piece of cable with identical fiber as is installed in the system. This is used as a jumper which facilitates splicing in a new section of cable. The kit must contain all tools, supplies and test equipment to perform routine and emergency maintenance procedures.

103 (2) **Cable Construction And Maintenance Base**

- Cable construction will impact the type of tools that will be required to perform repairs.
- Heavily armored cable will require special cutters and other tools while zipper cable will require minimal tools.

c. Connector

137 (1) **Connector Cleaning**

- The fiber endfaces can become contaminated with dust and dirt when the connector is unmated. This will cause the fiber to become scratched upon mating.
- This increases connector attenuation. Manufacturers recommend that connectors be cleaned after a specified number of matings and that protective covers be place on the connector when they are not mated.

Section II

Numerical Reliability Summary

SUMMARY OF NUMERICAL RELIABILITY DATA

In this section, numerical data is tabulated on operating hours, mean time to failure (MTTF) and mean time between failures (MTBFs) for the components and systems addressed in subsequent chapters.

The contents of this section are abbreviated which will facilitate its use as a quick reference guide. More detailed information can be found in each of the associated chapters. The detailed information includes the raw data from which calculations were made, calculations, any assumptions which were made, amplifying information on the conditions under which the active device data was acquired, and references. The chapter and section(s) in which this additional information can be found is given in parentheses following the device or system name/nomenclature in this summary.

Due to the available data, it was necessary to make several assumptions that have a significant impact on the meaning of the following calculated numbers. Therefore, it is highly recommended that the individual chapters be used in conjunction with this summary. This will give the reader a better understanding of the limitations on the usefulness of these numbers.

COMPONENTS

All calculated values were evaluated using 90% confidence limits.

Fiber
(Chapter 1 Sections VIII, IX)

MTTF, Predicted* (10^6 hrs) . . . . . . 0.18 - 0.26

Cable
(Chapter 1 Sections, VIII, IX)

MTBF, Calculated** (10^6 hrs) . . . . . 0.40 - 0.84

MTBF by Failure Type

Relevant*** (10^6 hrs) . . . . . . 0.56 - 1.09
Non-relevant **** (10^6 hrs). . . . 1.34 - 3.61

---

* Throughout this section, predicted values are taken from manufacturers' data, designers' data, or laboratory results.

** Throughout this section, calculated values are calculated using field data.

*** A failure expected to occur in field service which can be attributed to inadequate part design and poor workmanship. Also, a failure caused by and dependent upon an independent failure of government furnished equipment or induced by an external event, e.g., ice storm or back-hoe digup.

**** A failure not expected to occur in field service which can be attributed to conditions external to the part, e.g., installation damage or mishandling.

Splices
(Chapter 2 Section VI)

MTBF, Calculated (10^6 hrs)

Field Mechanical and Fusion Splices . . . . 14.0 - 54

Connectors
(Chapter 2 Section VI)

Mating Durability, Predicted

Connector Style	Number of Matings
MIL-T-29504	1000
MIL-C-28876	500
MIL-C-38999	500
MIL-C-83522	500
MIL-C-83526	1000
FC Style	1000
FOHC*	500
FOMC**	2500

---

* Fiber Optic Hybrid Connector (FOHC) not yet qualified to MIL-T-29504.
** Fiber Optic Multi-channel Connector (FOMC) not yet qualified to MIL-C-28876. Qualified to Magnavox Specification GRC-206 for U.S.A.F. which specified 200 matings.

COMPONENTS (Cont'd)

Light Emitting Diodes
(Chapter 3 Section VII)

AlGaAs/GaAs

Construction Type	Operating Temperature(°C)	Output Power (mW)	Predicted Operating Hours (10^6hrs) Median Value*	MTTF**
Surface	25***	--	--	1.0-10.0
INP****	50	--	--	1.0
Planar Double Hetero-junction (High Radiance)	86	--	0.0265	--

InGaAsP/InP

Construction Type	Operating Temperature(°C)	Output Power (mW)	Median Value	MTTF
INP	25	--	1.0	--

AlGaAs/Si

Construction Type	Operating Temperature(°C)	Output Power (mW)	Median Value	MTTF
Full Surface	70 (junction)	--	0.24	--

* Median Value - The time at which one half of the devices under observation failed.

** Mean Time To Failure - For non-repairable items, the average time that all of the devices under observation failed.

*** The source data indicated room operating temperature which has been interpreted to be 25 °C.

**** INP - Information Not Provided.

COMPONENTS (Cont'd)

Laser Diodes
(Chapter 3 Section VII)

AlGaAs/GaAs

Construction Type	Operating Temperature(°C)	Output Power (mW)	Predicted Operating Hours (10^6 hrs) Median Value	Predicted Operating Hours (10^6 hrs) MTTF
Constricted Double Heterojunction (Ridge-Guide)	25**	3-4	--	1.0
Oxide Defined Stripe-Constricted Double Heterojunction	22	10	--	0.08
Double Heterojunction	70	5	--	0.13
INP	50	1.6	--	0.4

Generic (1.3 μm Wavelength)

Construction Type	Operating Temperature(°C)	Output Power (mW)	Predicted Operating Hours (10^6 hrs) Median Value	Predicted Operating Hours (10^6 hrs) MTTF
Not Specified	20-25	0.7-1.0	--	0.22-50*
↓	20-25	1.5	--	0.5-1.0*
↓	25	3.0	--	50
↓	20	5.0	--	1.0
↓	20	1.0	1.4	--
↓	Same Device } 30	1.0	0.56	--
↓	Same Device } 50	1.0	0.06	--

* These figures reflect data on more than one device which were tested under the range of conditions specified. Detailed data can be found in Chapter 3.

** The source data indicated room operating temperature which has been interpreted to be 25 °C.

COMPONENTS (Cont'd)

Laser Diodes
(Chapter 3 Section VII)

InGaAsP/InP

Construction Type	Operating Temperature(°C)	Output Power (mW)	Predicted Operating Hours (10^6 hrs) Median Value	MTTF
Buried Heterojunction	10	5	> 20	--
Buried Heterojunction	20	5	>0.5	--
Buried Heterojunction	25**	5	--	0.3- 10.0*
Buried Heterojunction	50	5	1.0	--
Double Channel-Planar Buried Heterojunction	50	5	0.27	--
Double Channel-Planar Buried Heterojunction	70	5	0.14	--
Oxide Defined Stripe	50	4	--	0.04
Native Oxide	25**	INP	--	10
Self Aligned Structure	50	3/5	--	0.3
Buried Crescent	50	3	--	2.4
Buried Crescent	50	5	--	17
INP	25	1.6	--	0.4
INP	50	5	--	0.1

* These figures reflect data on more than one device which were tested under the range of conditions specified. Detailed data can be found in Chapter 3.

** The source data indicated room operating temperature which has been interpreted to be 25 °C.

COMPONENTS (Con't)

Laser Modules
(Chapter 3 Section VII)

Generic (1.3 μm Wavelength)

Construction Type	Operating Temperature(°C)	Output Power (mW)	Predicted Operating Hours (10^6 hrs) Median Value	MTTF
Not Specified	50	1.0	--	> 0.219
	20	1.5	--	0.35
	25	1.0	--	0.173
	20	5.0	--	≅0.72

Photodetectors
(Chapter 4 Section V)

APD

Material	Operating Temperature(°C)	Output Power (mW)	Predicted Operating Hours (10^6 hrs) Median Value	MTTF
INP	170	--	--	0.1
Ge	260	--	0.004	--

PIN

Material	Operating Temperature(°C)	Output Power (mW)	Predicted Operating Hours (10^6 hrs) Median Value	MTTF
InGaAs/InP	25	--	--	1.0

Unknown Construction (1.0 - 1.6 μm Wavelength)

	Material	Operating Temperature(°C)	Output Power (mW)	Predicted Operating Hours (10^6 hrs) Median Value	MTTF
Same Device	InGaAs/InP	85	--	--	>>0.02
	InGaAs/InP	≤ 30	--	--	>2.0

COMPONENTS (Con't)

Photodetector Modules
(Chapter 4 Section V)

APD

Material	Operating Temperature(°C)	Output Power (mW)	Predicted Operating Hours (10^6 hrs) Median Value	MTTF
Si*	25***	--	--	10
Ge**	25***	--	--	10

APD

Material	Operating Temperature(°C)	Output Power (mW)	Predicted Operating Hours (10^6 hrs) Median Value	MTTF
InGaAs**	--	--	--	408

* 0.85 μm Wavelength
** 1.3 μm Wavelength
*** The source data indicated room operating temperature which has been interpreted to be 25 °C.

COMPONENTS (Con't)

Couplers
(Chapter 5 Section V)

MTTF, Calculated (10^6 hrs)	2.16 - 3.30
MTTF by Failure Type	
Relevant	4.67 - 9.52
Non-relevant	2.64 - 4.26

Switches
(Chapter 6 Section IV)

Switch Durability, Predicted (10^6 On/Off Operations)

Switch Type	
Fiber Alignment	20-100
Prism/Mirror movement	1
Shutter	10

Enclosures
(Chapter 7 Section III)

MTBF, Predicted (10^6 hrs)	0.26 - 0.35

SYSTEMS

AN/FAC-3
(Chapter 8 Section V)

MTBF, (hours)

System, Predicted	8,000
System, Calculated*	24,000 - 30,000

AV-8B CNI SYSTEM
(Chapter 9 Section IV)

MTBF, (operational-hours)

Predicted	1,000
Calculated	25,065

LIFETIME,** (operational-hours)

Predicted	10,000
Calculated	57,712

FOTS-LH
(Chapter 10 Sections V, VI)

MTBF, System Predicted	770 hours
MTTR, System Predicted	0.25 hours

---

* This value is an estimate based on field reports.

** This represents the total number of accumulated operational-hours over the life cycle of the system.

Section III

Components

Chapter 1. Fiber and Cable

Optical fiber and optical fiber cable are used in all areas of communications such as voice, facsimile, data, process control, computer applications, sensing, military communications and other applications. Fiber optic systems offer significant advantages over conventional twisted pair or coaxial cable due to their immunity to electromagnetic interference (EMI), radio frequency interference (RFI) and electromagnetic pulse (EMP), high information transmission capacity, small size, and light weight. There are many different type of fibers and cables and each has its own advantages for individual applications. The mechanical and optical parameters of optical fiber and cable that affect reliability, maintainability and logistics are addressed in this chapter.

I. DESCRIPTION OF FIBER TYPES

1. MULTIMODE FIBER TYPES

Multimode fiber types are available with two main types of index profiles; step and graded. In a step-index fiber, the core has a uniform refractive index with a step or abrupt change at the cladding-to-core interface as shown in Fig. 1-1. This type of fiber is the simplest type and has core diameters in the range of 50 to 1000 micrometers (μm) with the 50 μm & 100 μm core sizes the most prevalent in the industry except for specialty applications, such as sensing, where the larger fiber sizes may be used. This type of fiber is subject to modal dispersion because the light reflects differently for different modes, resulting in some rays following longer paths than others and therefore taking a relatively longer time to reach the same destination. The 100 μm core step-index fiber is used in short-haul systems which are connector intensive such as Local Area Networks (LANs), aircraft or shipboard applications. Because of the relatively large core size, connectorization and splicing precision is not as critical as in single mode fiber which results in less time consuming installation and repair procedures.

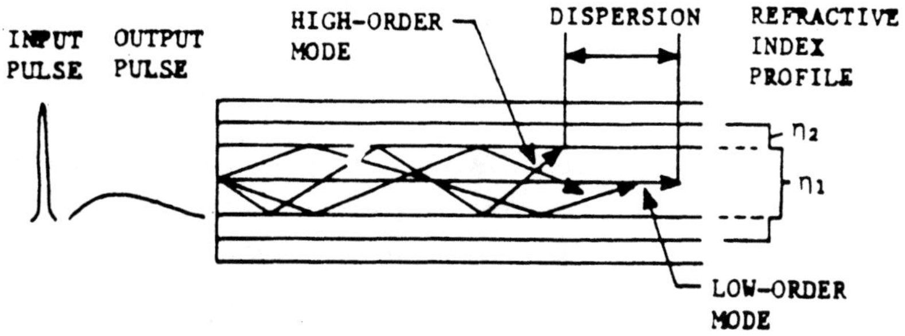

Figure 1-1. Multimode step index.[1]

In a graded index multimode fiber, the core is manufactured as a series of concentric rings with each successive ring having a lower refractive index. Due to the properties of refraction, light in graded index fiber which is farther from the axis travels faster due to its lower index. This causes all modes to arrive at the same destination at approximately the same time as illustrated in Fig. 1-2. This effectively reduces modal dispersion. Graded index multimode fibers are used in intermediate length systems ($\leq$ 10km) without repeaters.

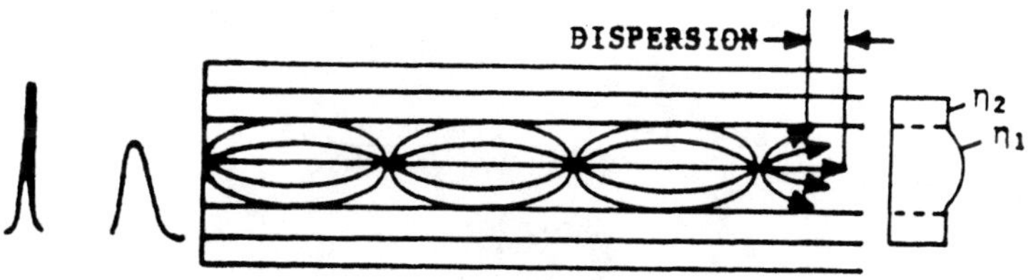

Figure 1-2. Multimode graded index.[1]

2. SINGLE MODE FIBER TYPES

Single mode fiber types are commonly available with the step index profile. Single mode fiber (see Fig. 1-3) has a very small core diameter, typically from 2 to 10 μm. The size of the core is small enough so that the fiber can only propagate one mode efficiently. This limits modal dispersion because only one mode is being propagated. The small size of the core has a disadvantage though, and that is that it is difficult to connectorize and splice due to the high degree of precision needed to align the small cores. This typically results in more time required for installation and maintenance procedures. Single mode fiber is used for high capacity/high speed data where long links are required. Repeater spacings are much greater than spacings in multimode systems, typically 30-40km. This affects maintenance in that less equipment is used, therefore, there is less maintenance to perform due to fewer parts. This makes the overall system more reliable, for the fewer parts there are to fail, the higher the reliability. The telephone companies use this type of fiber in their long trunk lines.

Figure 1-3. Single mode step index.[1]

There are several new types of single mode fiber that are being experimentally developed and are in use at government and private developmental installations. The first is a dispersion shifted fiber. Fig. 1-4 shows that this fiber has shifted the point of zero dispersion from 1300nm to 1550nm where attenuation is lowest for single mode fibers. The shift of the zero dispersion point is obtained by manufacturing the fiber with a core refractive index profile having a triangular shape. The core is surrounded by a depressed cladding index profile made up of a series of up to 24 rings which have been chemically deposited.[2] These rings contain up to four different dopant chemicals. The primary advantage of this type of fiber is that repeater spacings can be as great as 80km. This further improves system reliability and reduces maintenance as a consequence of there being fewer parts. The reduced parts count means fewer spares have to be stocked which alleviates logistics concerns associated with the cost, storing and tracking of parts. Disadvantages are that the fiber may only be operated at 1550nm, it can't be used at 1300nm. Fiber which operates at 1550nm is not as readily available which may impact the decision to utilize this fiber type for critical systems where logistics requirements are critical.

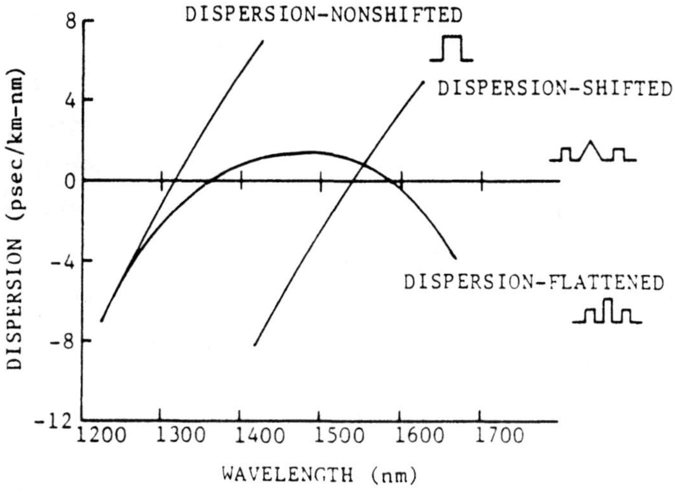

Figure 1-4. Dispersion characteristics of experimental fibers.[3]

Another way of increasing capacity at 1550nm is to flatten the dispersion curve as shown in Fig. 1-4 so that it is low between 1300nm and 1550nm. This leaves a large, high-capacity transmission window. This type of fiber is called dispersion flattened fiber and is still in the developmental stage.

Polarization maintaining fibers may play an important part in the development of fiber optic sensors and components.[4] Every single mode fiber has two propagation modes that are very similar except that their polarization planes are perpendicular with respect to each other. In normal operation of single mode fibers, the light signal is transmitted through only one of the polarization planes. The optical signal can easily be transferred from one plane to another by any kind of disturbance to the fiber; for example, a bend. For some applications, such as sensing, it is necessary to preserve the polarization state. Fibers are currently under development to preserve the polarization planes; that is preventing the alternations between polarizations.[5] Such fibers have been used in the fiber optic gyroscope and interferometric sensors. As fiber optic sensors come of age, the polarization maintaining fibers will be further developed and refined.

A comparison of the characteristics of the individual fiber types is given in Table 1-1. This information can be used by the system planner to narrow down the available types to those that meet the fundamental requirements of a given application. These requirements include permissible wavelength, attenuation, applicability in long-haul or short-haul systems, and fiber size.

TABLE 1-1. Optical Fiber Characteristics

FIBER TYPE	WAVELENGTH (nm)	SYSTEM LENGTH ≤10km SHORT	SYSTEM LENGTH LONG	INDEX PROFILE STEP	INDEX PROFILE GRADED	INDEX PROFILE UNIQUE/ SPECIALTY	MODAL DISPERSION	ATTENUATION (dB/km)	CORE/ CLADDING DIAMETER (μm)	EASE OF CONNECTORIZATION & SPLICING
Multimode	850,1300	o		o	o		o	3.0/1.0	50/125, 62.5/125 85/125, 100/140 and others	High
Single Mode	1300,1550		o	o				1.0	8/125	High
Dispersion Shifted	1550		o			o		1.0	8/125	Low
Dispersion Flattened	1300-1550		o			o		1.0	8/125	Low
Polarization Maintaining	1300-1550	o				o		1.0	8/125	Low

II. DESCRIPTION OF CABLE TYPES

There are several cable types that are currently manufactured. Optical cable encloses fiber within a protective structure in order to protect the fiber throughout its lifetime. A description of the common cable designs and typical applications and any advantages or disadvantages of each type will be discussed.

1. LOOSE TUBE

Loose tube cables are designed such that the fiber is protected within an outer protection device such as a tube or slot, but the fiber is still able to move freely. The fiber is free to move within the tube during temperature changes and physical stresses to the cable such as bending or clamping. This reduces the possibility of the fiber being subjected to microbends which can severely increase the attenuation of the fiber. Some designs employ a loose tube in which a single or multiple fibers are threaded as shown in Figs. 1-5a and 1-5b, respectively. Other loose tube designs employ a solid, star shaped core as can be seen in Fig. 1-5c. The fibers are laid in the slots which are formed around the circumference of the star. The star shaped core is then covered with a mylar tape. These loose tubes are then wound around a central strength member made of steel or of a dielectric material depending upon the given application. The voids in the fiber tubes or slots are generally filled with a filling compound, either a gel or powder, which prevents the ingress of water into the area surrounding the fibers. A layer of strength material may then be placed around the tubes and an outer jacket is then extruded over the assembly. This type of cable is suitable for many applications including aerial installations when used with a messenger wire and duct installation. If additional armoring is applied, these cables may be used for direct buried, or self-supporting aerial installation. Depending on their size, these cables can accommodate from 1 to 144 fibers. There are new developmental designs which can accommodate up to 600 fibers.[6]

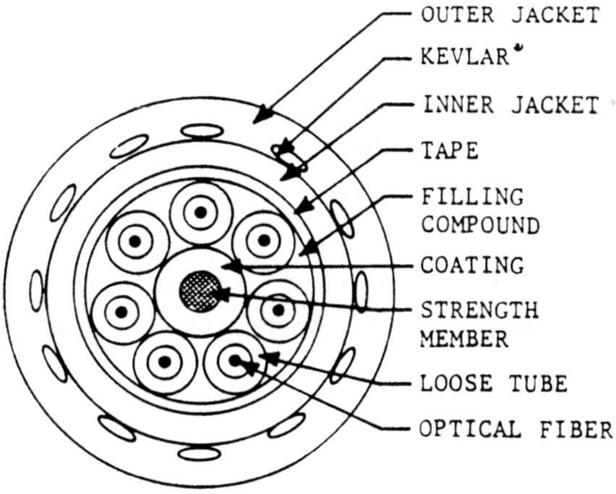

a. Single fiber per tube.

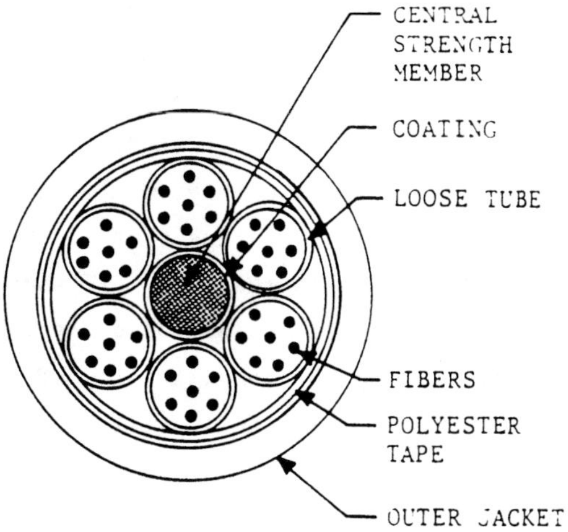

b. Multiple fibers per tube.

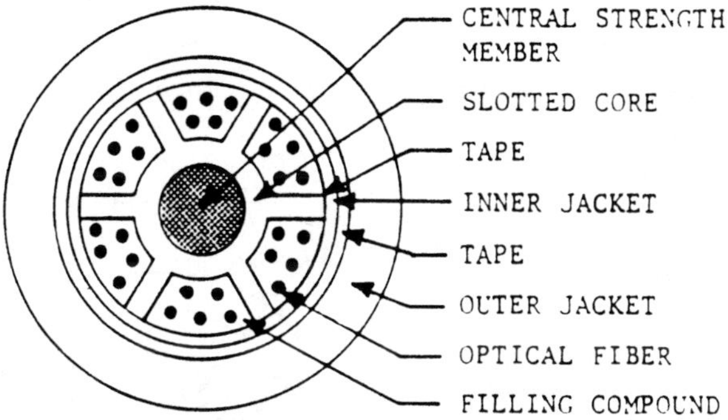

c. Slotted core design.

Figure 1-5. Loose tube cable constructions.[7]

The loose tube design can be somewhat more difficult to terminate than a tight tube design due to the tube material characteristics. The material must be pliable enough to allow the necessary movement of the termini when placing them in the connector, yet firm enough to provide adequate protection to the fiber. If the tube material is too rigid, it can kink and bend the fiber. Loose tube cable designs protect the fiber from microbends and allow the fiber to "float" within the tube. If these aspects of cable design are not considered in the selection criteria, the time to repair can be significantly increased and the compatibility of the cable with different connector styles and connector backshells will be greatly reduced.

2. TIGHT TUBE

Tight tube or tight buffer cables as they are commonly called, have a buffer coating extruded directly onto the optical fiber. The buffered fibers are then extruded with a jacket. For multiple fiber cables, two examples of which are given in Fig. 1-6, the jacketed fibers are then cabled together around a central strength member. A layer of aramid fibers is generally wound over the cabled fibers and an outer jacket is then placed over this strength material. Because the fibers are tightly encased, this type of cable is more compact, lighter in weight, smaller in diameter and more flexible than a loose

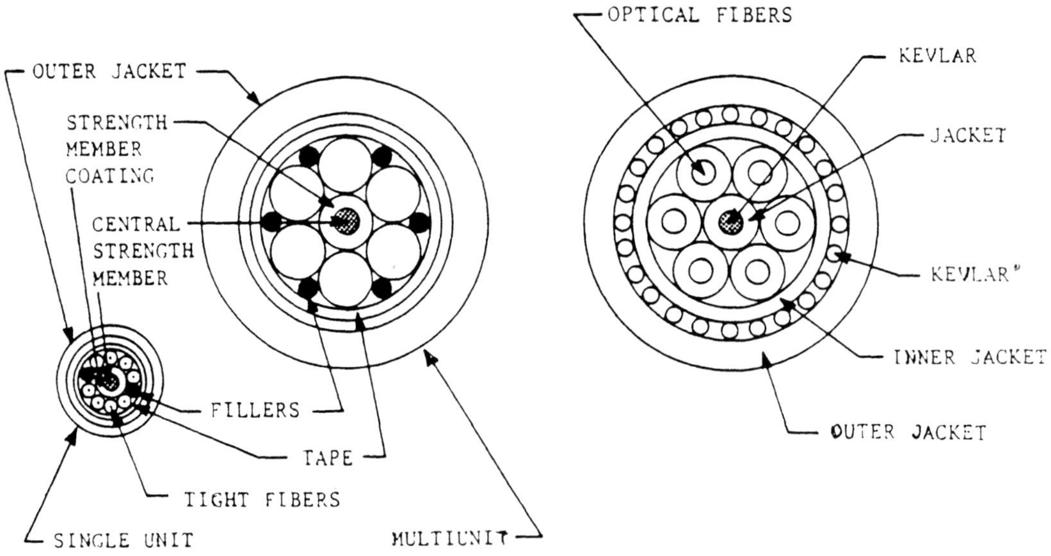

Figure 1-6. Tight buffer cable designs.[7]

tube cable. These characteristics make the cable more crush resistant and enable it to withstand high impacts. This cable design is used for tactical military use where crush resistance and small size and weight are important. Because the fibers are not free to float as they are in a loose tube cable design, this cable is more prone to experience microbends which will degrade the reliability of the system if the cable is used improperly. This type of cable should not be used in aerial or direct buried applications. Generally, this cable type has a higher attenuation than loose tube cable which is made with the same type fiber. This is due to the stresses placed on the fiber by a buffer and jacketing material being directly extruded onto the optical fiber itself. The removal of the cable sheath on a tight tube cable is generally more difficult than a loose tube cable due to the fact that the cable sheath is extruded directly onto the circumferential strength members of the cable. This usually makes preparation for termination of tight tube cable difficult though the termination itself may be easier because the buffer tube around the fiber is very flexible. This decreases installation and maintenance time. Some loose tube cables use the same extrusion method when a double sheath design is used. Therefore, there is no clear cut distinction between the two cable types with regard to the difficulty of sheath removal. When choosing the cable, the system designer should insure that the cable sheath is easily removed by analyzing the sheath construction methods.

3. PLENUM CABLE

A plenum cable is a cable that has been specifically designed for use in air handling or plenum spaces above suspended ceilings. Cables that are installed in these spaces must meet the requirements of the National Electric Code (NEC) for fire resistance and smoke producing characteristics. This cable can be either of tight or loose tube design although most cables manufactured to date are of tight tube construction (see Fig. 1-7).[8] The primary difference in this type of cable is the jacketing material. In order to pass stringent flame tests required by the NEC, the flame resistant jacket must be made of extremely flame resistant material such as a TEFLON[R] fluorocarbon resin or a TEFZEL[R] fluoropolymer material.[9]

[R] TEFLON and TEFZEL are registered trademarks of the Du Pont Company.

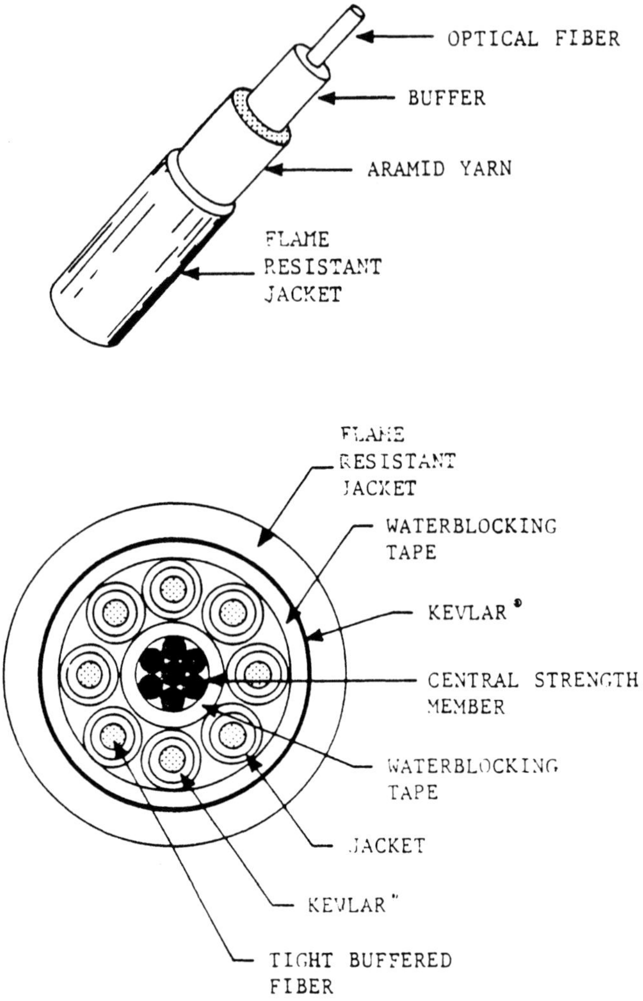

Figure 1-7. Plenum cable designs.[8,9]

For emergency repair situations, the same cable type should be kept in storage so that it may be spliced into the damaged area. The TEFLON[R] and TEFZEL[R] cable jacketing materials are difficult to remove from the cable which will affect maintenance crew requirements and the duration of the repair time. Heavy duty cutting tools are needed to repair this type of cable and they must be kept in the emergency repair kit to avoid extended downtime.

4. RIBBON CABLE

Fig. 1-8 illustrates ribbon cable which is a package of coated fibers having small outside diameters of approximately 0.24mm.[6] This cable is packaged into linear arrays of from 6 to 12 fibers. The fibers are generally held in place by a mylar tape which is extruded over the fibers. The ribbons are then placed inside a cable sheath. A typical telephone company ribbon cable has twelve ribbons of twelve fibers each. The ribbons are stacked on top of one another and surrounded by a sheath which is embedded with strength members. This type of cable is very space efficient and can contain a large number of fibers. Ribbon cable design is commonly used in telephone company trunk lines. The fibers are sufficiently protected for this type application but this cable type would not be suitable for tactical military environments or in conditions where the cable is going to be subject to impact, crush or excessive bending. The fibers are coated with a thin buffer material as opposed to the thicker and more protective buffers used on other cable types which makes them more prone to microbends. Connectorization can be accomplished individually or by mass-splicing the fibers. An array splice which simultaneously splices all of the fibers in the ribbon, is typically used to splice this type of cable. An impact to maintenance and logistics is that the special tools which have been designed for splicing ribbon cable must be kept on hand. Technicians must be trained in ribbon cable installation procedures and the special maintenance equipment. Ribbon cable termination is quite different from other cable termination techniques, especially with regard to the hardware.

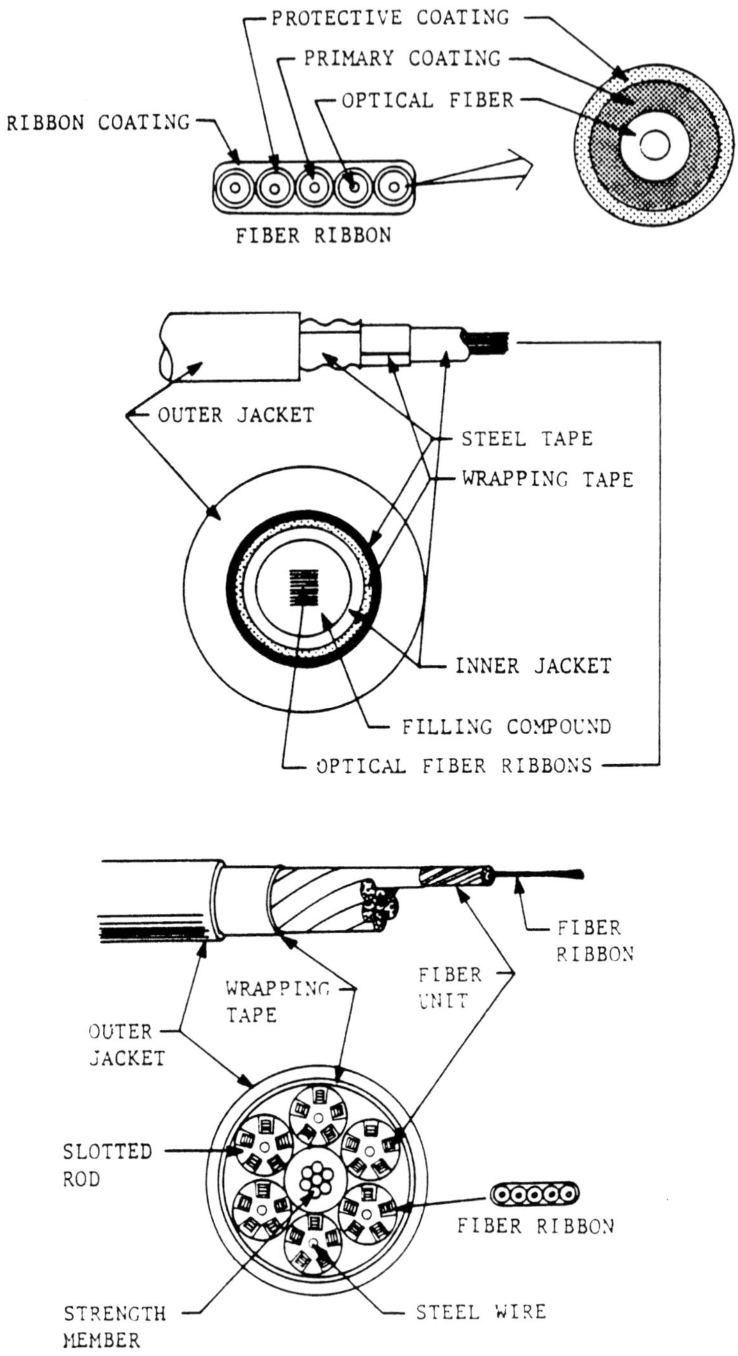

Figure 1-8. Ribbon cable designs.[6,10]

5. SUBMARINE CABLE

Submarine cable is laid in place under a body of water such as a lake, river, or ocean. The cable is generally of loose tube construction but is heavily armored to withstand the high pressures, environment, and fish and shark bites. The difference between thi, cable and the loose tube cable is two layers of armor as indicated in Fig. 1-9. Depending on the depth of the water and the length of the crossing, more or less armor may be needed. Generally, these cables are spliced to a regular loose tube type cable at a point very near the shore. The loose tube cables are then routed to the terminal equipment and connectorized. One application used a cable containing 48 fibers with an outside diameter of 18.5mm (≅ 0.73 in) for a 200 ft. deep, 3km lake crossing.[11] This type of cable is application specific and is only used for water crossings. Special tools are needed to remove the cable sheath for splicing and emergency repair. Underwater cable installations require considerable more time to repair than other cable types since the cable is usually brought to the surface prior to performing repairs. Underwater cable runs should have built-in redundancy, that is, spare fibers should be available for alternate transmission paths should one fiber fail.

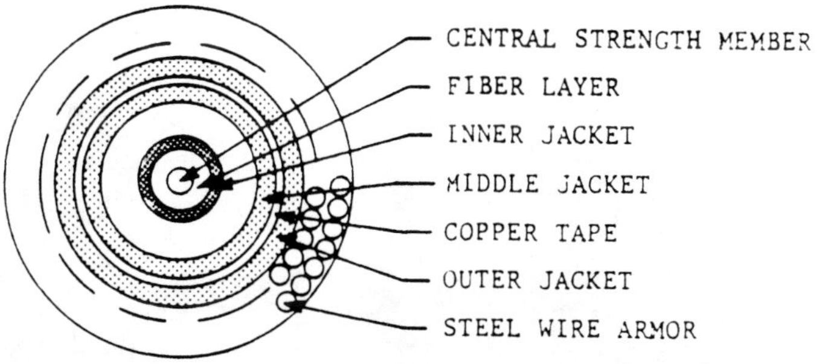

Figure 1-9. Submarine cable.[11]

Table 1-2 lists the characteristics of the cables which have been discussed. The applications in which the individual cable types can be used and the important mechanical properties associated with installation and durability are given. This information can be used by the system planner to narrow down the variety of cable types which are appropriate for use in a particular application.

TABLE 1-2. Optical Cable Characteristics

	APPLICATION					PROPERTIES					
CABLE TYPE	AERIAL	DUCT	DIRECT BURIED	SUBMARINE	TACTICAL	FIBER FREE TO MOVE	CRUSH RESISTANCE	FLEXIBILITY	IMPACT RESISTANCE	EASE OF CONNECTORIZATION
Loose Tube	o	o	o	o		Yes	Low	Medium	Medium	Medium
Star Core	o	o	o			Yes	Medium	Low	Medium	Medium
Tight Tube					o	No	High	High	High	High
Plenum		o				No	High	High	High	Medium
Ribbon	o	o	o	o		No	Medium	Medium	Low	Low

III. FILLING COMPOUNDS

Filling compounds are used in cables to prevent the ingress of water into the cables. Moisture around the fiber can cause existing microcracks to propagate which can cause degradation or even failure of the system over time. Filling materials are generally used in two different places in the cable. The first location is in the loose tubes or fiber enveloping areas. The second is in the interstices of the cable which includes all the areas not in direct contact with the fiber. The filling material in the loose tubes can be either a gel or powder compound. The filling material is used to block water from entering and to prevent wicking of the water along the fiber. The filling material used in the interstices of the cable is a water blocking material which is a very thick gel. Depending on the application of the cable, filling material may be used in both the buffer tubes and the interstices or only in one of the locations. For direct buried, aerial or duct installations, cables are generally manufactured with filling materials in both locations. More recently, most cable designs have employed filling materials in both locations. The chemicals that the filling materials are made of are very important as they may affect the optical parameters. There have been numerous studies in the past few years on hydrogen migration into the core of the fiber which causes an increase in attenuation. Silicon has proven to be one material which has experienced problems with hydrogen migration.

When hydrogen migrates into the core of the fiber, OH radicals cause serious degradation of the optical performance at wavelengths greater than 1050nm. Recent studies have indicated that hydrogen itself may cause optical signal losses of a significant nature.[12]

Submarine cables experience more hydrogen loss related problems than land cables due to the severe pressure which can drive molecular hydrogen into the silica fiber core. Hydrogen is generated within cables by outgassing from the plastics used in the cable construction and by electrolysis of the seawater. When water enters the cable, an electrochemical reaction occurs between the water and aluminum or iron armoring which causes hydrogen to be released. In a study performed by Kokusai Denshin Denwa's research laboratories in Japan, a phenomenon was discovered which related to the oxidation of metals in the cable when seawater reacted with conducting metals.[12] The study showed that a substantial increase in attenuation occurs when the cable contained aluminum or iron but no increase occurred when only copper or no metals were present in the cable.

The hydrogen absorbs light and converts the energy into vibration within the hydrogen molecule. The vibration occurs at resonance frequencies of 2240 nm and 1240 nm.[12] In order for the molecules to absorb light, they must be electrically polarized. Thus, if the cable contains metals which interact with the seawater, an electric field is produced which can cause the polarization.

The hydrogen from the seawater penetrates the cladding of the fiber in from two to three days. Then, the hydrogen molecules are polarized by electrolysis or outgassing. The molecules then bond to the fiber which causes the increase in attenuation. Cables must be designed with materials that do not generate hydrogen. When obtaining a cable with filling compounds, the fiber and compounds must be compatible.

IV. STRENGTH MEMBERS

Strength members are needed to provide mechanical strength to the cable so that it can tolerate environmental and mechanical conditions. Strength members can be divided into two broad categories: dielectric and non-dielectric. When using fiber optics, the transmission medium is dielectric. If the fiber and a dielectric strength member are enveloped in a dielectric sheath, a dielectric cable is developed. Dielectric strength member materials include epoxy reinforced glass, fiberglass, and impregnated aramid fibers such as Kevlar[R] reinforced with epoxy. This cable type is immune from EMI and does not sustain the severe damage that cable with metallic strength members sustain.

[R] Kevlar is a registered trademark of the Du Pont Company.

There are two strength member constructions; central and circumferential. Central strength members are usually a rather stiff material which is often designed to prevent the cable from bending smaller than the specified minimum bend radius of the fiber. (See Fig. 1-10a.) Circumferential strength members are used along the outer circumference of the cable and in a dielectric cable, are generally made of an aramid fiber (see Fig. 1-10b).

Non-dielectric cables are used in extreme or adverse conditions where armoring is needed. Applications for non-dielectric strength members include areas of heavy rodent population, aerial cables with an internal messenger wire, and submarine cables. The non-dielectric material generally used is steel in the form of strands or sheathing. The central strength member may be a large strand while the outer layers of the cable are protected by an inner sheath of steel, or several such sheaths, before the outer jacket is applied.

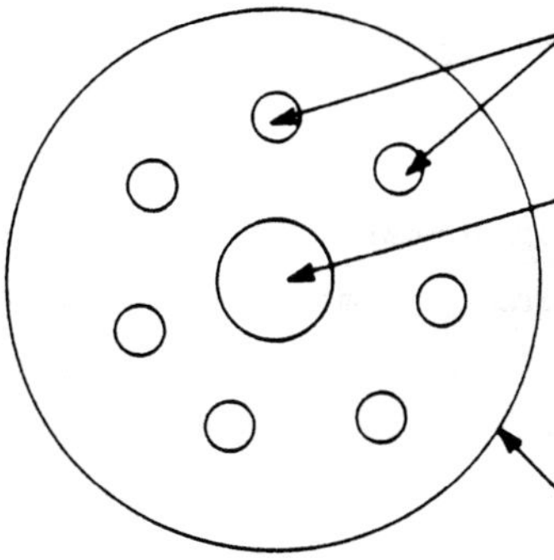

a. Central strength member.

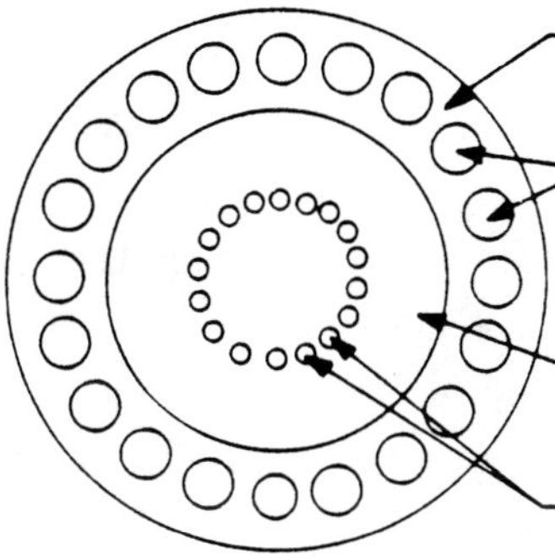

b. Circumferential strength members.

Figure 1-10. Strength member locations.

V. IMPACT OF RADIATION EXPOSURE TO OPTICAL FIBERS

The effects of radiation on optical fibers may be a potential problem in tactical environments during a nuclear scenario or in levels of low dose radiation such as in a nuclear reactor. The most common result of a fiber being exposed to nuclear radiation is a darkening of the fiber core which causes an increase in attenuation. When a fiber is exposed to radiation, the molecular structure of the glass changes. A fiber's response to nuclear radiation is a very complex non-linear function of total dose, dose rate, light level transmitted, and temperature, all of which are related to the core composition. Pure silica cores with a high OH^- content (200-1500 ppm) typically perform better than those with low OH^- ($<$ 5 ppm).[13] If dopants are used in the fiber, as they are with most multimode graded index and single mode fibers, the type and concentration of dopants has a dramatic impact on radiation sensitivity.

These disturbances to the molecular structure caused by nuclear irradiation can result from ionization, migration of electrons, migration of dopants, and the formation of OH^- from free hydrogen. When the structure of the fiber changes, it allows light to be absorbed in the ultraviolet or visible spectrum. This causes bonds of various energy levels to form which causes an increase in attenuation at the wavelength or band of wavelengths where the light was absorbed. The more the fiber is exposed to a particular dose rate, the more the fiber will darken.[13] Darkening is usually less severe at longer wavelengths.

Once a fiber is darkened due to radiation, it may return to its initial state, or recover, to some degree after irradiation stops. There have been many studies done on the length of time required for recovery. During recovery, the molecular structure of the glass rearranges, therefore, the darkness gradually disappears. The length of time for recovery to occur varies from microseconds to days to infinity depending on the composition.[13] Researchers have discovered that the recovery time may be accelerated by a process called thermal photobleaching which is the process of transmitting light through the fiber in ambient temperatures greater than 38°C. The higher temperatures cause a more rapid rearrangement of the molecular structure of the fiber.

The fiber and cable can experience mechanical changes as well as optical changes due to radiation exposure. Fibers and cables can experience thermal and mechanical shock when they are exposed to a high total dose ($\leq 10^6$ RAD) or to very high dose rates ($>10^6$ RAD/sec).[13] A typical mechanical change is the altering of the chemical bonds in the glass. Increased cross-linking can occur at high total dose levels which cause changes to the polymer properties and degrades the mechanical performance of the buffer and cable jackets.

There is still much research being performed on the effects of radiation on optical fibers. Results from preliminary studies vary widely and this makes it difficult to draw conclusions. One report indicates that the best fiber available for resistance to radiation at both 850nm and 1300nm is a low OH borofluorosilicate clad fiber.[14] Germanium-doped fibers perform well at long wavelengths but are not radiation resistant at the short wavelength. Conversely high OH fluorosilicate fibers have excellent radiation resistance at short wavelengths but not at long wavelengths. Plastic Clad Silica (PCS) fibers are used extensively in medical applications where the fibers are cleaned prior to use in surgery by irradiating them.[14] PCS fibers have fast recovery times and resist darkening when exposed to radiation. A relatively new fiber, Hard Clad Silica (HCS), is proving to be rather promising for use in radiation environments. The core of this fiber is pure silica with a high OH^- content which has a bonded cladding.

Fiber manufacturers can improve the radiation resistance of a fiber by pre-irradiating the fiber. Generally, this increases the loss of the fiber initially but improves the performance after future exposure to radiation. When the fiber is pre-irradiated, the areas most prone to darkening will react by changing their molecular structure. This results in a more stable and less reactive configuration.

There are no published standardized test procedures to date that will allow laboratories to test products with any degree of assurance. Data taken by various laboratories differ widely and there are many voids in the data.

Work is underway within the North Atlantic Treaty Organization (NATO) to develop a standard test procedure to measure the effects of fiber to nuclear radiation. Until such time that a standard exists, users of cables that must operate in radiation environments must make decisions based on the most recent literature published on the subject.

VI. MANUFACTURING TECHNIQUES

1. FIBER

Fibers are made by first preparing a glass blank or preform. The primary method in which preforms are prepared use variations of a doped deposited silica process. These methods include:

a. Outside Vapor Deposition (OVD)
b. Vapor Axial Deposition (VAD)
c. Modified Chemical Vapor Deposition (MCVD)
d. Plasma Chemical Vapor Deposition (PCVD)

In each of these processes, the raw materials are heated to a very high temperature (1150° - 1500°C). The soot particles from the flames are used to form the preform. The various processes develop the preform in different fashions but they all result in a fiber blank or preform. Before the OVD and VAD preforms can be used they must be consolidated or sintered. This is performed by dehydrating the preform. The MCVD and PCVD do not require sintering, but the preform must be collapsed at high temperature to eliminate the cylindrical void at its center before the fiber can be drawn.[15]

Once the preform has been developed, the fiber is drawn from the preform. The elements of a typical drawing tower are shown in Fig. 1-11. The preform is placed in a furnace where the tip is heated to 2100° - 2200°C. A capstan pulling apparatus is used to pull or draw the fiber into a specified diameter. While the fiber is drawn, the specified geometric relationship between the core and cladding is maintained. The diameter of the fiber is monitored by a laser gauge during the drawing process.

The fiber drawing process impacts the optical characteristics, dimensions and strength of the fiber. The optical properties are affected by the atomic absorption in the glass structure; contaminants that become trapped in the fiber during the draw process; density fluctuation of the glass; and microbending loss due to improper coating design and application.[15] Cleanliness during the drawing process is very important.

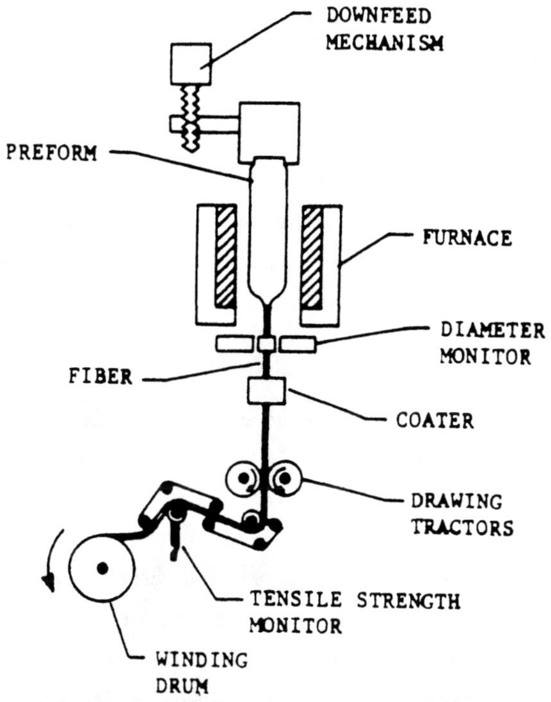

Figure 1-11. Schematic of a fiber drawing system.[15]

2. CABLE CONSTRUCTION

Cables are made to contain optical fibers for use in often harsh environments. The principal objectives of fiber optic cable design is fiber survivability and retention of fiber transmission characteristics. In order to meet these objectives, the cable must be well designed.

Glass fiber has very little elongation before it breaks. The typical value is 1%, whereas copper can be elongated more than 20% before it breaks.[16] This places constraints on the cable manufacturer who must insure that the fibers are not elongated during construction of the cable. The fibers also must not be under any residual tension after they are cabled. If the fibers are under stress, any microcracks in the fiber may elongate. The optimum situation is for the fiber to be under zero stress. However, it is satisfactory for the stress to be below 33% of the rated proof tested tensile strength of the fiber.[17] Under these conditions, any microcracks will propagate so slowly that the fiber will most likely never be subject to static fatigue which is the severing of a fiber due to crack growth caused by excess stress. However, the fiber can still fail due to outside mechanical and environmental forces that the cable is subjected to even if the cable is not under any residual tension.

In order to maintain the transmission characteristics of the fiber, during cable construction the fiber axis must not be unduly bent which could cause microbends or macrobends. The stranding process of the cable must insure that no added loss results from bending the fiber. This is accomplished by process controls of the manufacturer. The cable must also be designed so that any thermal or mechanically induced forces generated during the cable's lifetime do not cause significant transmission degradation. This is accomplished by designing the cable for an intended environment and specifying to the purchaser the conditions that the cable can withstand. Or conversely, the customer can specify the requirements needed for a specific application and the manufacturer then designs a product to meet the customer's needs.

Fibers cannot be the primary load bearing elements of the cable whereas copper can bear the primary load. The entire fiber length must be placed inside the cable but the length of the fiber is greater than the length of the cable, this is called fiber overfill or overlength. The coefficient of elongation for the cable is calculated at different tensile strengths and then trigonometric calculations are performed to insure that the average position of the fiber is in the center of the loose tube.[18] For these loads, if the loose tube inside diameter is too narrow, the fiber will not be able to "float". Thus the inside diameter of the loose tube must be adjusted for each cable design dependent upon the elongation properties of the cable. Therefore, when the cable is under an axial load and elongates, the fiber is not strained, its geometry changes.

The fibers must be near the cable central axis or have provisions to allow them to move to positions of reduced stress when the cable is bent. If all fibers can't be near the central axis, the fibers furthest from that axis can be subject to significant strains when the cable is bent. In order to prevent this, the fibers must be stranded around the central axis of the cable. The fibers are generally helically laid around the axis. The lay length must be determined by a trade-off between choosing a length short enough to provide protection against bending strains and long enough to prevent fiber breaks due to static fatigue.

The load bearing materials of the cable and the cable jacket material must be carefully selected. The materials used will affect the cable's weight, strength, frictional characteristics and thermal expansion characteristics. The most common tensile strength member is Kevlar^{R}, with epoxy impregnated glass, steel and graphite also used. All of these materials have a high tensile modulus. The Kevlar^{R} provides a great deal of radial compressive strength if the yarns are mechanically coupled to the rest of the cable structure. This is usually accomplished by embedding the Kevlar^{R} in the jacketing material. The additional compressive strength provided prevents strain due to polymer shrinkback of the cable jacket and also limits compressive strain due to thermally induced compression.

VII. DESIGN TESTS

1. FIBER TESTS

Fiber can be affected by environmental and mechanical conditions which may affect the attenuation, strength and resistance to losses caused by microbends. Environmental tests are run on fiber by the manufacturers to predict performance. Optical properties of the fiber are measured before and after the environmental stress testing is performed and a comparison between the two readings is made. The tests examine the effectiveness of the coating material for if it is not adequate, the strength and/or microbend resistance of the fiber will be degraded. This will become apparent during the environmental testing.

Common tests that are run on optical fiber are listed in Table 1-3. The primary mechanical test that is run on optical fiber is the tensile strength test in which a tensile load is applied to the fiber to cause flaws to fail. All fiber is proof tested by the manufacturer to a specified value prior to cabling so that existing flaws larger than a given size are forced to fail. A typical proof test value is 50 kpsi. For special applications where greater strength is required, fibers have been proof-tested up to 400 kpsi. Proof testing does not weaken flaws that do not fail at the applied tensile load.

TABLE 1-3. Typical Fiber Tests.

Test	Condition
Acidic and Basic Water Soaks	pH1, pH7, pH12 salt & fresh water
Temperature/Humidity Cycling	-10 °C @ 4% RH* to 65 °C @ 98% RH* 24 hour cycles for 30,60 or 90 days
Fungus	ASTM G21
Flammability/Toxicity	IEEE-STD-383
Extended Heat	65 °C for 87 days 125 °C for 91 days 200 °C for 99 days 300 °C for 0.0035 days
Abrasion	FOTP-66
Tensile Strength	Proof Tester 50 kpsi - 400 kpsi

* Relative Humidity

Ref. 19

2. CABLE TESTS

Tests are run on the optical fiber cable to ensure that it will withstand the anticipated mechanical and environmental conditions while maintaining the specified optical parameters. The manufacturer goes through a qualification procedure to initially qualify the cable. This procedure requires extensive testing on many samples and is very lengthy and costly. Once the manufacturer has qualified his product, the remaining lots are tested on a periodic basis for mechanical and environmental performance (see Table 1-4). The optical performance of the fibers is tested on 100% of the cables (see Table 1-5).

TABLE 1-4. Typical Mechanical and Environmental Cable Tests.

Tests	Procedure
Temperature Cycling	FOTP-52
Thermal Shock	FOTP-160 *
Humidity	REA PE-90
Tensile Strength	FOTP-33
Fluid Immersion	FOTP-40
Flexibility	FOTP-104
Crush Resistance	FOTP-41 DoD-STD-678/2040
Twist-Bend	FOTP-91
Impact	FOTP-25
Ultraviolet Resistance	FOTP-97
Flammability	IEEE-STD-383
Shock	FOTP-71 *
Vibration	GTE GTS 8542
Radiation Resistance	FOTP-49
Smoke Index	NES-711
Toxicity Index	NES-713
Compound Drip	FOTP-81
Hot/Cold Bend	FOTP-37
Water Penetration	FOTP-82
Jacket Tensile and Elongation	ASTM D638, D2633
Environmental Stress Crack (PE Jackets)	ASTM D1693

* FOTP in development

Ref: 18, 20, 21

TABLE 1-5. Typical Optical Performance Cable Tests.

Tests	Procedure
Attenuation (cutback method)	
Single mode	FOTP-78
Multimode	FOTP-46
Bandwidth	FOTP-30, 54
Numerical Aperture	FOTP-47
Core Size	FOTP-58
Mode Field Diameter	FOTP-166
Cutoff Wavelength	FOTP-80

Ref: 22

VIII. COMMON FAILURE MECHANISMS AND THEIR CAUSES

Optical fibers are rarely used in their pure state except for laboratory and testing conditions. Most optical fibers are assembled into an optical cable. During the cabling process, the fiber is subjected to stresses which can result in failures during cabling or premature operational failures. The most prevalent failure mechanism in optical fiber is fracture of the fiber due to stress corrosion or fatigue which is the propagation of existing microcracks while under stress.[19] If the residual or threshold tension is less than 33% of the rated proof tested tensile strength, the fiber should last 10 - 20 years. This lifetime range assumes the cable does not experience any other mechanical failures which would degrade the fiber strength.[19] If the cable sheath is damaged so that fluids are able to enter the cable, the fiber lifetime will be significantly reduced. One fiber manufacturer states that the fiber should last forever provided the residual strain is as stated above.[17] Generally, the cables are designed so that the fiber will not experience undue stress, thus the likelihood that the fiber will fail due to residual stress is minimal. Other flaws may exist in the surface of the fiber in the form of chemically or mechanically induced defects. Whether or not these flaws will fail during installation or during the cable service life is dependent on the residual stress the fiber is under, the length of the fiber and the duration of stress. When moisture is present in the cable, surface defects will propagate in the fiber.[23] This reduces the level of stress that can be tolerated before fractures may occur. Design the cable to prohibit moisture from contacting the fiber and to allow the fiber to be in a totally relaxed state (no residual stress) before and after installation.

Attenuation of the optical fiber can be affected in the cabling process due to mechanical stress placed on the fibers. Excessive attenuation can also be caused by hydrogen diffusion into the core of the fiber. This situation is most evident at wavelengths greater than 1050 nm and when materials used in the manufacturing of the cable react and outgas hydrogen. Also, the cabling process can introduce microbends into the fiber causing significant losses.[18] The coating on the fiber is designed to minimize these microbend induced losses but if the coating is damaged or cracked, the bare fiber is exposed which will degrade the strength of the fiber and make it highly susceptible to further microbends and moisture penetration.[24]

Stress, attenuation, and microbending can all cause premature failure of the fiber. Here, failure is defined as the fiber degrading to the point that system performance is compromised. The failure can be a gradual increase in attenuation until the power margin is exceeded or the fiber could break causing a catastrophic failure in that segment of the system. Failures with the cable are primarily associated with the installation of the cable and the integrity of its jacket. During installation, the cable is subject to damage by friction, improperly operated burial equipment, excessive tension or bend radius limitations being exceeded. The cable jacket can be physically damaged by excessive tension therefore reducing the protection provided.[25,26] Manufacturers specify a maximum installation tension and a maximum residual tension for their cables. A typical value for maximum installation tension is 600 pounds, whereas, 100 pounds is a typical value for residual tension. If the residual tension level is exceeded, the jacketing materials will be under stress which can cause the jacket to weaken and eventually tear or separate.

Cable jackets are generally very strong and no reports of tearing of jackets have been made in reports published by two major users of optical cable.

Environmental factors such as excessively low or high temperatures, cycling temperatures, ultraviolet exposure and water or fluid immersion can affect the cable jacket integrity. Design the jacket to prevent its shrinking, cracking, swelling, or splitting during long term exposure to such harsh conditions. If the jacket becomes brittle or cracked, moisture will enter the cable and eventually increase attenuation. If the jacket shrinks or swells, it will cause mechanical problems at the connector or splice and cable interface. This could produce high losses or complete failure due to microbends or broken fiber which would affect overall system reliability.

Another environmental problem that exists with external cables is rodent or shark/fish bites. For cables that are going to be buried directly underground or under water, there must be adequate armored protection to allow the cable to withstand these bites. There have been numerous studies performed which have measured the force that rodents (groundhogs and gophers primarily) exert on cables. A cable needs to be a minimum of 0.75" in diameter with two layers of armoring to resist destruction due to rodents. Submarine cables are subject to shark and fish bites and are generally very heavily armored. Research is in progress to determine the force and the time it takes sharks to damage cable. Preliminary studies show that if the cable can withstand the first 72-96 hours, the sharks/fish will cease attacking Researchers attribute this phenomena to shark/fish protecting their environment and when the cable doesn't attack them, it is considered to be a friend rather than a foe.

There have been few published papers concerning reliability data of installed fiber optic cables. Table 1-6 details one installer's data on types of failures and the average restoration time for each type of failure.[27] The failure rate has been calculated by a major telecommunications company to be one incident per year per 146 sheath-km of cable. This number is comparable to coaxial transmission cable.[27] If construction practices could be improved, a reduction in failures would be realized. Improvements would require that contractors working in the area where a fiber optic cable is installed be informed of the exact location of the cable and given a set of plans which depict the depth and location of the cable. The National Underground Contractors Association (NUCA) has requirements for marking the area and providing plans/drawings for the local and state governments for electrical contractors. Plans are in the development stage within the Electronics Industry Association to implement the same type of requirements for the fiber optic industry.

TABLE 1-6. Fiber Optic Cable Failure Types (Dec 84 - Jan 86).

Cause of Failure	Number of Failures	Average Restoration Time (minutes)
Parallel Construction	6	518
Contractor Activity	1	450
Backhoe, Digging	3	567
Gunshot	2	260
Hurricane	3	489
Flood Damage	1	10,080
Ice Crush	1	840
Vandalism	1	53
TOTAL	18	

Failure rate is one incident per year per 146 sheath-km of fiber optic cable.

Ref. 27

In another study, three types of failures were defined: non-relevant, relevant nonchargeable, and relevant chargeable. A relevant nonchargeable failure is a failure which is caused by an uncontrollable incident such as accidental dig-up or the breaking of a supporting pole. Non-relevant failures are those failures which are a result of improper installation. Relevant chargeable failures are fiber failures that occurred while the cable was handled, installed and operated within specifications. These handling and installation problems only surface when a failure occurs. They can be minimized by training and making sure the proper procedures are followed. This also applies to repair procedures.

Studies on the reliability of installed cables have revealed that 83 incidents of failure were observed in 5612 sheath-km of installed cable.[27,28,29] Of these 83 incidents, only one was a relevant chargeable failure and that was a factory splice point made during the manufacture of the cable to produce a longer continuous cable. The vast majority of the failures were due to parallel construction; backhoe cuts, digging, other activity, rodents, gunshots and vandalism (see Fig. 1-12).[27,29] The causes of failure due to weather included hurricane, flood damage and ice crush.

The authors have calculated reliability parameters for this cable data. In making these calculations, it was assumed that the distribution is continuous and random. This allows the use of the Chi Square distribution to determine the confidence limits on the Mean Time Between Failure (MTBF). The two-sided truncated time Chi Square equation given in equation 1-1 was used to calculate the 90% confidence interval. The source data and results of the calculations are included in Table 1-7.

$$\text{MTBF Confidence Limits} = \left(\frac{2T}{\chi^2\left(\frac{\alpha}{2}, 2r+2\right)}, \frac{2T}{\chi^2\left(1-\frac{\alpha}{2}, 2r\right)} \right) \qquad \text{Equation 1-1}$$

T = sheath-km-hours = 27,731,263
α = risk of error = 0.1
r = number of failures
χ = Chi Square distribution

TABLE 1-7. Source Data and Calculated Reliability Parameters For Fiber Optic Cable.

SOURCE DATA

Installed Cable

Sheath-km	5,612
Sheath-km Hours (Est)	27,731,263

Number of Incidents

Relevant Non Chargeable		69
Non-relevant		13
Relevant Chargeable		1
	TOTAL	83

CALCULATED DATA

Reliability Calculations For 69 Relevant Non Chargeable Failures

Calculated MTBF (10^6 hours)	between 0.622 and 1.096
Calculated Failure Rate ($/10^6$ hours)	between 0.912 and 1.607

Reliability Calculations for 13 Non-Relevant Failures

Calculated MTBF (10^6 hours)	between 1.34 and 3.61
Calculated Failure Rate ($/10^6$ hours)	between 0.277 and 0.745

Reliability Calculations For 1 Relevant Chargeable Failure

Calculated MTBF (10^6) hours	between 5.85 and 541
Calculated Failure Rate ($/10^6$ hours)	between 0.0019 and 0.171

Reliability Calculations For All 83 Failures

Calculated MTBF (10^6 hours)	between 0.53 and 0.884
Calculated Failure Rate ($/10^6$ hours)	between 1.13 and 1.89

Ref. 27, 28, 29

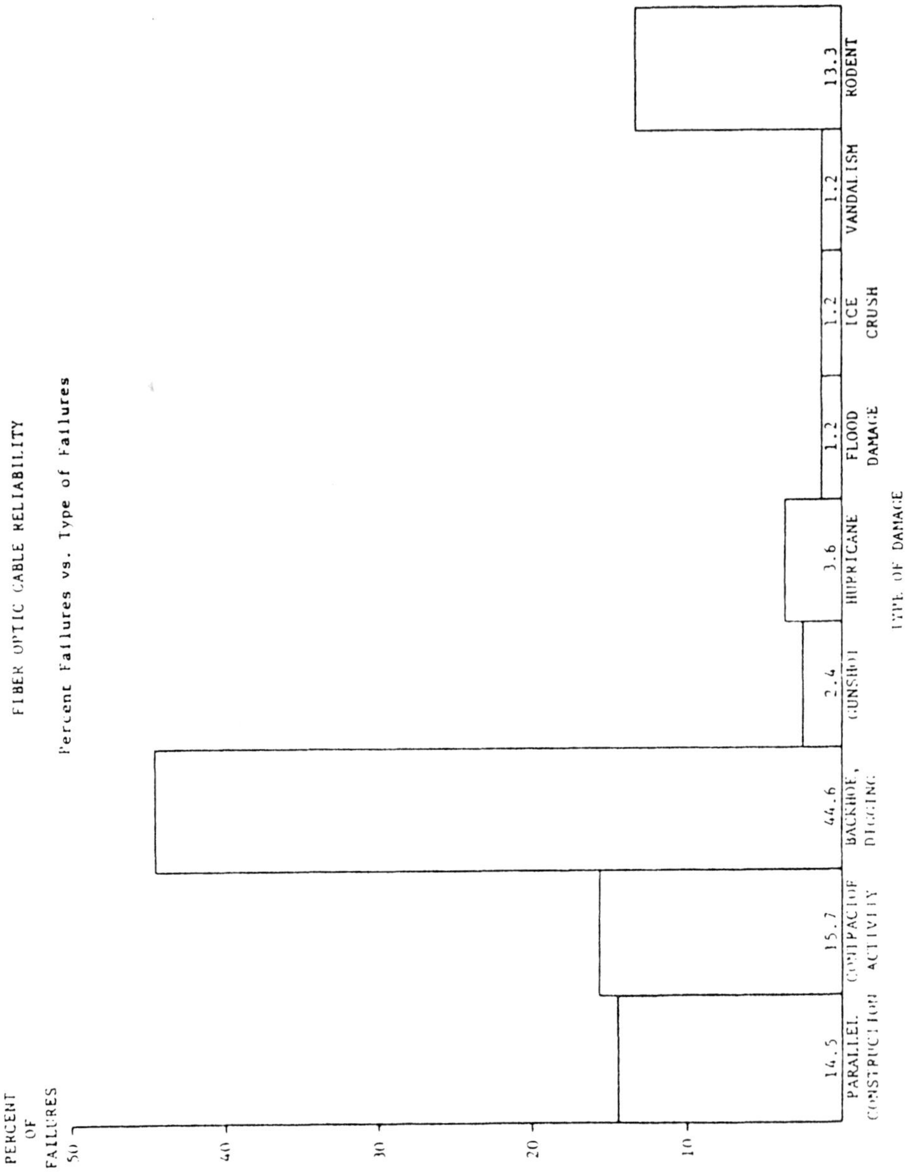

Figure 1-12. Histogram for the distribution of causes of fiber optic cable failures.

The reliability calculations were performed individually for the failures of each category of failure and then as a cumulation of the total number of failures. The MTBF and failure rate was calculated for each group of failures. Performing calculations in this manner provides a measure of the cable reliability for each cause of failure individually.

IX. CRITICAL DESIGN CRITERIA AND APPROACHES TO MINIMIZE COMMON FAILURES

Fiber design is critical to the reliability of optical systems for it must be able to withstand the mechanical and environmental conditions in which it will be operated and stored. The fiber must be proof-tested at the proper level and the fiber coating must be sufficient to preclude microbending losses. The material used in the construction of the fiber should not cause excessive attenuation.

Fiber has to be stored in a temperature and humidity controlled environment prior to cabling. The typical storage environment for optical fiber is -60°C to +55°C at less than 98% RH and the typical rated operating temperature is -60°C to +85°C. With the necessary precautions taken, fiber is expected to have a 30 year operating lifetime.[19]

Cable storage and operating temperatures from several manufacturers are listed below:

Manufacturer	Storage °C	Operating °C
A	-40 +80	-40 +80
B	-50 +70	-40 +70
C	-40 +70	-40 +70
D	-46 +71	-46 +71

Military Fiber Optic Cable Specification, DoD-C-85045, requires a storage temperature range of -55°C to +85°C and an operating temperature range of -46°C to +71°C. Generally, storage temperatures cover a broader range than operating temperatures. Fiber optic cable manufacturers have experienced problems with the low end temperatures due to brittleness of the cable jacket materials and shrinking of the jacket. At high temperatures, the jacket material elongates and the filling compounds tend to run. The fiber itself can withstand a broader range of temperatures.

Most cables have a life expectancy of 10 to 20 years.[18] The cables on the lower end of the lifetime scale tend to be made for tactical applications and those at the high end of the scale are for telecommunications applications. Military cables are generally required to be a tactical cable which requires a tight tube construction. These cables do not allow the fiber to float within the cable and are more prone to suffer from increased attenuation due to microbends and macrobends.

During design of the cable, attention to cable trigonometric constraints, such as fiber overfill or overlength, is vital.[18] Cabling process parameters must be closely monitored and measurements must be made on the cable after completion to insure that correct fiber overfill is achieved. The fiber parameters generally do not change during cabling unless the cabling process is not tightly controlled. The fiber length generally is longer than the cable length due to the fact that the fiber is typically stranded in a helical fashion around the central members. Each cable construction type is a custom product with very different requirements, thus it is difficult to make generalizations about the parameters which impact the optical performance of the completed optical cable.[18] Some of the parameters that usually vary with the application are:

a. tube size
b. number of tubes per cable
c. fibers per tube
d. fiber type and packaging method (single mode, multimode, index profile, loose tube, tight tube)
e. cable type (dielectric, non-dielectric, armored, plenum, aerial, submarine)
f. attenuation levels (long-haul, local distribution)
g. bandwidth

These parameters can be combined to design a multitude of cable types with each type having different operating parameters.

To avoid failures of the fiber within the cable, proper installation techniques must be practiced. This includes monitoring the tension placed on the cable to ensure the manufacturer's specified tension level is not exceeded and providing enough slack cable at manholes, poles and termination points such that the tension in the cable and fibers can decrease to a level below the long-term tensile rating. To minimize other non chargeable failures, cable must be designed to withstand lightning, rodents, shotgun blasts and severe environmental conditions.

X. MAINTENANCE

Most fiber failures occur due to a non-relevant failure of the cable itself. For this reason, emergency restoration procedures are needed. When a failure occurs, the first step is to determine if the failure is due to the cable and if it is, then the next step is to identify the fault location. One method of doing this is to use an Optical Time Domain Reflectometer (OTDR) to identify the fiber that is not transmitting properly and the exact location of the fault. In some cases, the fiber will break due to tensile forces applied at a point other than the cable break point. When this occurs, it is called an offset break (see Fig. 1-13). Offset breaks generally occur in cables that have the fibers in the center of the cable as opposed to being helically wound around the central strength member and which also have high tensile ratings.[27] In loose tube cables using filling compounds commonly used by the phone companies, no offset breaks have been reported. In cases where the cable sheath has been partially severed and not all of the fibers have been broken, it is possible to repair the fibers without interrupting the operation of the other fibers. For cable breaks where all fibers are broken, the cable must be restored to service rapidly with a temporary repair. This allows permanent restoration to be made to one fiber at a time without disturbing the operation of the temporarily repaired fibers. To provide for rapid restoration, an emergency restoration kit should be available at every installation site. Having an emergency restoration kit on site is imperative in order to prevent unnecessary delays in bringing the system back into operation.

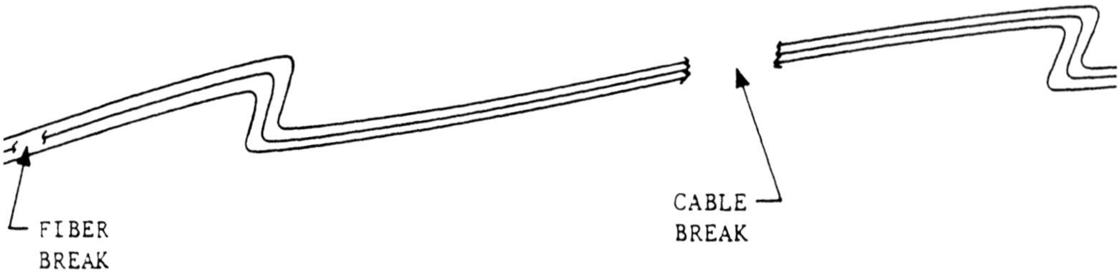

Figure 1-13. Offset fiber break.

Experience has proven that being prepared saves hours of downtime as was discussed in an article by Szentesi.[27] In one case, an eight-fiber cable was back in service in 5 hours when an emergency repair kit was available. In another case, a six-fiber cable required over 24 hours to restore, the primary delay was locating proper equipment. In other studies[27], it has been shown that fiber-cable troubles are a mere 10% of the total number of failures, yet 88% of the restoration time was spent on fiber/ cable repair. The average time was 2145 minutes (almost 36 hours) per fiber-cable failure. The excessive repair times were attributed to the newness of the technology and therefore to a lack of adequately trained and experienced maintenance personnel. The study also indicated that experience should improve these figures. Another study on single mode systems indicated an average restoration time of 479 minutes (almost 8 hours). This study was done after the previous study with a fourfold improvement in the repair time. The marked improvement is due to an additional 2-3 year experience base that personnel have been exposed to and the fact that the technology is coming of age. This is not to say that technicians require 2-3 years of training and experience before they are competent. Because failures of fiber optic systems do not occur very often, the technicians do not get much field repair experience.

Preventive maintenance is not generally required for an optical cable once installed and often is discouraged to prevent maintenance induced failures. Provided the cable has been installed according to the manufacturer's specifications, including securing the cable(s) to its resting place properly, maintenance on a routine basis is not needed.

XI. LOGISTICS CONSIDERATIONS

Logistics is a primary factor in repair time of fiber/cable systems. For this reason, each installation (aircraft, ship, etc) should use a standard cable/fiber type throughout the installation in order to facilitate installation and maintenance and reduce logistics concerns. Tool kits and repair kits can be standardized to reduce the number of different kits that must be kept in stock. As was mentioned earlier, the expedient repair of fibers/cables is heavily dependent on having the proper equipment readily available. The emergency repair kit contains all of the supplies needed to repair optical fibers and the cable sheath. Depending upon the type of failure, an emergency repair may be temporary to restore service quickly or there may be sufficient time to initially perform a permanent repair. Adequate lead time should be allowed when procuring items to assemble an emergency repair kit due to the equipment involved. This is important when the kit is being assembled piece meal. See Table 1-8 for a list of equipment. The materials must be compatible with the fiber and cable; e.g., proper fiber core/cladding size and attenuation characteristics, and materials that will bond to the existing cable jacket. Cable manufacturers generally have these kits available and the materials are compatible with their cables. Personnel must be trained to use the equipment within the emergency repair kit including power meters and OTDRs and also to interpret their results. Troubleshooting skills must be emphasized and the training should be given by a recognized training organization within the fiber optics industry. Initial courses for technicians are generally one week long with refresher courses lasting 1-3 days.

TABLE 1-8. Emergency Repair Kit Contents (Typical).

1. Ruggedized Optical Time Domain Reflectometer
2. Ruggedized Optical Power Meter
3. 30 meter preconnectorized cable (Same cable type and fiber content as installed at site.)
4. Bare fiber connectors
5. Elastomeric splices
6. Fiber cutter
7. Cable stripping tools
8. Fiber stripping materials
9. Splice tray or patch panel

XII. SUMMARY

Reliability of installed optical cable in operating non-tactical systems, such as telephone installations, has proven to be very good. Cable problems represent approximately 10% of the reported failures for entire systems and of these failures, 95% are not due to design nor workmanship flaws but are due to unrelated construction and severe weather influences.[27] The rate of cable cuts is comparable to that of electrical cable. Preventive maintenance of installed cable is not needed nor recommended. Depending on the criticality of the system, repair procedures must allow for an immediate, temporary restoration of the link followed by a permanent repair. An emergency repair kit should be at every site which utilizes fiber optics and the kit must contain the proper tools and materials to fix the type of fiber/cable that is at the site. This repair kit is critical, as it significantly affects the overall downtime of a system.

Fiber optics should be considered whenever the benefits of fiber systems is needed to enhance the transmission quality or capacity of the system. If the benefits of fiber outweigh the other choices; copper, microwave, or satellite, the system designer should not hesitate to choose fiber optics for the transmission medium due to reliability considerations. Optical fiber, if cabled and installed properly, will last for years without any preventive maintenance. Reliability of optical cable is very good, designers of systems should seriously consider using fiber optic technology when it is appropriate for the given application.

Chapter 2. Splices and Connectors

Splices and connectors are used to interconnect optical fibers. Splices are designed to be used where mating and demating is not anticipated though some designs do allow mating and demating. Connectors are designed for repeated matings and dematings.

In most fiber optic systems, connectors are used at the end equipment and splices are used to join links of cable together. Fusion splices or mechanical splices may be used, although fusion splices seem to be used more often. During emergency restoration, mechanical splices are generally used as a temporary fix then at a later date permanently replaced with a fusion splice.

I. OPTICAL LOSS MECHANISMS

The overall insertion loss for a splice or connector consists of both intrinsic and extrinsic losses. Intrinsic losses are dependent upon fiber parameters, and can't be reduced by splice or connector design. Most of the intrinsic losses that occur result from differences in core diameters, and numerical aperture (NA) as shown in Fig. 2-1. These losses occur only when a signal is propagating from a larger core, larger numerical aperture to a smaller one of these properties.

For single mode and multimode fibers, if light is coupled from a fiber with a larger core to one with a smaller core, optical power will be attenuated. However, light coupled in the other direction will not experience intrinsic attenuation. Light propagating from a fiber whose NA is larger to one whose NA is small will experience a power loss. Light propagating in the opposite direction will not experience attenuation. If the core size and the NA are equal, no loss attributable to intrinsic losses will occur.

The design of mechanical splices and connectors may not compensate for intrinsic losses, but their design does impact losses caused by extrinsic factors. Extrinsic losses are associated with the accuracy with which the

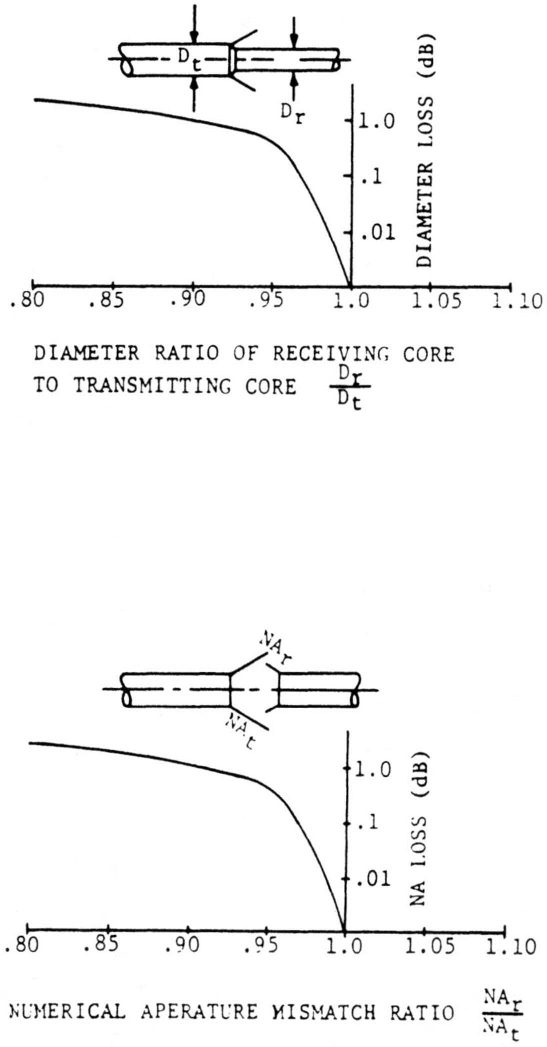

Figure 2-1. Intrinsic losses.[1]

device aligns the fiber cores. Connectors experience these losses regardless of whether they are the cylindrical sleeve/ferrule, the biconic sleeve/conical ferrule or lensed type. Mechanical splices also experience the extrinsic losses that are discussed below. These losses include lateral or axial alignment of the fiber cores, angular alignment of the cores and the size of the gap between fiber ends. Fusion splices generally do not experience extrinsic losses but they do experience loss due to the deformation of the core when it is heated to its softening point.

One of the most troublesome extrinsic losses is caused by lateral misalignment as shown in Fig. 2-2a. To minimize this loss, the fiber ends to be joined must be plane colinear about the aligned core axes.

Another alignment problem that affects optical loss is angular misalignment as shown in Fig. 2-2b. To avoid this, the fiber ends must be perfectly parallel to each other. Angular alignment is much easier to control than lateral alignment during connector and splice design.

Another alignment problem results from the gap that normally exists between the transmitting and receiving fibers of connectors and mechanical splices as shown in Fig. 2-2c. The fiber ends should be separated by a small gap to prevent the fibers from becoming scratched or cracked during mating or demating. The fibers may also come into contact with one another due to shock or vibration in mechanical splices or connectors which may cause damage to the fiber endfaces. The loss associated with this gap is caused by Fresnel reflection which is due to different indices of refraction of the fiber cores and the gap medium. For a glass to air interface, the differences in refractive index causes a Fresnel loss of 0.3dB to 0.4dB and varies with the gap size and relative indices of refraction. Fresnel loss can be reduced if not eliminated by applying an index matching fluid between the fibers. An advantage in using index matching fluids other than reducing Fresnel loss is that any losses caused by imperfectly prepared fiber ends are reduced since the ends are no longer discontinuous, but effectively flow continuously into each other via the fluid. Disadvantages of using index matching fluid includes its ineffectiveness when used with graded index fibers since the refractive index varies with the core radius. Also, it is difficult to use index matching fluid in field terminations due to the increased probability of contamination from dust and dirt. Over time, contamination will accumulate in the fluid after repeated matings and dematings which increases connector or splice loss. Some connectors have built-in reservoirs that when filled apply the index matching fluid automatically which eliminates the need to apply the fluid every time the connector is demated.

The effects of extrinsic losses become more tolerable by choosing a fiber with large core and cladding diameters. A large cladding makes it easier to align the working surfaces of the fibers and a large core will allow greater deviation from tight connector alignment tolerances without inducing additional losses. Also, large core and cladding diameters will make coupling easier and lessen the possibilities that the fiber will suffer from microbending since large core fibers are not as susceptible to microbends whether they are induced by coiling the fiber in a backshell of a connector for storage or by pressure exerted by the piece parts of the connector. However, the use of fiber with large core and cladding diameters has a higher attenuation per unit length and lower bandwidth thus restricting its applications.

Fiber end preparation significantly affects the coupling efficiency of a fiber-to-fiber junction. A smooth, scratchless, flat surface, perpendicular to the core axis will provide the best conditions for coupling light between fibers. There are two methods that are commonly used, the "grind and polish" method and "score and break" method.

The grind and polish method yields a high percentage of quality finished ends but requires a lot of practice to get repeatable results when performed manually. Also, this is a time consuming process which is more difficult to successfully complete in the field although many connector styles require this method of end preparation. The score and break method does not consistently produce acceptable surfaces, but is very quick and easy to perform and learn and is a common method of preparing fiber ends for testing. Medium to high loss values can be achieved with connectors that utilize the score and break method of endface preparation.

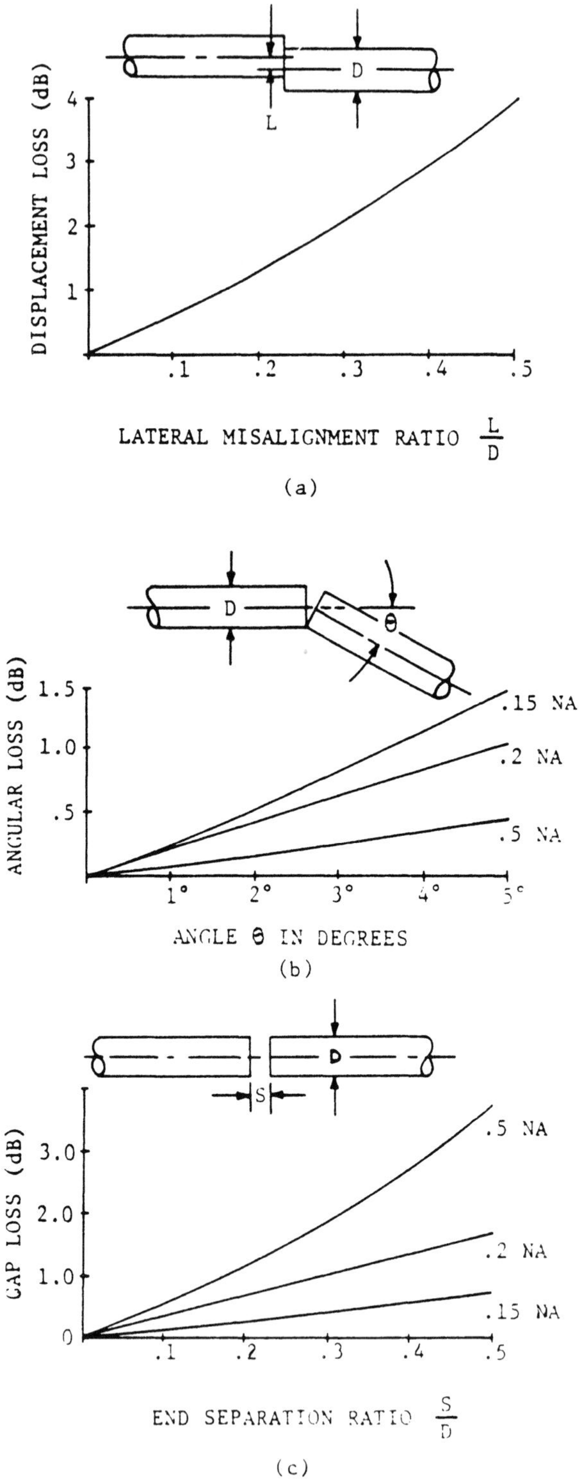

Figure 2-2. Extrinsic losses.[1]

II. DESCRIPTION OF SPLICE TYPES

1. MECHANICAL SPLICES

The splices described in this section are used with single fibers and multi-fiber ribbon cable. The single fiber splices which will be discussed are the glass tube or hollow tube splice, the three rod splice, the elastomeric splice, and the rotary splice. The ribbon cable splices discussed will be the silicon chip array splice and the rapid ribbon splice.

a. Loose tube splice -- In the glass tube or hollow tube splice, cleaved fiber ends are inserted into a capillary tube and butted together. The capillary tube has a square cross section and inside dimension which is greater than the outside diameter of the fiber as shown in Fig. 2-3. The

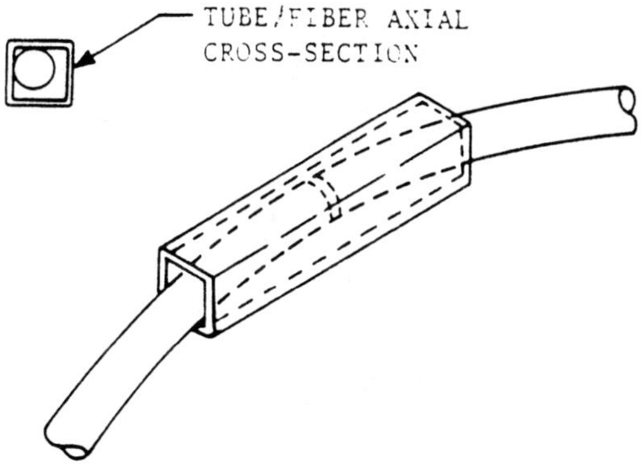

Figure 2-3. Loose tube splice.[2]

fibers are then bent slightly which forces the fiber ends into one of the interior corners of the tube. Index matching epoxy is wicked into the tube and cured forming a permanent splice.

b. Three rod splice -- The three rod splice is similar in principle to the glass tube splice. Three precision rods are bound together in parallel and the interstice they form is used to control the position of the fibers as shown in Fig. 2-4. Index matching epoxy is injected into the interstice, the cleaved fiber ends are inserted and butted together, and the epoxy cured.

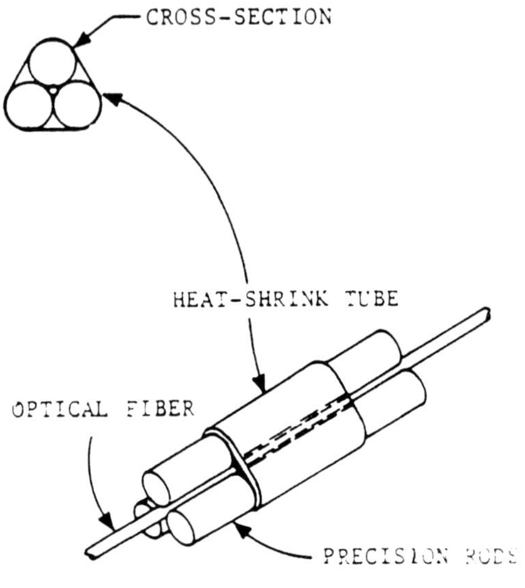

Figure 2-4. Three rod splice.[2]

A common problem shared by the glass tube and three rod splices is the core offset caused by splicing fibers having different outside diameters. To illustrate this problem, consider the result if two fibers of different outside diameters were inserted into the hollow tube splice. Since the tube aligns the fibers with respect to their outside diameters, the cores would not be aligned as shown in Fig. 2-5. This results in a splice attenuation above optimal. A similar problem could occur if the outside diameters were the same but the fibers had poor core/cladding concentricity. Core/cladding concentricity is the degree that the core is geometrically centered in the cladding or how much the core and cladding align axially. Thus, in order to minimize the attenuation, the fibers used with these splices must be compatible. This will impact logistics planning for the system by limiting the useful fiber sizes that need to be kept in stock and will affect the maintainability of the system if compatible fiber is not available when needed. The replacement splice must also be compatible with the spare coiled fiber that is existing in the system unless a new section of cable is installed. In the latter case, the splice must be compatible with the new fiber size. It is important that the proper splice sizes be available should a replacement be required.

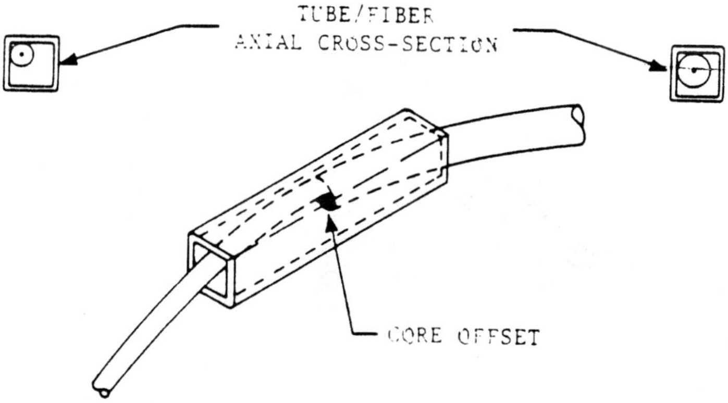

Figure 2-5. Core offset.

c. Elastomeric splice -- The elastomeric splice is molded in three parts to achieve fiber alignment. The two inner parts are elastic and identical except that one half has a 60° V-groove for aligning the fiber while the other half has a flat surface. To hold the inner pieces together, they are inserted into a rigid sleeve. The V-groove is tapered at both ends for easy fiber insertion and its cross-section is smaller than the diameter of the fiber. This causes the walls of the molded parts to elastically deform as the fiber is inserted. Then, the elastic restoring forces automatically center the fiber's outside diameter with the center of the V-groove as illustrated in Fig. 2-6. This provides alignment for fibers with different outside diameters but does not reduce losses resulting from poor core/cladding concentricity.

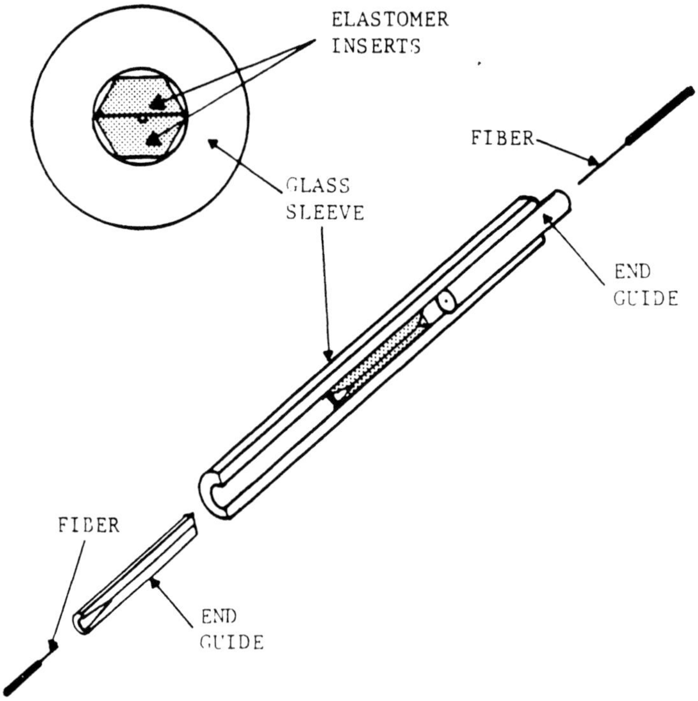

Figure 2-6. Elastomeric splice.[2]

d. Rotary splice -- The rotary splice consists of two precision drilled glass ferrules whose cores have a built-in eccentricity. The cleaved ends of the fiber are cemented into the ferrules which are then inserted into an alignment sleeve which is designed to allow them to be independently rotated. Rotating one ferrule with respect to the other causes the fiber cores to come into close axial alignment as shown in Fig. 2-7. The amount of rotation needed is determined by monitoring the fiber's output with an Optical Time Domain Reflectometer or an Optical Power Meter in order to obtain the minimum loss. This splice practically eliminates problems resulting from core/cladding concentricity and differences in fiber outside diameters.

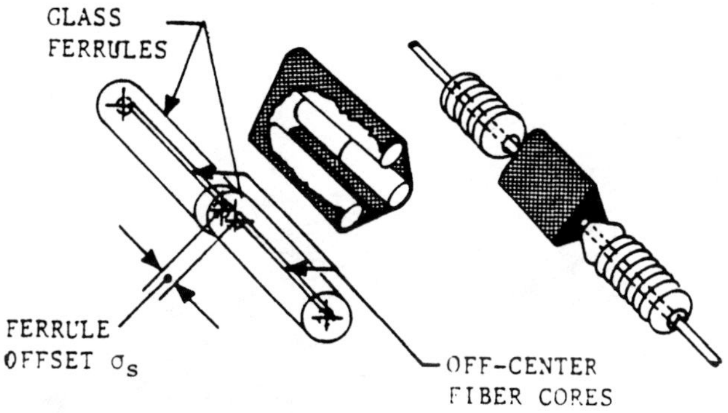

Figure 2-7. Rotary splice.[3]

e. Silicon chip array -- The silicon chip array splice is used with ribbon cable. In this type of splice, the fibers (which have had their ends stripped simultaneously by a special stripper) are laid in V-grooves which have been etched into a securing silicon chip. A chip with identical grooves is laid over the fiber ends and epoxied to the first chip. The ends protrude from the chips. This same procedure is performed on the mating fibers. The fiber ends are then simultaneously ground and polished while encased in the chips. The chips are then butted against each other and grooves on the top and bottom of the assemblies are forced into alignment by alignment chips which straddle the interface as shown in Fig. 2-8. The entire assembly is then clasped together with a metal clip.

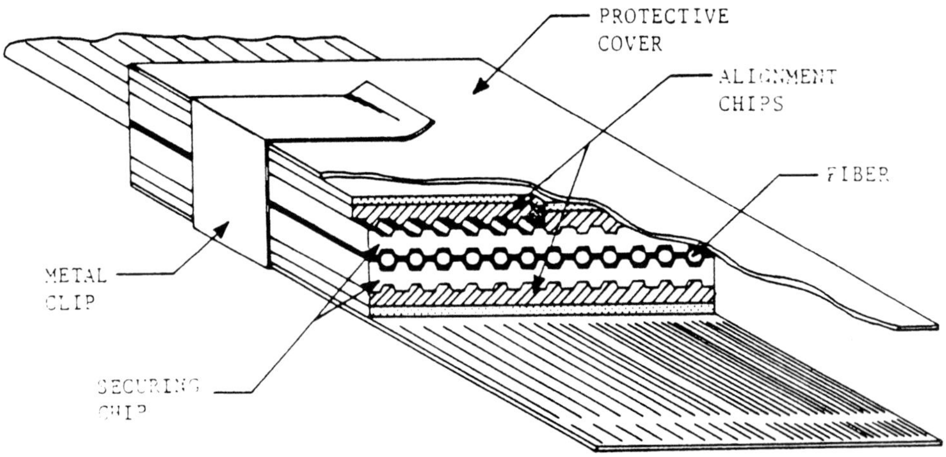

Figure 2-8. Silicon chip array splice.[2]

f. Rapid ribbon splice -- The rapid ribbon splice shown in Fig. 2-9 differs from the silicon array splice in that no epoxy is used during assembly. Instead, the fibers in the ribbon cable are stripped, polished, and ground simultaneously using special tools and fixtures. The fibers are placed in a tray which has grooves etched into it and aligned using a vacuum chuck assembly and microscope. A cover plate with identical grooves is then placed over the first plate and clips are applied to hold the assembly in place. Strain relief clamps are clipped into place and index matching fluid is injected into the splice. A suitable cover is placed over the hole in which the index matching fluid is injected to prevent contamination.

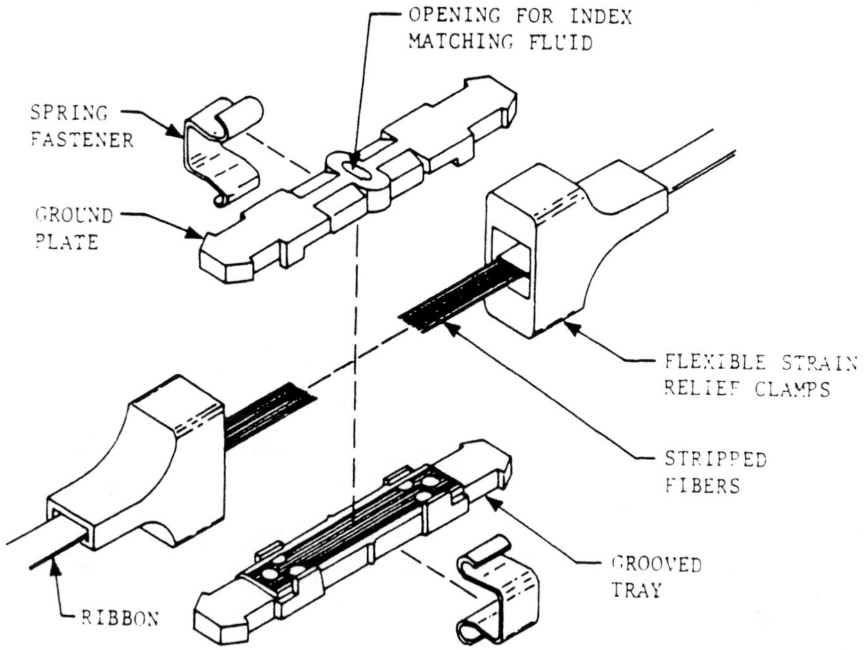

Figure 2-9. Rapid ribbon splice.[3]

2. FUSION SPLICES

Fusion splicing is done with a relatively expensive and sophisticated fusion splicing machine. The fiber ends are butted together and simultaneously melted with a brief electric arc. This fuses the fibers together which retains their alignment. Before the fiber ends are fused together, a protective sleeve is loosely placed over one fiber. The sleeve is slipped over the splice after fusing and is held in place with an adhesive or epoxy. Fusion splicing aligns the fibers with respect to their core diameters resulting in very low splice losses, typically less than 0.1dB.

There are two general methods of making a fusion splice: Profile Alignment System (PAS) and Local Injection and Detection (LID). The PAS method aligns the cores rather than aligning the outside diameter as most mechanical splices and connectors do which results in an ultra-low loss (0.05dB). This method was developed primarily for single mode fibers where a large loss can result from a minor offset between the cores of the fibers. Collimated light from an illumination lamp is shined onto the stripped fibers from a radial direction. This light enters the fiber and is refracted by the fiber core and cladding through an angle dependent on the change in refractive index between the core and cladding (see Fig. 2-10). A camera located along the "x" axis picks up the image of the light in the fibers. A computer analyzes the image, locates the core centerline of each fiber and then automatically aligns their axes using micro positioners. The computer then receives the image from a "y" axis camera and again aligns the cores, thus providing core alignment in both the "x" and "y" axes. Once aligned, the fibers are ready to be fused. PAS can be used with graded index multimode fiber but the centerline of the <u>fiber</u> rather than the centerline of the core is aligned since the core image is not distinct. This is due to the fact that graded index cores do not have clearly defined index of refraction profiles with respect to the cladding. Multimode splice losses average 0.05dB.

Local injection and detection (LID) systems eliminate the need for remote monitoring of the splice loss. LID systems use the principle of microbending to inject light into the core of the fiber through the cladding and buffer material. The light is detected also using the microbending

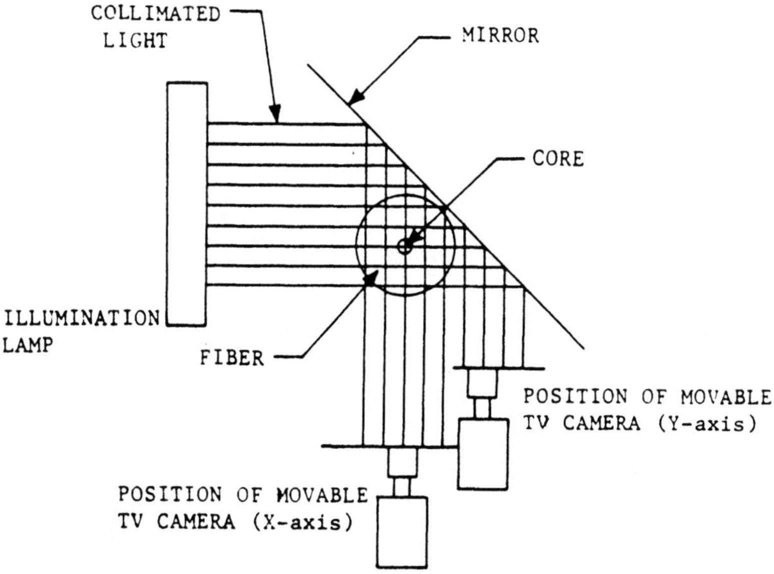

Figure 2-10. Profile alignment system.[4]

principle after passing through the splice point. When the detected light is maximized by adjusting the position of the fibers, the splice is ready to be fused. There have been problems with injecting sufficient light into fiber which has colored buffers. The opposite problem also occurs where the fiber cores may be so large that enough light is detected for the alignment indicator to give a good alignment reading before the fibers are actually aligned. This has occurred when multimode fiber is being spliced with a machine designed only for single mode fiber.

Since the fibers have to be bent into a small radius to inject and detect the light, opponents of the LID system claim that this can introduce microbends into the fiber which result in surface cracks and premature failure due to stress corrosion of the fiber. Machines are made that can be used with both multimode and single mode fibers and aligns the fibers in both the "x" and "y" axes. Once a fusion splice is completed by the PAS or LID method, it is almost impervious to variations in temperature and humidity and is extremely stable provided the sleeve and epoxy can withstand the environmental conditions. Fusion splices are also physically the smallest of the splices mentioned in this discussion.

III. DESCRIPTION OF CONNECTOR TYPES

There are many different connector types which are used within the industry for various applications. Although different in design, they all provide varying degrees of precision alignment between optical fibers. To achieve this, the connectors are composed of several pieces of which the ferrule and sleeve are the primary fiber positioning elements. To maintain this alignment, the fibers are held in position within each connector half by using either epoxies, adhesives or mechanical clamps. Epoxies have the disadvantage of needing proper mixing, filling, and curing time. However, epoxies take longer to set up which allows more assembly and alignment time. Although adhesives and epoxies provide mechanical security for delicate fibers, they are the principle reason connectors which require their use are not easily terminated in the field. This is due to misalignment that can occur while the adhesive or epoxy is setting.

1. BUTT JOINT

Perhaps the most widely used connectors have a cylindrical ferrule which butt joins to another cylindrical ferrule within a sleeve. This is basically a precision aligned butt joint with ground and polished ends.[3] Connectors that utilize the cylindrical ferrule include the ST, FC, D4, SMA and PC style connectors (see Figs. 2-11 and 2-12). Fig. 2-13a shows a generic design of a cylindrical ferrule style connector.

2. BICONIC SLEEVES

Another style of connector which has become the defacto standard for phone company long haul applications is the biconic sleeve connector with conical ferrules as shown in Fig. 2-13b and Fig. 2-14. This design was developed to minimize lateral misalignment of the fiber ends. The two ferrules are centered perfectly in the sleeves which makes this design susceptible to core eccentricity problems. In order to eliminate losses due to eccentricity, the conical face of the plug end of the connector must be precision ground so that the fiber core is precisely centered within the

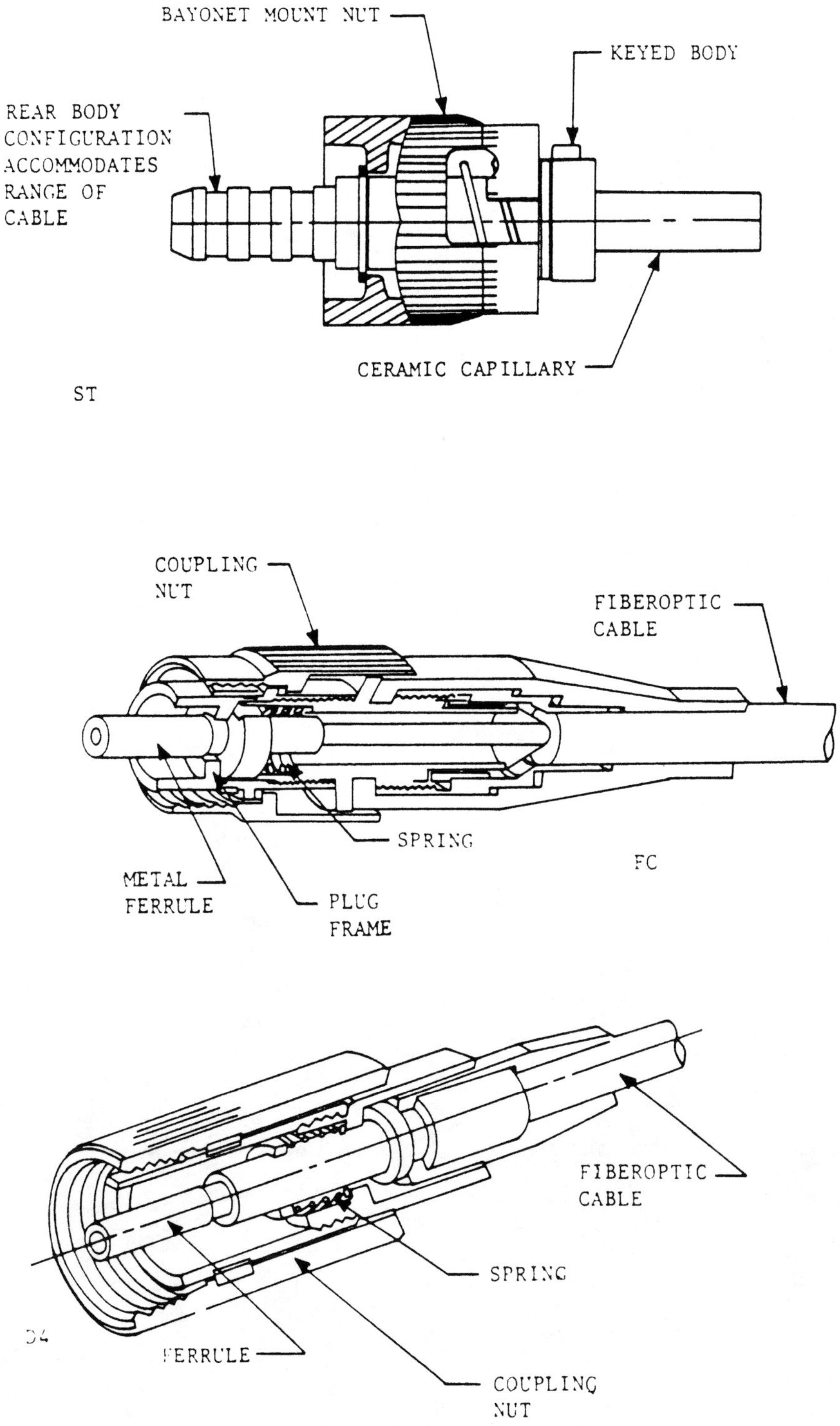

Figure 2-11. Cylindrical Ferrule Connectors.[5]

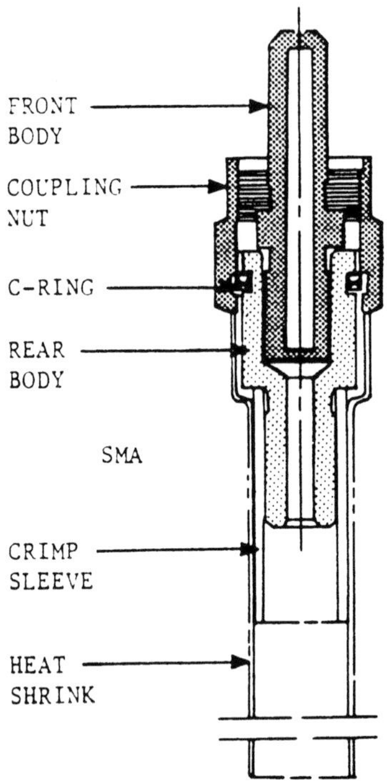

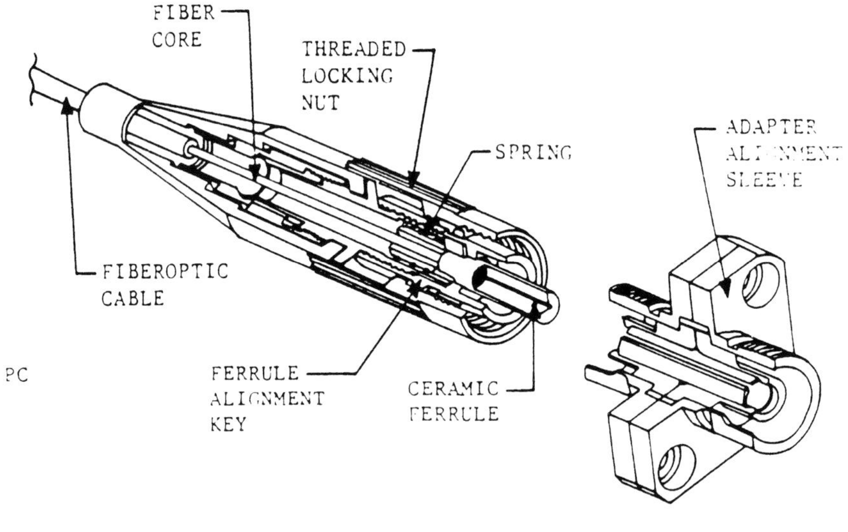

Figure 2-12. Cylindrical Ferrule Connectors, Con't.[5]

conical face of the plug. This process precludes the biconic connector from being installed in the field. The connectors are preterminated on a short, generally 1-3 meter, cable commonly called a pigtail and then fusion spliced to the system cable in the field.

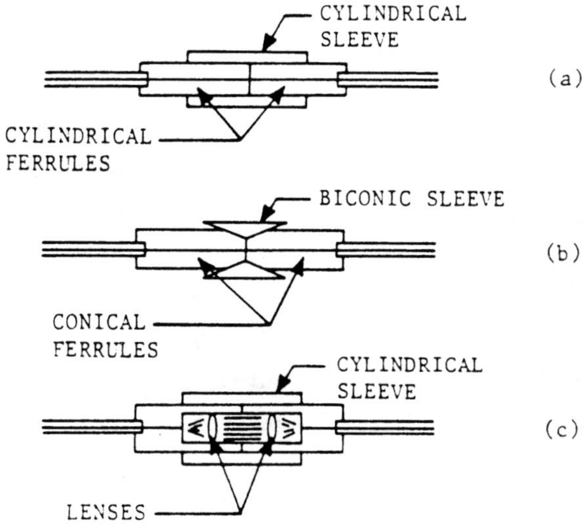

Figure 2-13. Typical connector styles.[3]

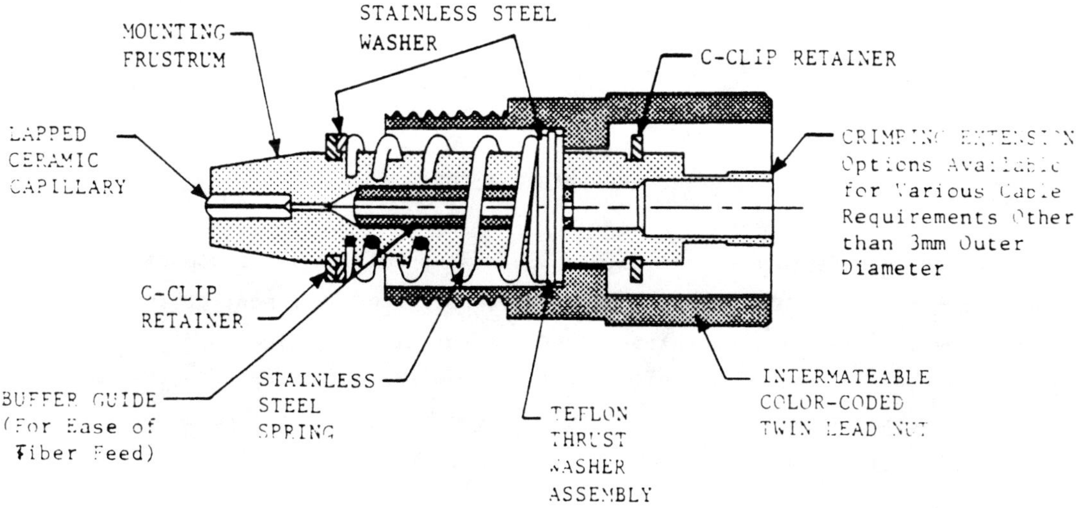

Figure 2-14. Biconic connector.[5]

3. LENSED

The lensed connector, as shown in Fig. 2-13c, uses lenses to increase the cross-sectional area and to collimate the optical output of the sending fiber and to focus this light onto the receiving fiber. Because the output is enhanced in this manner, the concentricity requirements are not as stringent but the angular requirements become much more important. This style connector is less susceptible to damage during multiple matings and its performance is less sensitive to debris due to the magnified optical diameter. However, if the connector uses an index matching fluid between the two lenses to minimize Fresnel reflections, then the fluid can trap debris which will increase the loss. Lensed style connectors are used less frequently than the cylindrical and conical style connectors because they are newer connectors and have not captured as big a market as the original SMA style connectors. The advantages of using a lensed connector have not warranted system designers to retrofit their systems with lensed connectors. Some users claim that there are no optical advantages to using lensed connectors.

4. FUSED LENS

A new connector type which melts the fiber end thus forming a precise ball lens onto and from the glass of the fiber itself was introduced to the market in 1987.[6] This type of connector is not as sensitive to lateral misalignment. No reliability information is available on this connector since it has not been used in the field yet.

IV. FABRICATION

In the manufacture of optical connectors, precision is of utmost importance to minimize extrinsic losses. To have good alignment of the fiber cores, the diameter and concentricity of the ferrules and sleeves must be precisely controlled for the cylindrical type connectors and precise taper length and concentricity control of the conical parts is required for the biconic connectors.

1. MATERIALS

The function of a particular part and the application in which the connector is used will largely determine the materials from which the part should be fabricated. A connector that is repeatedly mated and demated is prone to wear out fast unless it is fabricated using abrasion resistant materials like stainless steel and ceramics. Soft materials should not be used for this type of application. Also, exposure to severe environments, such as military field applications, requires that connectors be constructed using ruggedized materials such as stainless steel. This will provide protection against the crushing forces, high impacts and deterioration from the elements experienced in the field. For benign environments, and applications where matings and demating is infrequent, such as offices, no special allowances need to be made for the connector material.

A list of common ferrule and housing materials used within the connector industry is given in Table 2-1. This list, although not all inclusive, can be used by system designers to determine whether the materials that a particular application requires is readily available or if special production is needed. For example, aircraft or shipboard applications may require the use of certain materials and restrict the use of others. Also, losses of the same style connectors vary greatly from manufacturer to manufacturer as well as within one manufacturing company due to the use of different ferrule and housing materials. It should be noted that not all manufacturers use all of the listed materials.

TABLE 2-1. Ferrule And Housing Materials.

Material	Ferrule	Housing
Thermoplastic	X	X
Copper alloy plated	X	X
Brass	X	X
Stainless steel	X	X
Nickel plated zinc		X
Ceramic	X	
Thermoset	X	
Plastic	X	X
Polyethersubonef		X
Glass filled epoxy	X	
Aluminum	X	X
Silica filled epoxy resin	X	
Arcap	X	
Polycarbonate		X
Epoxy	X	
Copper	X	
Prevex		X
Silica compound	X	
Nickel silver	X	X
Tungsten	X	

Ref. 7

2. FABRICATION TECHNIQUES

Generally, ferrules and sleeves are manufactured using one of the following methods:

- o Machined
- o Molded
- o Extruded

Most connector styles use a combination of these methods. For example, the ferrule may be machined and the sleeve may be molded. Molding is one of the cheapest means for producing parts in quantity, but it generally does not provide as much part precision as a machine process.[7] However, there has been a recent development of a transfer-molded biconic sleeve whose dimensions can be held more closely than those of machined sleeves.[3] There is no particular method of manufacturing connectors that can be considered the best method. Each connector must be rated on its performance characteristics which will indicate the quality of the manufacturing process.

V. DESIGN TESTS

1. SPLICE TESTS

Since splices are affected by environmental and mechanical conditions which may increase their optical loss, environmental tests are performed on the splices by their manufacturer. Generally the attenuation of a completed splice is measured before and after the environmental testing is performed and a comparison between these two readings is made. The difference is used as a measure of the splice's sensitivity to the test condition and as a prediction of its performance in the field. Common tests and test methods that optical splices are subjected to are listed in Table 2-2.

TABLE 2-2. Optical Splice Testing.

Test	Method
Insertion Loss	EIA-455-20
Salt Spray	MIL-STD-1344 Method 1001
Fluid Immersion	MIL-STD-1344 Method 1016
Impact	DoD-STD-1678 Method 2030
Operating Temperature	EIA-455-72
Thermal Shock	MIL-STD-1344 Method 1003
Humidity	MIL-STD-1344 Method 1002
Temperature Life	DoD-STD-1678 Method 4010
Tensile Loading	2,000 Newtons/Minute
Mechanical Shock	MIL-S-901
Vibration	MIL-STD-1344 Method 2005
Temperature Cycling	-40°C to +70°C for 50 cycles

Ref. 7

2. CONNECTOR TESTS

Connectors are also affected by environmental and mechanical conditions which may increase the optical loss. The attenuation of a terminated connector is measured before and after the environmental tests are performed and a comparison between these two readings is then made. Common tests and test methods that are run on optical connectors are listed in Table 2-3.

TABLE 2-3. Optical Connector Testing.

Test	Method
Insertion Loss	EIA-455-34 or EIA-455-20
Temperature	High temp 48 hours/ Low temp 48 hours
Humidity	MIL-STD-1344 Method 1002 or MIL-STD-202 Method 106
Submersion	1 hour at 1.83 meters or 96 hours at 5 inches
Tensile Loading	18 Kg axially for 1 minute
Cable Strain Relief	Visual Inspection
Flex Life	5 Newton Force, ±90° Flex, 250 cycles
Twist	180° twist for 1,000 cycles
Mating Durability	300 cycles/hour for 2 hours
Temperature Cycling	MIL-STD-1344 Method 2008
Temperature Shock	MIL-STD-202 Method 107
Vibration	MIL-STD-202 Method 204

TABLE 2-3. Optical Connector Testing. (Con't.)

Test	Method
Shock	MIL-STD-1344 Method 2004 or MIL-STD-202 Method 213
Salt Fog	MIL-STD-1344 Method 1001
Dust	MIL-STD-202 Method 110
Fluid Immersion	MIL-STD-1344 Method 1016 or MIL-C-81511
Cable Retention	MIL-STD-1344 Method 2009

Ref. 9

VI. COMMON FAILURE MECHANISMS

1. SPLICE FAILURES

Failure mechanisms, which may cause long term degradation in splices, include breakdown of the index matching fluid, separation of the fiber endface and the index matching fluid and fiber movement in the lateral and axial directions.

The breakdown of index matching fluid can be prevented by using the proper type of fluid for a particular environment. For example, some types of fluid become cloudy or stiff when exposed to cold temperatures, others may become too thin to stay in the splice when exposed to high temperatures. For this reason, the index matching fluid should be carefully analyzed to insure that it can withstand the anticipated environment.

Separation of the fiber endface and the index matching fluid will cause additional attenuation due to the newly created glass-air/air-gel interface. To prevent this, an index matching fluid which will remain soft and pliable in the anticipated environment should be used. This will allow the fluid to adapt to changes in fiber position resulting from external forces while still reducing Fresnel reflection. The most popular type of gel used is a silicon gel which is inserted as a liquid and gelled by heating with a heat gun.

Splices and coiled spare fiber can move even when stored within the splice closure. The splice itself is held in place by a securing mechanism inside of the splice closure. Spare fiber is coiled up inside of the closure for future maintenance purposes. This fiber is also subjected to movement. The splice closure and organizer must be chosen such that there is ample room to coil the fiber with respect to its minimum bend radius and the fiber must be secured at several points to prevent movement. Failures within the splice cases have occurred but the failures have not been at the splice point itself but rather in the spare fiber. Installation and maintenance crews must be instructed on the possible effects of exceeding the minimum bend radius and the proper method of securing the splices within the splice closure. These instructions can be given through training courses to prevent installation and maintenance actions from adversely affecting the reliability of the installed system.

In response to survey questionnaires[10,11], splice manufacturers anticipated the most prevalent failure mechanisms to be:

Bad cleaves
Fiber breakage
Fiber endface separation due to improper assembly
Dirt
Vibration
Improper assembly techniques.

Predicted operational failure rates reported in the questionnaires ranged from 1 to 15%[10,11] and estimated operating life ranged from 25 years to permanent[10,11] for the V-groove and the glass alignment rod splice, respectively. The manufacturing production yield is the percentage of splices which have passed the mechanical and environmental production tests run by the individual manufacturers. This yield ranged from 85 to 99.5%.[10,11] This information was reported in surveys without specific conditions being stated.

Reports published on mechanical and fusion splices indicate both of these splice types are extremely reliable. Most system designers prefer a combination of fusion and mechanical splices with fusion splices often being the choice for long-haul and network installations and mechanical splices selected for specialized and emergency situations.[12]

Reliability indicator figures have been calculated from factory and field splice data received from major manufacturers. For the individual splice types (mechanical, fusion and linear array), the source field data was combined from two manufacturers and reflects splice performance in various environmental and mechanical conditions. The need to combine the data is a result of the lack of detail with which the source data was provided. That is, no information about the applications in which the splices were used was available.

To calculate the reliability figures, it was assumed that the splice failures are not time dependent and that when a failure did occur, it was replaced with an identical splice. Also, the literature does not indicate that the quality of a splice is something other than good or bad. Therefore, it was assumed that when a splice fails it does so catastrophically, otherwise it performs as specified.

The <u>field</u> splice failure data[13,14] reported by manufacturers for mechanical and fusion splices was assumed to be continuous and random which allows the Chi Square distribution to be used. The data included the total number of mechanical and fusion splices, the total number of failures for each splice type, and the total splice-hours for both splice types combined. The Mean Time Between Failure (MTBF) at a 90% confidence interval was calculated using equation 2-1. The source data and calculated results, in which a standard Chi Square table was used, are given in Table 2-4.

$$\text{MTBF Confidence Limits} = \left(\frac{2T}{\chi^2\left(\frac{\alpha}{2}, 2r+2\right)}, \frac{2T}{\chi^2\left(1 - \frac{\alpha}{2}, 2r\right)} \right) \qquad \text{Equation 2-1}$$

T = total splice hours
α = risk of error = 0.1
r = total number of failures
χ = Chi Square distribution

The failure rate was determined by calculating the reciprocal of the MTBF.

TABLE 2-4. Field Splice Data Summary.

Total Number of Splices	5,982
Number of Fusion Splices	5,721
Number of Mechanical Splices	261
Total Number of Failures	7
Number of Fusion Failures	4
Number of Mechanical Failures	3
Total Splice-Hours	177,447,966
Calculated MTBF (10^6 hours)	between 13.5 and 54
Calculated Failure Rate (10^{-6} hours)	between 0.0185 and 0.0741

Ref. 12,13

The mechanical splice failures given in Table 2-4 were reported as failing due to improper handling, whereas the fusion splices failed due to improper cable anchoring in the closure. These latter failures did not occur at the actual splice itself. All of these failures are considered avoidable.

Cable manufacturers perform factory fusion splices, to concatenate fiber within cable to increase the cable length. These splices are considered to be so reliable that they no longer constitute a failure mode. Using data from a study which tracked factory fusion splices performed between 1979 and 1984, as shown in Table 2-5, it can be seen that no splice failures occurred during this time frame.

The MTBF at a 90% confidence interval and failure rate were calculated for factory fusion splices using equation 2-2. The data was assumed to be continuous and random which allows the Chi Square distribution to be used. The source data and calculated results, in which a standard Chi Square table was used, are given in Table 2-5.

$$\text{MTBF Confidence Limits} = \left(\frac{2T}{\chi^2 (\alpha, 2r + 2)} , \infty \right) \qquad \text{Equation 2-2}$$

T = total splice hours
α = risk of error = 0.1
r = total number of failures
χ = Chi Square distribution

TABLE 2-5. Factory Fusion Splice Summary.

Total Number of Splices	1,027
Total Splice-Hours	11,363,352
Total Number of Reported Failures	0
Calculated MTBF Lower Limit (10^6 hours)	4.94
Calculated Failure Rate (10^{-6} hours)	0.2

Ref. 13

A major cable manufacturer supplied field fusion splice reliability data[13] as presented in Table 2-6. No time intervals concerning the date of installation, number of splices installed per installation, nor the service life of the splices is available, thus no reliability data can be calculated. The information was obtained from a major cable manufacturer and was dated in 1986. The manufacturer has installed systems which have been operational for several years.

Table 2-6. Field Fusion Splice Data Summary.

Number of Splices	500,000
Number of Failures	0

Ref. 13

Source data on field linear array splices is given in Table 2-7 along with the associated reliability indicators.

The estimated minimum mean life of field linear array splices was determined by using the partial histogram shown in Fig. 2-15. Only the first 228,000 splices were included in the calculation. These splices were installed in the four year period from 1979 to 1983. It has been assumed that these splices have been operational four years after the 228,000th installation which means that the first splice has been in operation for eight years. The assumption was also made that an equal number of splices was installed in each of the first four years.

Referring to Fig. 2-15, the area of triangle ABC is determined and the area of triangle DCE is subtracted from that value which results in an area for ABDE of 1.19×10^{10} splice hours. Dividing this value by the number of splices (228,000) gives a mean life of 5.256×10^{4} hours. This is a conservative figure for the minimum mean life of the field linear array splice due to the fact that there have been no failures reported to date.

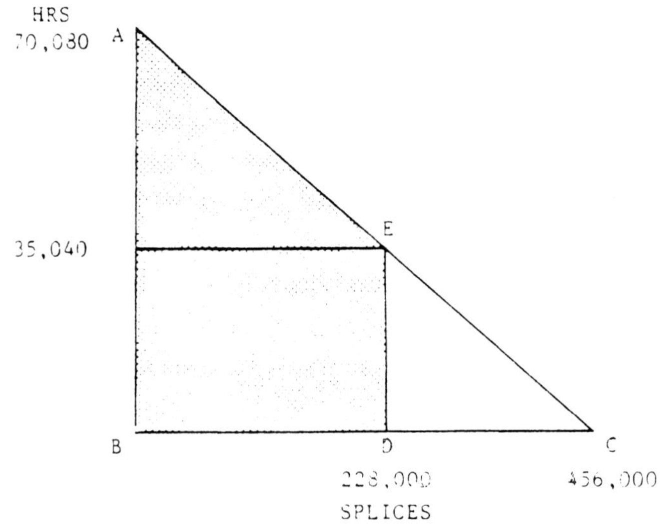

$$ABDE = \triangle ABC - \triangle DCE$$

$$\overline{BC} = \frac{x}{70{,}080} = \frac{228{,}000}{35{,}040}$$

$$\overline{BC} = 456{,}000$$

Figure 2-15. <u>Field Linear Array Splice Reliability Calculations</u>.

Table 2-7. Field Linear Array Splice Data Summary.

Total Number of Splices	728,000
Total Number of Failures	0
Calculated Minimum Mean Life (10^6 hours)	.05256
Calculated Minimum Failure Rate (10^{-6} hours)	19.022

Ref. 3,13

2. CONNECTOR FAILURES

The primary failure mechanism expected by manufacturers as reported in surveys is dust or small particles of dirt which enter the connector when unmated.[15,16] For this reason, most manufacturers have recommended that the optical contacts be periodically cleaned. Their recommendations range from every time the connector is demated to every 100th demating. Manufacturers were unable to predict a failure rate for their connectors,[15,16,17] however, one manufacturer indicated the life cycle of one style of connector to be 500 matings and another style is 2500 matings.[17] The difference in the number of matings the connector can withstand is due to the different design objectives, materials and applications for each connector style. The manufacturing production yield of connectors which passed the mechanical and environmental production tests run by individual manufacturers ranged from 96 to 98%.[15,16]

A connector is considered to have failed when the loss of the connector increases to the point where it is above the maximum value specified for the system in which it is operating. When a connector fails catastrophically, it causes loss of optical transmission in the terminated cable at the connector. A connector can also degrade slowly or wear out, but continue to transmit optical power. This degradation is in the form of a gradual increase in loss at the connector interface. The system will continue to operate until the connector loss causes the system power nargin to be exceeded which will degrade data transmission. The primary causes of connector loss increase stem from repeated matings and dematings. The most frequent cause is a result of contamination blocking the light at the fiber endfaces. If contamination remains after mating, the fiber endfaces will become permanently scratched. This severely reduces the coupling efficiency between mated fibers. After repeated mating and dematings, the mechanical tolerances on the connector parts begin to degrade which causes optical misalignment between fibers. This also reduces the coupling efficiency which increases loss. Misalignment can also occur to the ferrules of the mating connectors if the epoxy has not cured properly, if the connector is exposed to extreme temperatures, or if the connector is subjected to impacts.

The authors found no published reports on connector reliability nor could any of this data be obtained from manufacturers. Generally though, reliability is based upon the number of times the device can be mated before the attenuation increases above the specific acceptable performance level. Individual connectors are designed to endure a specified number of matings. The mating durability test which is performed during factory environmental testing verifies this design specification. The mating durability of a connector is extremely important since repeated matings result in wearout of individual piece parts in the connector, misalignment problems, or contamination or damage to the fiber endface. All of these conditions increase the loss of the connector which may impact the overall reliability of the system. Since connectors can only be mated a finite number of times before system performance is degraded, the mating durability of connectors should be compared to the number of times the connectors are expected to be mated when selecting a connector. A list of connector styles, their design mating durability requirements, and test method is given in Table 2-8.

A problem that has been reported is "pistoning" of the fiber in epoxy style connectors. Pistoning is defined as fiber movement within the ferrule of the connector. This problem usually occurs at temperatures exceeding 150°C. This phenomenon has been attributed to epoxies which vary due to lot differences and from inconsistencies in the uniformity of the mixing and application. This causes a variation in the epoxy lap shear strength which can cause improper curing or adhesion qualities. This may result in pistoning if the epoxy is then exposed to high temperature or temperature cycling

TABLE 2-8. Fiber Optic Connector Mating Durability.

Connector Style	Rated Number of Matings	Test Method
MIL-T-29504 Termini, Fiber Optic Connector	1,000	MIL-STD-1344 Method 2016
MIL-C-28876 Circular, Multiple Removable Termini	500	MIL-STD-1344 Method 2016
MIL-C-38999 Circular, Miniature, Environment Resistant	500	Specification Test
MIL-C-83522 Single Terminus	500	Specification Test
MIL-C-83526 Circular, Environment Resistant, Hermaphroditic	1,000	MIL-STD-1344 Method 2016
FOHC Fiber Optic Hybrid Connector	500	MIL-C-38999 Specification Test
FOMC Fiber Optic Multi-Channel Connector	2,500	MIL-STD-1344 Method 2016
FC Full Contact	1,000	Not Available

Ref: 9, 15, 16, 17, 18, 19, 20, 21

conditions. The Electronic Industries Association Working Group FO-6.3.1 on Connector/Cable Interface problems reported in January 1986 that using epoxy as a potting agent does not result in as strong a bond as when it is applied in several thin layers. When used at extreme temperatures, some epoxies will lose mass or shrink at temperatures greater than 150°C after fill cure. There have been instabilities reported at temperatures greater than 200°C which have caused a breakdown of the epoxy lap shear strength. The glass transition temperature for most epoxies is 100°C - 110°C. As the epoxy is exposed to those temperatures, it becomes more plastic and loses its ability to hold an object securely. These temperatures are not uncommon in engine compartments of aircraft and vehicles, space applications and environmental testing requirements for military applications. The industry is working together with the Department of Defense to develop a better epoxy to withstand these extreme temperatures.

A common area of concern for both connector and cable manufacturer's is the interface between the connector and cable when a termination is made. The problem that can occur at this interface is breaking of the fibers at the interface due to flexing. The problem is addressed by the connector manufacturer rather than the cable manufacturer. Strain relief mechanisms are designed into the cable end of the connector. The purpose of the strain relief is to provide a material that will limit the degree of the bend radius to which the cable can be subjected. Single fiber connectors generally use a piece of heat-shrink tubing installed at the cable/connector interface while multi- channel connectors usually use a rubber or elastomeric boot. Both the heat shrink tubing and the boot serve to stiffen the cable at the interface point which limits the degree of bending that the cable can be subject to and it allows the rubber to take some of the strain off of the cable. Manufacturer's and users can test the quality of the strain relief by subjecting the completed assembly to flex tests. There are many variations of flex tests used throughout the industry but one of the most common is the cyclic flexing test.

The ultimate connector design will eliminate the need for mixing and applying epoxy by the technician in the field. During 1986, industry began researching a crimp and cleave style connector which uses no epoxy and a connector style where the epoxy is an internal part of the connector thereby eliminating the application step. The latter design would have a shelf life before the epoxied parts would become unusable due to the setting of the epoxy. The military and the industry have recognized the reliability problems that the use of epoxies can cause and both are expending considerable effort to develop a solution.

VII. FACTORS IMPACTING SELECTION AND PERFORMANCE

1. SPLICE SELECTION

Table 2-9 lists design characteristics which should be considered before using any of the splices discussed. This data represents typical values taken from manufacturer's data sheets. The actual time required to complete the splice may vary depending on the type of epoxy used, the skill of the technician and environmental constraints.

TABLE 2-9. Splice Critical Design Criteria.

Splice Type	Typical Loss (dB) Multi Mode	Single Mode	Operating Temperature(°C)	Average Assembly Time (Min.)
Hollow Tube	0.2	0.2	-40 to +70	5-7
Three Rod	0.2	0.2	-40 to +70	5
Rotary	0.035	0.035	-40 to +70	10-19
Elastomeric	0.2	0.2	-40 to +70	3-5
Silicon Chip Array	0.15	0.3	-40 to +70	10
Rapid Ribbon	0.25	0.3	-40 to +70	20-30
Fusion Splice	0.1	0.1	*	5-40

* Limited by the fiber or the protection sleeve.

None of the splices listed in Table 2-9 should be subjected to any tensile load and they should be protected in a suitable enclosure with respect to the environment in which they will be used. The three-rod splice is rarel' used in the field since a slight variation in the fiber's outside diameter will either make the fiber impossible to insert or too loose to properly align with the other fiber. The silicon chip array and rapid ribbon splices provide strain relief by clamps which grip the cable. The rapid ribbon splice is equipment intensive since it requires a vacuum chuck and a special microscope which attaches to the chuck for alignment purposes.

The quality of a fiber splice is extremely sensitive to the cleanliness of the fiber end, alignment of the fibers and retention of the fibers. Splices are sensitive to unfavorable environmental conditions in which they are performed. Average field-splicing times range from 3 to 40 minutes per splice depending upon environmental conditions and the type of splice being used.

2. CONNECTOR SELECTION

When selecting an optical connector, the ability of the connector to align the optical fiber cores is the most important feature. The fiber endfaces must be held flat and perpendicular to the alignment surface axis. Other selection criteria and factors that contribute to loss are listed in Table 2-10.

TABLE 2-10. Typical Connector Dimensional Values For Critical Parameters.

PARAMETER	MULTIMODE	SINGLE MODE
Eccentricity between bore and alignment surface axis (See Fig. 2-16a) (μm)	5	1
Diametral clearance between the largest bore I.D.and the smallest fiber O.D. (See Fig. 2-16b) (μm)	5	2
Angular misalignment between bore and alignment-surface axis (See Fig. 2-16c) (°)	1	1/2

Ref. 3

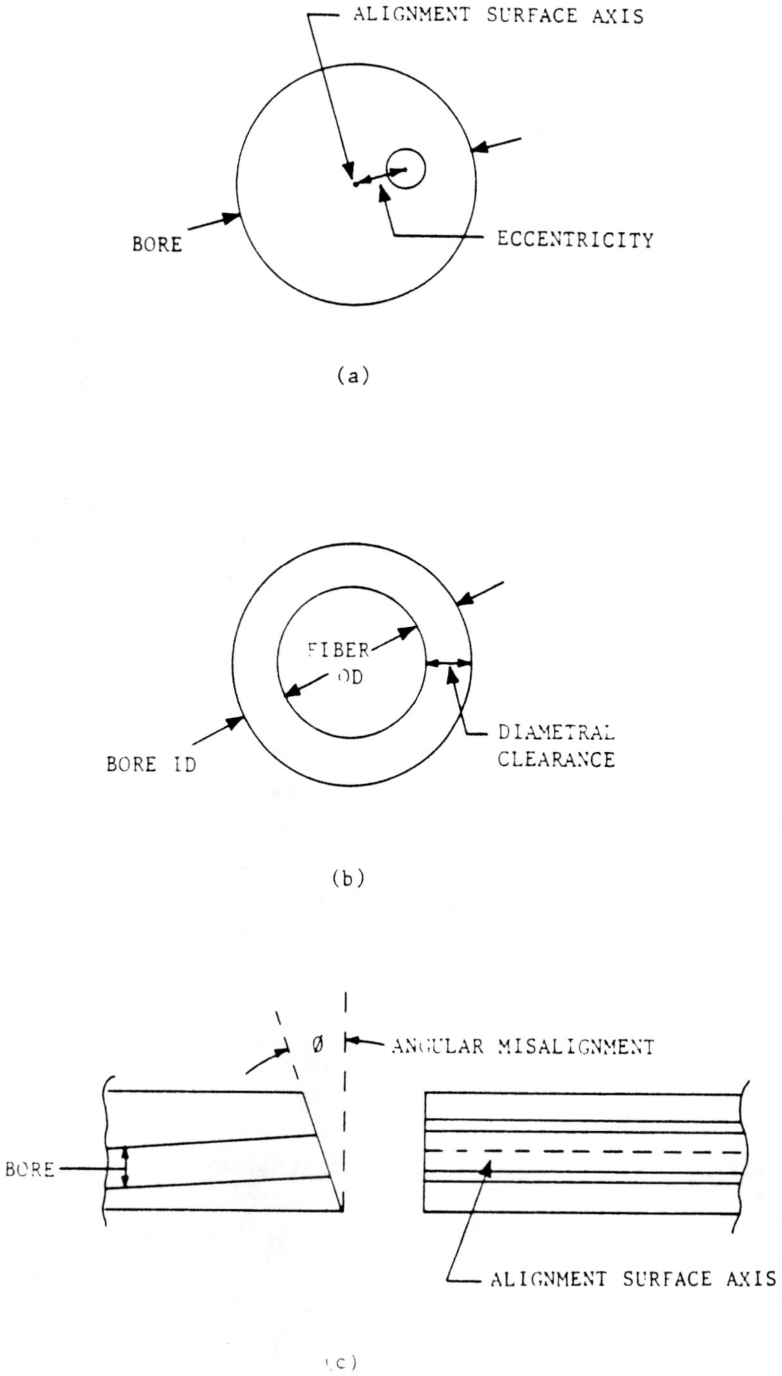

Figure 2-16. Fiber optic connector critical design parameters.

3. CONNECTOR PERFORMANCE

Connector styles are available in both multimode (MM) and single mode (SM) versions. The alignment of fiber cores is the main purpose of any connector whether it be MM or SM. The differences in core size, 50 μm for MM and 8-10 μm for SM, make the alignment of the cores much more difficult for SM connectors. Connectors generally have greater losses than splices because fiber alignment is not fine-tuned as it is in the rotary splice and index matching fluids are not used as frequently in connectors. Both MM and SM connectors suffer from wear and endface damage due to repeated matings and dematings. The damage to the endface does not occur during mating because the design allows for adequate endface separation but it may occur when the connector is unmated due to bumping or dropping the unmated connector. This is where the importance of protecting the endface with an endcap comes in. Not only are the above parameters extremely important but the connector design plays a significant role in the connector loss. The range of losses for each connector style is given in Table 2-11.

TABLE 2-11. Connector Loss Ranges.

Style	Nominal Attenuation - First Use (dB)	Nominal Attenuation After 1,000 matings (dB)
Cylindrical sleeve with cylindrical ferrules (Fig. 2-13a)	0.2 - 2.0	0.2 - 2.5
Biconic sleeve with conical ferrules (Fig. 2-13b)	0.5 - 0.7	0.6 - 1.0
Cylindrical sleeve with lenses (Fig. 2-13c)	0.6 - 1.5	0.6 - 1.7

Ref. 18

4. CONNECTOR ASSEMBLY

The time required for connector installation also varies due to the termination method. Epoxy, polish and crimp terminations generally require more time than the crimp and cleave type terminations. Epoxy curing time and the polishing step can be quite lengthy depending upon the operator's skill. The time needed to assemble a single fiber terminus can range anywhere from less than 5 minutes to more than 30 minutes. Manufacturers do not indicate average assembly times on their data sheets due to the many variables possible. The skill of the operator is the main driving factor behind the assembly time. Generally, the assembly of an optical termini requires a good deal of finesse.

VIII. MAINTENANCE REQUIREMENTS

1. SPLICES

Splices require no maintenance. To ensure long life, splices should be placed in closures which prevent tensile stresses from being applied to the splice and also prevents the fibers from being bent beyond their minimum bend radius. This necessitates proper strain relief mechanisms to be used with the protective closure. The splices must be kept dry since water could be absorbed by the fibers and/or epoxy which will degrade the transmission of light. Also, water and dirt could contaminate the index matching fluid which would cause high attenuation at the splice.

2. CONNECTORS

Connectors must be designed to be mated and demated repeatedly to minimize degradation of their performance due to wear. However, when connectors are demated, the potential for contaminating the fiber ends exists. This requires a maintenance action of routinely cleaning the connectors after dėmating by simply wiping the fiber endfaces with a swab dipped in acetone. If the connector's fiber endfaces are damaged or scratched, the connector may have to be replaced or the damaged fibers repaired if the loss degrades overall system performance.

The expedient repair of severed cables is generally performed by splicing in a new section of cable or by attaching connectors to the severed ends and installing a jumper cable. A temporary repair is generally performed as soon as the break is located and then at a later date, a permanent repair is made. The most common method of temporary repair is to use elastomeric splices or bare fiber connectors to install a piece of cable.

IX. LOGISTICS CONSIDERATIONS

Logistics is a primary factor in the repair time of optical systems. Expedient repair is heavily dependent upon having the proper equipment immediately available. Typical emergency repair kit contents are given in Chapter 1, Table 1-6. Personnel must be trained in the use of the equipment and must be given refresher courses in order to facilitate quick and reliable repairs.

Splices must be stored and operated at the manufacturer's specified temperatures. An average of three manufacturers' storage temperatures specifications ranged from -30°C to +79°C and operating temperatures ranged from -30°C to +70°C.[10,11]

Connectors also must be stored and operated at the manufacturer's specified temperatures. An average of four manufacturers' storage temperature specifications ranged from -51.25°C to +117.5°C. The average operating temperature specifications ranged from -49.25°C to +102.5°C.[15,16,17]

Splices must be kept on location for emergency repair. Depending upon the maintenance philosophy of the governing agency, emergency repairs may be made with elastomeric splices on a temporary basis and later repaired with the original system configuration splice, or the permanent splice may be made immediately. The maintenance philosophy will impact logistics in respect to what is kept in stock.

Most splices are sized to the fiber outside diameter with a small degree of tolerance. The typical multimode fiber outside diameter can vary ±3 μm throughout its length. This means that several splice sizes must be kept in stock because of the varying fiber sizes. The stocking personnel must know the specified size and tolerance of the system fiber so that the proper spares can be ordered and stocked. Each splice type also requires specific tools and assembly equipment. Equipment may be needed for both permanent and temporary repairs. The tools that are used for stripping the fiber are also very tightly controlled with respect to tolerance. If the system uses multiple fiber sizes, all the proper tools will have to be stocked.

Connectors also may be needed for emergency repair situations. They may be utilized when an existing connector has failed, or when a new section of cable is being installed to repair a damaged section.

Connectors are much like splices in that they too are sized to the fiber outside diameter with a small degree of tolerance. Therefore, due to the variation in fiber diameter, a large number of connectors, or ferrules if it is a multiple channel connector, must be available.

The stocking of equipment is a concern due to the different installation techniques. Epoxy, heat guns, ultraviolet lamps, grinding wheels, polishing disks, special cleavers, and many other tools may be needed depending upon the connector style being used within the system.

Standardization of the connector and splice type within a particular system is imperative to minimize logistics considerations. The industry has developed specifications for most of the connector types mentioned in this chapter. This results in intermateability between different manufacturers' for the same type products. These products meet the requirements of the same industry specification. The connectors may be designed with a different number of piece parts, but if it meets the intermateability requirements of the specification, it is a suitable connector. Use of a standardized connector in the system will simplify stocking considerations and insure that more than one source is available.

Specifications on splices have not been standardized within the industry. The Electronic Industries Association is currently developing a specification which will cover most splice types. The military has one specification for a single fiber splice which is very similar in style to the SMA connector.[8] The telecommunications industry has standardized the use of the biconic and linear splices. The telecommunications corporations develop their own specifications for these splice types and then form licensing agreements with manufacturing companies. Manufacturers then produce the splices according to the telecommunications specification.

Use of standardized components in any system design will insure that replacement components will be readily available from multiple sources at a reasonable cost. The system designer should analyze all standard components available before arbitrarily designing a specialized device.

X. SUMMARY

1. SPLICES

Reliability of installed optical splices, both mechanical and fusion, have proven to be very good. The failures are not intrinsic but rather they are failures occurring in the splice case due to improper storage of the splice and excess fiber. Generally, failures have not occurred at the splice point itself. Preventive maintenance of installed splices is not needed nor recommended.

2. CONNECTORS

Reliability of installed connectors, is very dependent upon the quality of their installation. If the connector is installed properly, the chances for fiber pistoning due to improper preparation of epoxy or improper cure time are reduced. Generally, if the connector is subject to less than its rated mating durability and the environmental conditions are within those tested and specified by the manufacturer, the connector should provide reliable service.

Repair procedures generally allow for an immediate, temporary restoration of a damaged link followed by a permanent repair. An emergency repair kit should be at every site and it should contain the proper tools and materials for the installed system. Splices and connectors are essential elements of this kit for their installation significantly affects the overall downtime of a system.

Chapter 3. Emitters

Fiber optic emitters are one of the core components of fiber optic systems. They are imbedded in and driven by the electronics of transmitters and transceivers. Their performance will impact the entire system. If the emitte fails to deliver an adequate amount of power, the received power will decrease thus increasing bit error rate until it becomes unacceptably high. If the emitter becomes unstable with temperature variations, then the signal quality will be degraded. A multitude of problems can result from improper emitter performance. All of these potential problems must be considered and methods developed to reduce their occurrence with a high degree of confidence over a 20-25 year system lifetime.

The causes of unreliable emitter performance, techniques to screen those devices and predicted lifetimes are discussed. Also, a brief review of common emitter constructions is given.

This discussion is limited to lasers and Light Emitting Diodes (LEDs) fabricated only from semiconductor materials since these types of devices are the most extensively used in fiber optic systems.

I. FUNDAMENTAL PHYSICAL CONCEPTS

1. ABSORBED AND EMITTED ENERGY

The operation of semiconductor lasers and LEDs are governed by the rules of quantum mechanics. These rules state that the internal energy of semiconductor crystals can only have discrete values or quantization levels. In semiconductor materials, the levels of interest are called the conduction band and the valence band. When free electrons are available in the conduction band, then current can flow. These bands are separated by an energy band gap, Eg. To transition this band gap, an amount of energy equal to the difference between the conduction and valence bands must be expended. When an electron makes this transition, energy can either be absorbed or emitted with a frequency that is proportional to Eg. This freouency is given by equation 3-1.

$$f_o = \frac{E_g}{h}$$

Equation 3-1

where h = Planck's constant

Eg is in eV

Or, using the relationship $f_o = \frac{c}{\lambda}$, equation 3-1 becomes

$$\lambda = \frac{ch}{E_g}$$

Equation 3-2

where c = speed of light

Since the value of Eg in a semiconductor is dependent upon the materials used, the wavelength of emitted energy can be controlled by careful selection of the materials. This has allowed the production of active devices using chemical compounds with operating wavelengths anywhere from the visible region to the near-infrared. Some of the more common compounds used in LEDs and lasers are shown in Fig. 3-1 with their respective operating wavelength ranges.

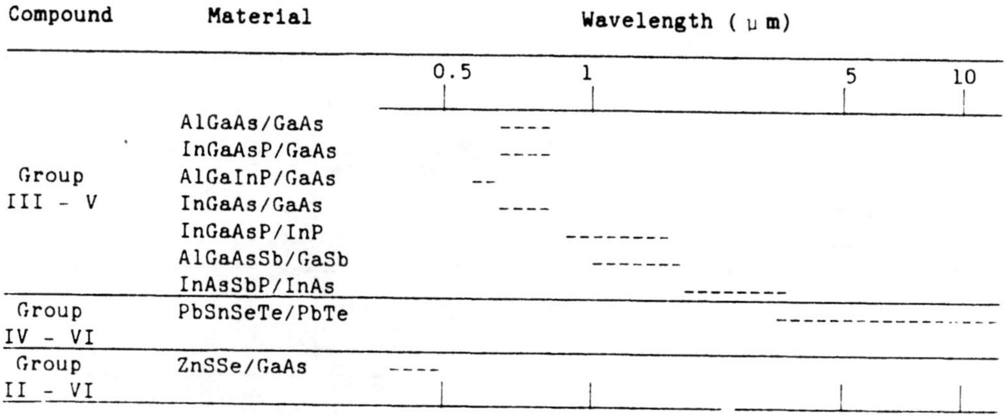

Figure 3-1. Common compounds used for LED and laser active layers.

Not only is the selection of the materials important, but the fraction of each element used in the compound must be considered. Varying the percentage of each element used per group determines the index of refraction of the individual elements. Subscripts are used in the compound nomenclature to designate the percentage of the particular elements used per group in the compound. For example, if x = 0.47 in the compound $Al_xGa_{1-x}As$, this would indicate that the compound is composed of 47% Al and 53% Ga from group III elements and the only group V element is As. (Table I-1 in Appendix I lists the common elements used in semiconductor device fabrication and their abbreviations). Generally, AlGaAs lasers are approximately one fourth as sensitive to ambient temperature fluctuations than comparable InGaAsP lasers. Also, single mode lasers operating in the 0.8 - 0.9 μm wavelength range have output powers which are roughly an order of magnitude greater than lasers that operate in the 1.2 - 1.7 μm range.[1]

2. DIRECT AND INDIRECT TRANSITIONS

Different elements can have different probabilities of emitting light. This is due in part to there being two types of transitions that can occur; direct and indirect. These transitions are shown in Fig. 3-2 where two different band gap structures are shown in E versus K (energy versus wave vector) diagrams.

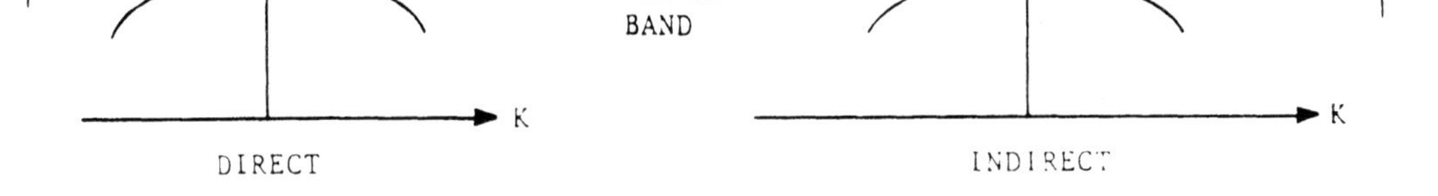

Figure 3-2. Direct and indirect transitions.

Direct transition materials have a high probability of emitting energy and a quantum efficiency approaching unity. This transition can happen when the lowest energy of the conduction band and the highest energy of the valence band occur at the same value of wave vector as shown in Fig. 3-2a. GaAs is a direct transition compound. An indirect transition can happen in a material when the lowest energy of the conduction band and the highest energy of the valence band occur at different wave vector values as shown in Fig. 3-2b. For this transition to occur, the transitioning electrons must collide with the atoms of the crystal to accomplish the change in the wave vector value. This results in the generation of heat and a very low probability that optical energy will be emitted. As a result, the quantum efficiency is often less than 0.001.[2] Silicon and germanium are indirect transition materials and are commonly used as isolation layers in emitters.

3. SPONTANEOUS EMISSION, RESONANT ABSORPTION AND STIMULATED EMISSION

Electrons can transition between energy bands through three methods: spontaneous emission, resonant absorption or stimulated emission.

a. Spontaneous emission -- Electrons can spontaneously transition from the conduction band to the valence band without the application of an external energy source. The rate at which these electrons transition is directly proportional to the number of electrons left in the conduction band. This results in an emitted energy which has a frequency equal to f_o (see equation 3-1) and is produced randomly and spontaneously as shown in Fig. 3-3a. This is called spontaneous emission and is the basis on which the LED operates.

b. Resonant absorption -- This process, illustrated in Fig. 3-3b, occurs when an electron in the valence band absorbs energy from an external source at a frequency equal to f_o and recombines with a hole in the conduction band without emitting energy. This is called resonant absorption and is considered to be a loss mechanism which decreases LED and laser efficiency.

c. Stimulated emission -- An electron in the conduction band can be stimulated to transition to the valence band by the application of an external energy source. If the external energy source is a field of photons with individual energies equal to that of the photon in the conduction band and in phase with each other, then stimulated emission can occur. The electron is induced to transition the band gap to the valence band, thereby emitting a photon. This photon is in phase with the applied field. This is called stimulated emission and is the basis on which the laser operates. Fig. 3-3c depicts this process. The emitted light output is coherent in nature due to the phase relationship and resonance between the individual photons.

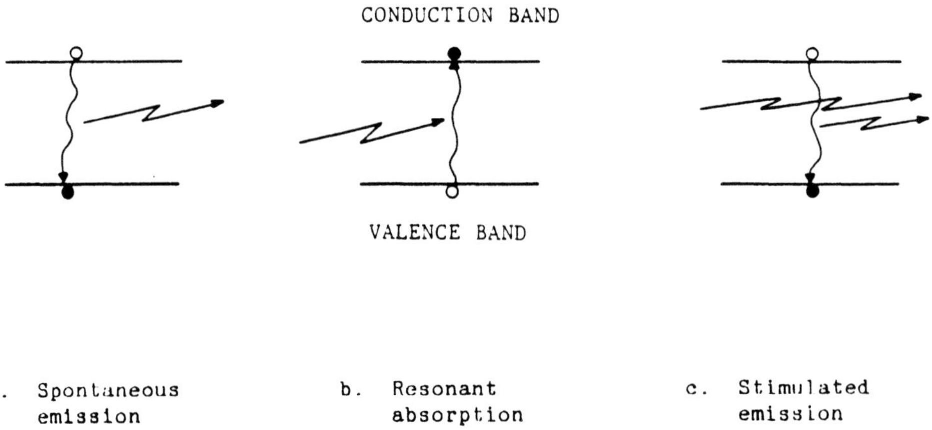

Figure 3-3. Methods by which electrons can transition energy bands.

The external energy source is supplied in the form of electrical energy for semiconductor emitters. This energy source produces the necessary field of photons which stimulate conduction band electrons through other spontaneous and/or stimulated emissions within the semiconductor material

II. BASIC LED OPERATION

An LED can be formed from a properly fabricated p-n junction as shown in Fig. 3-4.

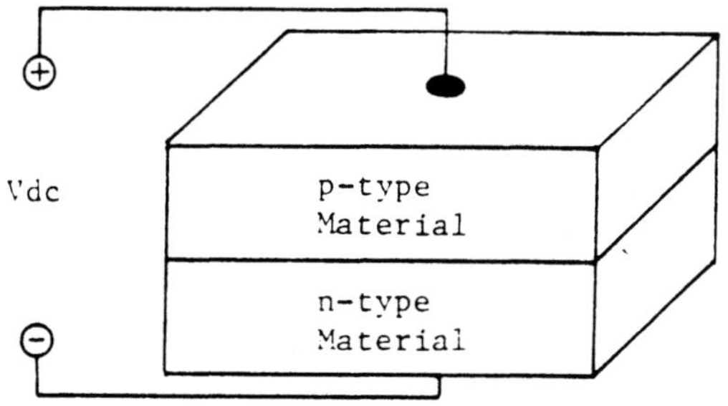

Figure 3-4. Basic p-n junction LED.

All LEDs operate on the principle of spontaneous emission which can only produce incoherent light. At thermal equilibrium, the occurrence of spontaneous recombination is low and the amount of energy emitted is insignificant. However, if electrons are injected into the p-n junction by applying a forward voltage as shown in Fig. 3-4, then electron-hole recombination increases. This causes a corresponding increase in spontaneous recombination and likewise, spontaneous photon emission.

III. BASIC LASER OPERATION

The word LASER is an abbreviation for Light Amplification through Stimulated Emission of Radiation. As the name implies, light is amplified. As previously described, spontaneous emission results when electrons which have been excited into the conduction band randomly and spontaneously transition to the valence band thus emitting a photon. There is a mean time that these electrons spend in the excited state before transitioning. Laser operation is based on stimulating these already excited electrons to transition before spontaneous transitioning occurs. Once stimulated, the resulting photons have the same energy, are in phase, travel in the same direction and have the same polarization. These are the characteristics of coherent light.

At thermal equilibrium, resonant absorption and spontaneous emissions far outnumber stimulated emissions. This is due to the number of electrons in the valence band, N_v, being much greater than the number of electrons in the conduction band, N_c. A good approximation to this ratio is given by the Boltzmann distribution given in equation 3-3.

$$\frac{N_c}{N_v} = e^{-\frac{hf_o}{kT}} \qquad \text{Equation 3-3}$$

where h = Planck's constant

k = Boltzmann's constant

T = temperature in Kelvins

To increase the number of stimulated emissions, the distribution of electrons must be reversed, that is, $N_c > N_v$. This is achieved by injecting electrons at a sufficient rate into the device (which is called pumping) through the application of an external current. When N_c is made larger than N_v, which is referred to as a population inversion, stimulated emissions outnumber spontaneous emissions. However, it is still necessary to overcome the losses associated with resonant absorption. This is primarily accomplished by proper construction of the laser diode. A basic laser diode is constructed using a Fabry-Perot resonant cavity as shown in Fig. 3-5. The "cavity" is solid, made up of the semiconductor materials. Its ends are formed by cleaving along parallel natural cleavage planes defined by the crystal lattice structure. The resulting flat and smooth endfaces form the front and rear facets of the laser which act as partially reflecting mirrors to stimulated photons. These reflected photons will pass back and forth through the cavity and are amplified by producing additional stimulated emissions in each pass. As more and more photons stimulate more and more electrons in the population inverted conduction band, fewer electrons remain to produce spontaneous emissions.

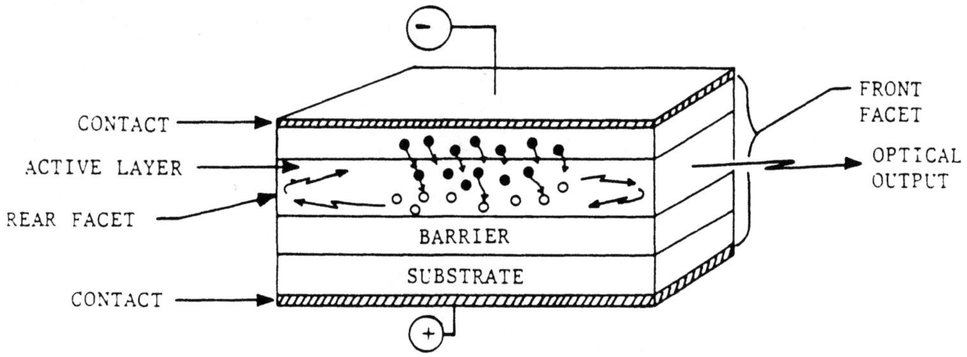

Figure 3-5. A basic laser diode structure.

The photons produced through stimulated emission add to specific electromagnetic fields or modes within the cavity at a rate which is proportional to the intensity of the radiation in each mode. The predominant mode or modes are determined by the dimensional characteristics of the individual laser. Optical amplification increases when the radiation reflects between the two facets of the cavity. During this reflection process, photons are being added to the individual modes thus increasing their total energy. Those modes having a sufficient gain to exceed the cavity losses will continue to be reflected and further amplified. When the gain of one or more modes exceeds the absorption incurred during one round trip of the cavity, then lasing is said to occur. At this point, the drive current associated with this condition is at a level called the lasing threshold current, I_{TH}. The change in optical output of a laser as the drive current increases is shown in Fig. 3-6.

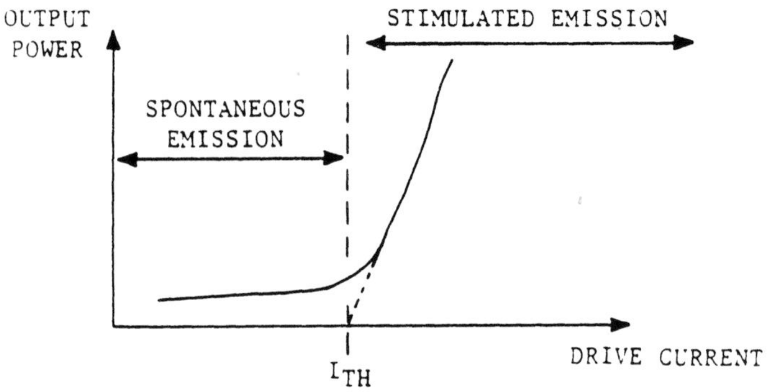

Figure 3-6. Change in laser output power as drive current increases.

IV. EMITTER CONSTRUCTION

There are a multitude of constructions that have been used in the fabrication of LEDs and lasers with some of the construction techniques used for both devices. The length, width and thickness dimensions are commonly referred to as the longitudinal, lateral and transverse dimensions, respectively, as shown in Fig. 3-7.

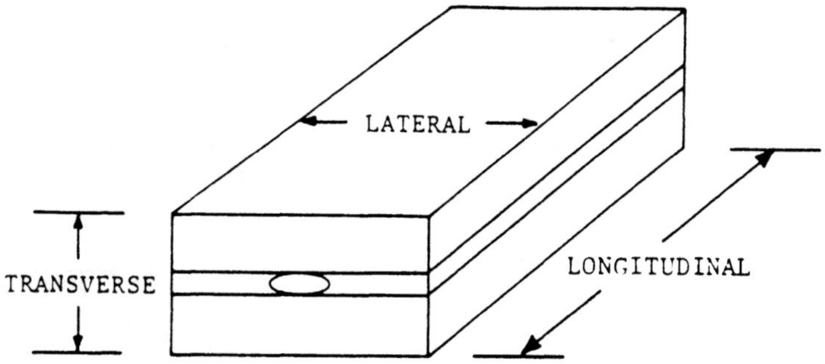

Figure 3-7. Identification of emitter dimensions.

1. LED CONSTRUCTION

The three basic LED constructions are surface emitter, etched-well and edge emitter. The construction of each will be discussed in turn.

a. Planar surface emitter - Fig. 3-8a shows the construction of a typical AlGaAs/GaAs planar heterojunction surface emitter and Fig. 3-8b shows a cross section at the emitting area.

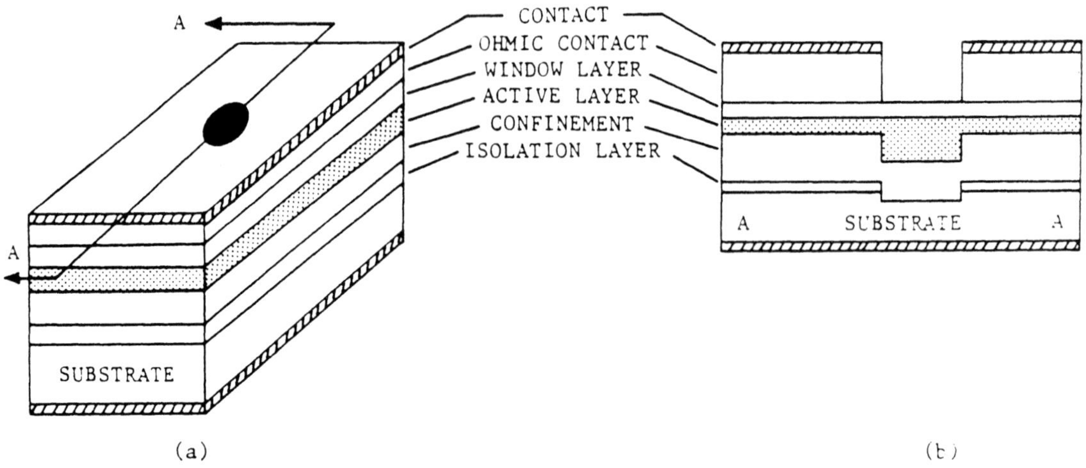

Figure 3-8. Planar surface emitting LED construction.

The device is fabricated by starting with a Zn-doped substrate with an isolation layer grown on top of it. A hole the diameter of which will define the emitting area of the final device is then etched through the isolation layer and partially into the substrate. Next a high resistivity layer is grown as shown which provides lateral injected carrier confinement. This helps to provide a higher concentration of injected electrons in the emitting area which are available for recombination and it restricts the occurrence of luminescence outside of this area. Both of these conditions improve efficiency. The active layer is then grown with the mole fraction of the individual compounds determining the wavelength of the emitted light. Next, the p-n junction is formed such that the band gap of the n-material is greater than that of the p-material. This makes the n-layer transparent to the light emitted and allows it to be directed out of the device. This layer is often referred to as the 'window' for not only does the light exit this layer but it also serves to protect the active layer from contamination. The next layer is provided to help establish good ohmic contact between the device and the metal contact which is applied last.

The surface emitter has a Lambertian radiation pattern. That is, the radiant intensity in any direction is equal to the surface normal intensity multiplied by the cosine of the angle between the direction of radiation and the surface normal.

The thermal impedance path from the active layer to the heat sink is the limitation on power dissipation before additional, degrading internal heat is generated. Since, for a given current density, the emitting area size is constrained by the rate at which heat can be conducted away from the active area, a trade-off must be made between the current density and the emitting area size. The larger the area, the lower the current density becomes which reduces the requirements on heat dissipation. However, a large emitting area means poorer optical power coupling efficiency into a fiber. On the other hand, a small emitting area requires a reduction in current density or device reliability will suffer due to excessive heat. Therefore, if a smaller emitting area is to be used to provide better coupling efficiency, then the current density will have to be reduced which will correspondingly decrease the maximum peak output.

Since this construction does not permit adequate power output with a small emitting area, it typically has an emitting area which is 50 to 100 μm wide. This makes the device inappropriate for use in long haul systems or for coupling into small sized fibers. Also, this relatively large emitting area results in a high device capacitance which precludes its use in high data rate systems. This is true for any large emitting area device. However, considering these limitations, this low cost device is widely used in digital applications operating up to 10 Mb/s over a maximum distance of several kilometers.

b. Etched-well emitter - The etched-well emitter was developed by Burrus and Dawson of Bell Telephone Laboratories[3] and for this reason is often referred to as a Burrus LED. Fig. 3-9 shows a typical etched-well emitter.

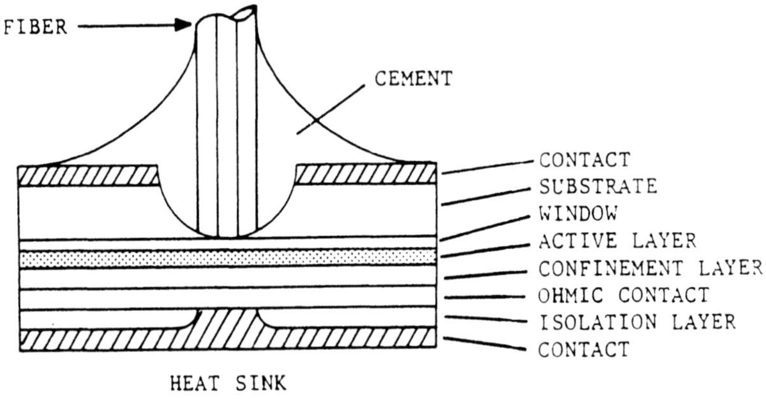

Figure 3-9. Typical etched-well LED construction.

This device is constructed similar to an upside down surface emitter with the optical output passing through a hole in the substrate. This provides improved thermal impedance characteristics which allows the device to be operated at higher current densities.

On the substrate are grown window, active and confinement layers which serve the same functions as was described for these layers in the discussion on surface emitters. Following the confinement layer is a contact layer which provides good ohmic contact for the metal contact. Next comes an isolation layer through which a small opening is etched to define the emitting area.

The metal layer is then applied over the isolation layer which touches the ohmic contact layer through the isolation layer opening. Finally, a well, which must be located directly above the opening in the isolation layer, is etched through the substrate. In this well a fiber is usually attached using an organic cement.

This construction allows the active emitting area to be extremely close to the heat sink which provides very good thermal conduction. Consequently, a higher current density can be maintained through a smaller emitting area. Therefore, even though this device also has a Lambertian radiation pattern, its optical power coupling efficiency is higher than the surface emitter. Also, this lower thermal resistance translates into a smaller peak wavelength shift with changes in temperature.[4] Another benefit to the small emitting area which is typically 50 μm wide, is the ability to operate at improved data rates of 10-20 MHz. This bandwidth can only be increased by making the active layer thinner or by having better control over the material impurity concentration. However, increased bandwidth will be accompanied by an output power penalty.[5]

Unlike the surface emitter in which all layers are grown from one side, the fabrication of the Burrus emitter requires that work be performed on both sides. All layers of the Burrus emitter are grown from one side, then, as explained, the region through which emission occurs is defined from the opposite side. This process requires dual-sided alignment which increases the cost of production over the surface emitter.

c. Edge emitter -- The edge emitting LED gets its name from the way in which it emits light; from its edge instead of through its top surface. Fig. 3-10 shows the construction of a typical double heterojunction edge emitter.

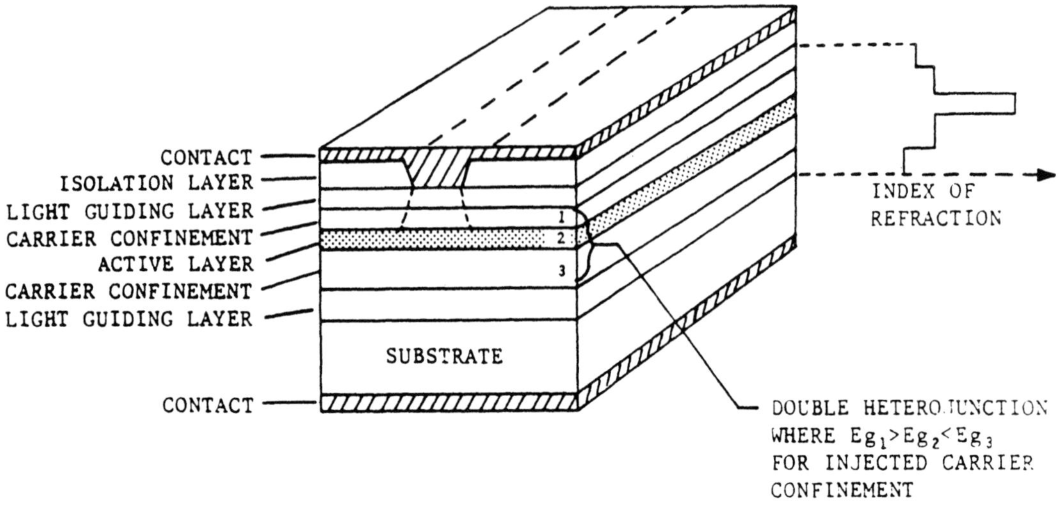

Figure 3-10. Edge emitter double heterojunction LED.

The edge emitter is designed to be more powerful and to emit a more directional beam than the surface or etched-well emitters. To accomplish this, confinement of both injected carriers and the optical field is achieved.

By definition of heterojunction, the carrier confinement layers and the active layer make up the double heterojunction because the band gaps of these confinement layers are greater than that of the active layer. Due to the size of the relative band gaps, the injected carriers are unable to diffuse into the surrounding layers. This provides two benefits: 1) a higher concentration of injected carriers which are available for recombination and, 2) the occurrence of luminescence only in the active layer. Also, the higher energy band gaps of the carrier confinement layers do not absorb the active layer radiation. This allows more light to be coupled into the fiber.

Transverse optical field confinement is accomplished by making the active layer extremely thin (≅ 0.05 μm)[6] relative to the 2-3 μm thick carrier confinement layers. The transverse radiation is unable to remain in the active layer. Instead, it leaks into the adjacent layers where the absorption is very low compared to the self-absorption of the active layer due to careful mole fraction selection. The leaked light remains in these layers since the index of refraction of the optical confinement layers is less than that of the active layer as indicated in Fig. 3-10. This gives the carrier confinement layers waveguide properties resulting in the radiation being guided and not further diffused. Therefore, most of the light perpendicular to the active region is channeled to the endfaces in a narrow transverse pattern. The light generated parallel to the active region is not effected by the waveguide characteristics.

Lateral optical field confinement is determined by the dimensions of the metal contact stripe which is defined by the isolation layer. This stripe runs the full longitudinal dimension of the LED. The effect is to laterally limit injected carrier recombination to just beneath the stripe as indicated by the dashed line in Fig. 3-10.

The benefit of this optical confinement is a much smaller emitting area which produces a directional radiation beam pattern comparable to that of a laser diode. This provides a much better coupling efficiency into a smaller sized fiber and allows operation at higher data rates, up to 100 MHz. However, lasers can still typically couple 10-15 dB more power into a small core fiber.[7]

A higher radiance can be achieved if light is emitted from only one end facet of the LED. This is done by using a dielectric reflector at the rear endface and an antireflection coating at the output endface. Fig. 3-11 shows the typical beam pattern for the edge emitter.

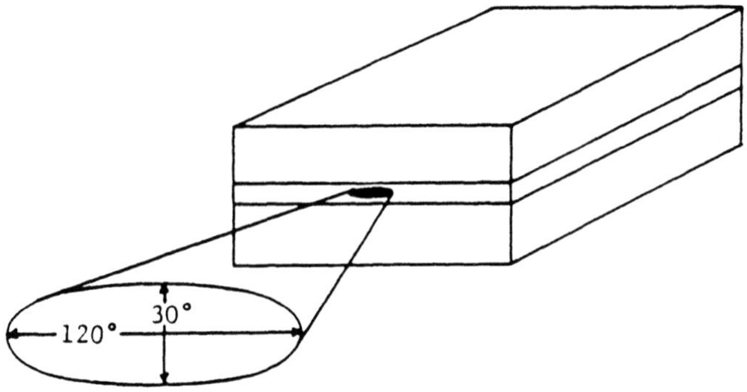

Figure 3-11. Typical beam pattern of an edge emitter LED.

2. LASER CONSTRUCTION

Many laser constructions exist today with the objective of a particular design often being single mode operation. The electromagnetic fields or modes set up in the semiconductor laser cavity can be described in terms of the longitudinal modes, lateral modes and transverse modes. The laser characteristics that these modes determine is given in Table 3-1.

TABLE 3-1. Laser Modes and Their Determining Characteristics.

Longitudinal	Lateral	Transverse
Determines basic spectral response of optical output.	Determines lateral shape of optical output.	Determines transverse shape of optical output.
Typically many modes exist due to cavity longitudinal dimension being much greater than the operating wavelength.	The number of modes depends on the surface of the cavity side walls and lateral dimension of the cavity.	Determines radiation pattern and threshold current.

Ref. 8.

All laser diodes discussed here emit their radiation pattern from one end facet like the edge emitting LED. These semiconductor laser diodes can be separated into two distinct classes: gain-guided and index-guided. Gain-guided lasers establish lateral mode confinement by controlling the current flow through device geometry and typically operate with multiple longitudinal modes. Index-guided lasers have lateral waveguide confinement as a result of the lateral channels having a relatively lower refractive index and they typically operate with only one or a few longitudinal modes. However, this type of device can only be used with single mode fiber since modal noise can result if used with multimode fiber.[7] Specific constructions of each type will be discussed in detail.

a. <u>Homojunction lasers</u> - This device is made from a single semiconductor material as shown in Fig. 3-12.

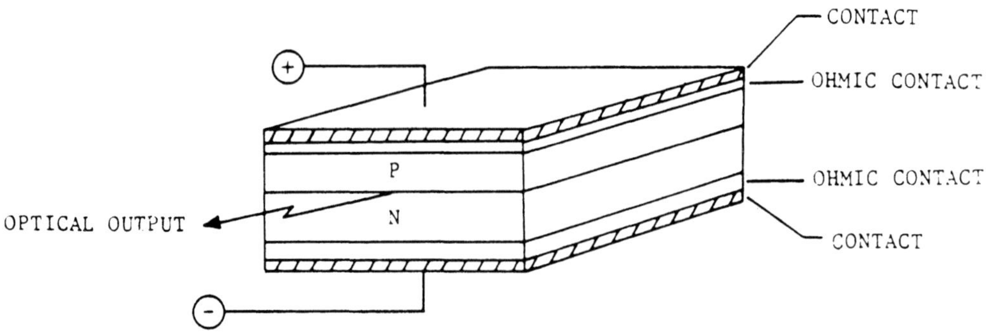

Figure 3-12. Homojunction construction.

A p-n junction is fabricated and metal contacts applied to supply the injected carriers in order to achieve a population inversion. The endfaces are cleaved along natural planes such that they are extremely smooth, flat and parallel. These serve as the reflective surfaces needed for optical feedback which is required for sufficient gain to be achieved.

The lateral modes are suppressed by roughening the vertical side faces, thereby making this a gain-guided device. Since there is no mechanism to confine the transverse modes, the energy density associated with these modes extends beyond the population inversion region. This energy is not available to support stimulated emission and is wasted. As a consequence, the gain of the laser suffers which means a significant increase in current density before lasing current is reached. Therefore, this is a low efficiency, low power device. These devices have a low tolerance to elevated temperatures and are best operated in the pulsed mode. However, the maximum pulse width specified must not be exceeded due to undesirable junction heating. Peak pulsed power out can be as high as 10 watts.[2]

b. Single heterojunction laser -- The single heterojunction (SH) laser is the easiest heterojunction laser to fabricate. A Zn-diffused p-n junction is formed in the n-type material which makes up the active layer. This is shown in Fig. 3-13.

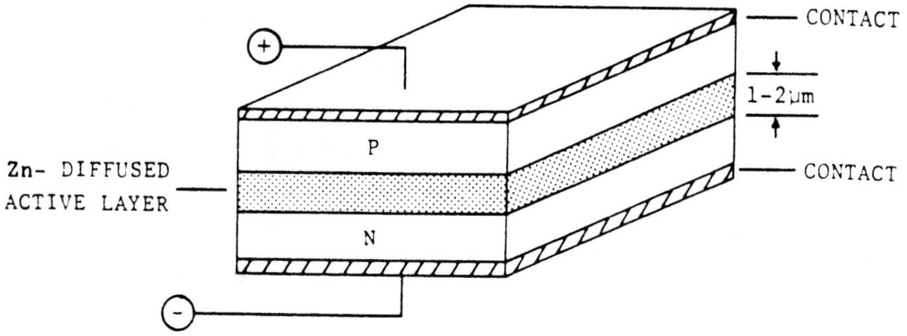

Figure 3-13. Single heterojunction laser construction.

The band gap of the p-type material is greater than that of the n-type while the band gap at the p-n junction interface is equal to that of the n-type, thus, the name single heterojunction.

For highest efficiency and lowest threshold current density, the thickness of the active layer should be between 1-2 μm.[2] This is due to the inability of the population inversion region to extend beyond a 1 μm thick layer adjacent to the p-n junction interface.

Transverse optical field confinement occurs only at the active layer and the p-type material heterojunction due to the significant difference in refractive index. This is an improvement over the homojunction; however, the SH laser must be operated in the pulsed mode at room temperature due to its high current density.[2] The maximum duty cycle must be observed or junction heating will occur. This device is not useful for room temperature continuous wave operation due to the detrimental effects of heating.

The lateral field confinement is achieved as in the homojunction, by roughening the side walls, therefore, this is a gain-guided device.

c. Double heterojunction laser - The double heterojunction (DH) laser is the most commonly used laser construction. Due to the double heterojunction on both sides of the active layer, this laser is very efficient and has a very low threshold current density. Fig. 3-14 shows the construction of this device in its basic design along with the relative indices of refraction of the different layers.

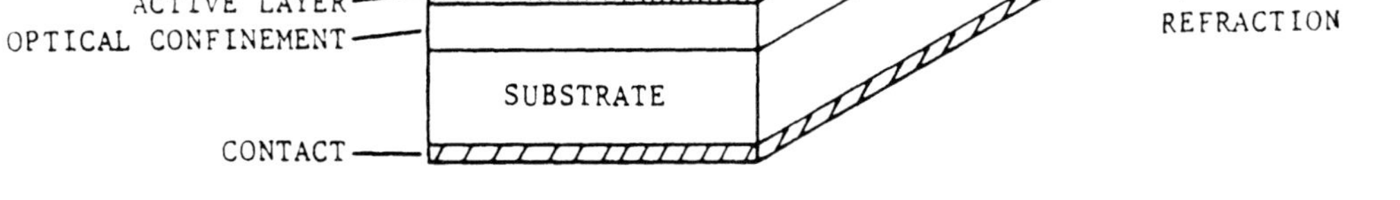

Figure 3-14. Double heterojunction laser construction.

The DH laser has the benefits derived from optical field confinement and injected carrier confinement as described in the discussion on edge-emitting LEDs. The principles are the same and they apply equally to lasers. The result is the availability of a higher number of photons in the active layer to contribute to stimulated emission which gives this device a higher gain, higher efficiency and a lower threshold current density. The energy density distribution for a DH laser is much more confined to the active layer relative to that of the homojunction or SH lasers.

The basic DH laser provides no additional lateral mode control over the homojunction and SH laser constructions. Therefore, the generated light can spread over the entire width of the chip. To accomplish lateral mode control the stripe geometry is employed as was described for the edge emitter LED.

d. <u>DH planar stripe geometry</u> - This gain-guided device has the same optical field and injected carrier confinement mechanisms as does the basic DH laser. There is an additional benefit of lateral active layer excitation confinement through the use of a stripe geometry. This device is shown in Fig. 3-15.

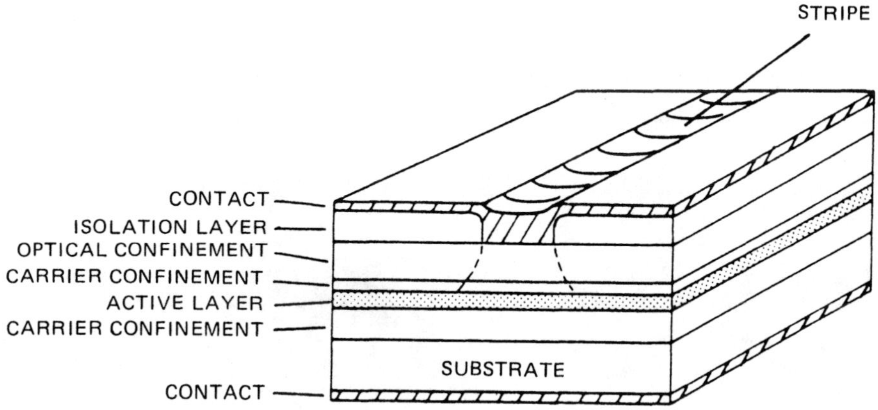

Figure 3-15. <u>DH planar stripe laser construction</u>.

The ability to decrease the lateral distribution of the population inversion area results in a decreased threshold current density. In turn, this will reduce the number of lateral modes and make possible single mode operation which can occur with stripe widths up to approximately 29 μm.[2] The wider the stripe, the greater will be the output power, however, single mode operation will be less stable and the drive current will be higher. Conversely, the narrower the stripe width, the more stable single mode operation will be and the lower drive current will be, but, optical output power will suffer.

This device provides no means to control lateral diffusion at the active layer edges. Therefore, if the stripe is made too narrow in an effort to reduce threshold current density or to achieve stable single mode operation, lateral mode instability will occur as a result of this diffusion process thus defeating any attempt to achieve single mode operation.

Unlike the homojunction and SH lasers, the stripe DH laser can be operated in both continuous wave and pulsed modes at gigabit/sec data rates. Output power levels in excess of 25 mW continuous wave is achievable. Due to the narrow spectral width of 1-2 nm (as compared to 35-90 nm for LEDs)[7] pulse dispersion is not a significant problem.

Three more gain-guided designs which have been developed to significantly reduce lateral mode instability will be discussed. The first of these uses a Zn-diffused p-type material as shown in Fig. 3-16.

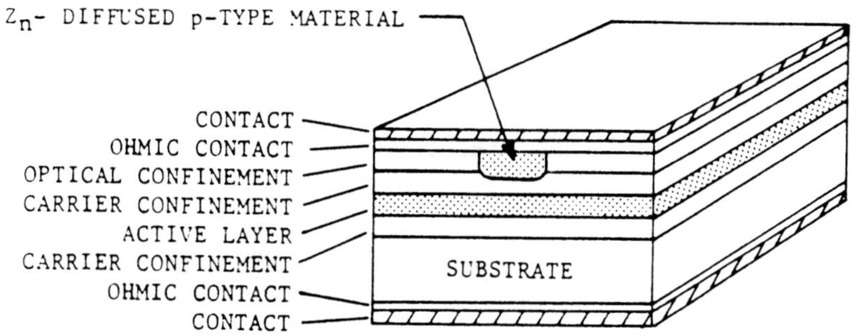

Figure 3-16. DH planar Zn-diffused laser construction.

This design does not establish definite lateral edges within the laser cavity, it only provides better control over the spreading of the current as it travels vertically. The same is true for the DH planar etched stripe laser shown in Fig. 3-17.

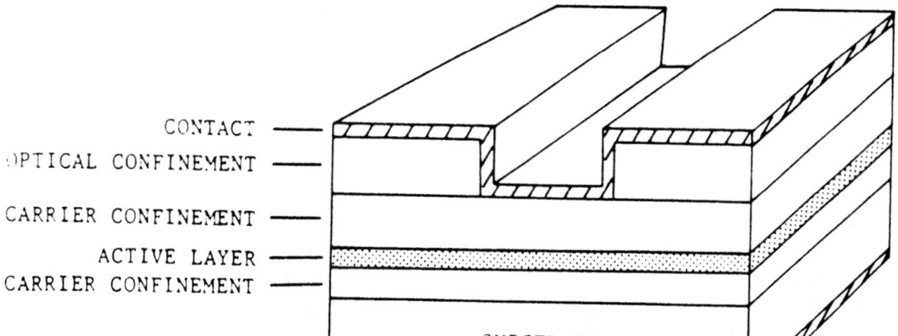

Figure 3-17. DH planar etched stripe laser construction.

The DH proton implanted laser shown in Fig. 3-18 provides very good lateral mode control. This is done by establishing very high resistivity regions everywhere in the optical and carrier confinement layers except for the region directly beneath the stripe. The high resistivity regions are developed by proton bombardment.

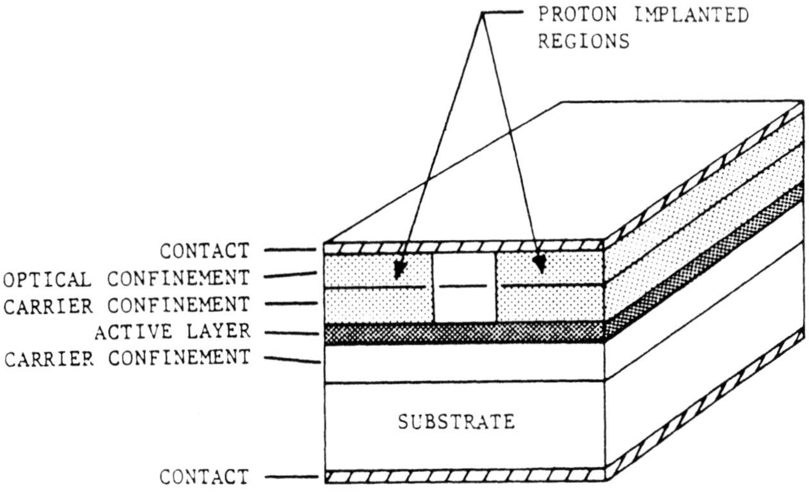

Figure 3-18. DH proton implanted laser construction.

This design allows the formation of narrow (10 μm) stripe widths which will allow stable single mode operation.

e. Buried heterojunction (planar substrate) laser -- The buried heterojunction (BH) laser is also known as a buried mesa device. Fig. 3-19 shows a typical BH laser construction.

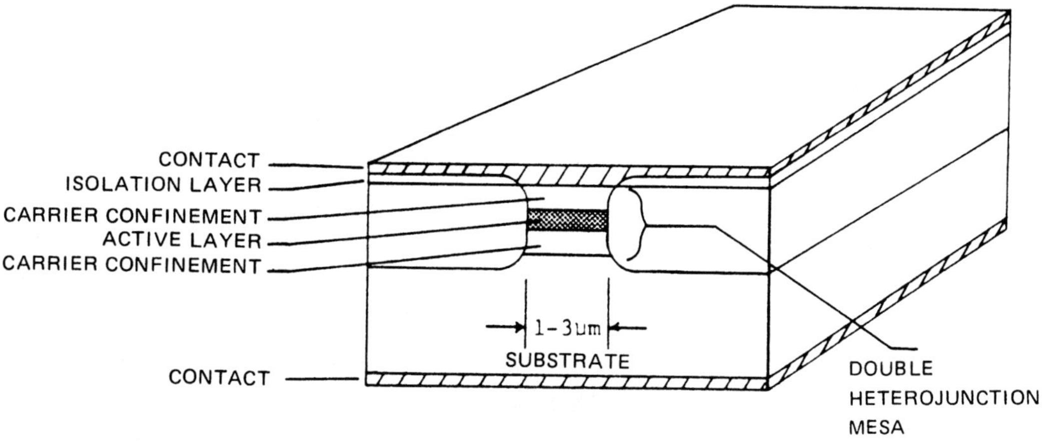

Figure 3-19. Buried heterojunction laser construction.

This index-guided device is fabricated by etching a slab of planar DH material such that a mesa is formed the entire longitudinal dimension of the laser. This constitutes the stripe through which injected carriers will flow. The longitudinal sides of this stripe are then passivated (coated for protection from contamination and chemical reactions) and then buried by growing material in the regions that were removed during the etching process. These interfaces form vertical heterojunctions which effectively block the lateral diffusion of current. However, due to impurities introduced during the two-step fabrication process, leakage current through the buried region in the vicinity of the active region is the primary cause of degradation.

The BH laser construction is capable of providing one of the lowest threshold currents ($\cong$ 10 mA), however, the average value is on the order of 20 - 25 mA.[1]

Considering that this device has double heterojunctions providing strong confinement to both lateral and transverse modes, the development of more than one lasing mode is very unlikely. Therefore, this design can easily operate in a stable, single mode. Because of these characteristics, this device is considered to be a strong candidate for use in underwater applications.[9,10,11]

Many more laser constructions exist, however, the majority of them are still in the developmental stage or have only been available for a very short time. Therefore, they will not be discussed in detail, however, this study does address reliability statistics on their lifetimes. Additional constructions include: oxide defined stripe (ODS), ridge waveguide (RWG), double channel-planar double heterostructure (DC-PDH), distributed feedback (DFB), inverted rib waveguide (IRW), crank-transverse junction stripe (crank-TJS), native oxide (NO) and the self-aligned structure (SAS).

V. EMITTER DEGRADATION CHARACTERISTICS

When addressing the subject of emitter lifetime, the degradation of the device over time, whether the device is an LED or laser, is the primary factor to be considered. Although catastrophic failures do occur, their occurrence early in the devices' lifetime has been drastically reduced through the implementation of careful screening procedures.

A tremendous amount of work is being conducted in an effort to understand and reduce the failure rate of emitters and great progress has been made. However, a word of caution is in order when correlating experimental results developed at separate facilities due to a lack of standardization of criteria and technique. A major problem is the definition of failure. The most commonly used definitions of failure are: a 50% reduction in optical output power for LEDs and a 50% increase in threshold current for lasers. However, these definitions are not consistently used. Also, when numerous emitters are being tested, their combined lifetimes may be statistically calculated using the mean or the median. The mean being the average lifetime of all devices tested. The median being the time at which 50% of the devices have failed. These are two very different concepts. Another parameter whose value varies widely is the activation energy, E_a. The activation energy is a coefficient used in an exponential function to characterize a device's lifetime as a function of temperature. Values assigned to the activation energy vary with the researchers and range from 0.3 to 1.0 eV. Small changes in this parameter have a significant impact on predicted lifetimes. Also, the testing techniques employed vary widely in an effort to obtain a greater understanding of new and existing phenomena. For example, similar tests may be independently conducted to analyze the same phenomenon but one set of devices may have been through a 100 hour burn-in and the other set may not. Therefore, the performance of these devices is expected to be different at the outset of testing. This is important since devices are prone to fail in this burn-in period which allows their removal before testing. If burn-in is not performed before testing, the results will be skewed.

These variations must be taken into consideration when statistically analyzing the results and when applying the analysis to a given situation.

When an emitter fails catastrophically, it ceases outputting optical power at the onset of failure. Failure is immediate and final, not gradual.

An emitter can also degrade slowly, or wear out, but continue to provide optical output power. The degradation is typically in the form of a reduction in output power. This can be countered by an increase in drive current which acts to raise the output power to the proper level. When wear out is the degradation mode, failure occurs when the drive circuitry can no longer raise the output power to the specified level or if the increased drive current overheats the device leading to catastrophic failure. In any event, failure due to wear out is usually dictated by the system's specifications and drive circuitry. This makes it necessary to know detailed system specifications before comparisons between individual device lifetime performances can be made between systems. However, if the degradation rate of an emitter is known along with the minimum acceptable output power for proper system operation, the useful lifetime of the emitter can be estimated for a given system.

1. LED DEGRADATION

There are two general LED degradation modes, rapid and slow.[5] Rapid degradation is caused by the formation of dark line defects (DLDs) and dark spot defects (DSDs). Both DLDs and DSDs are areas of nonradiative recombination in the active region caused by material impurities or crystal lattice defects generated during the fabrication process. Their formation is dependent upon the square of the current density, temperature and the impurity concentration in the active region. They appear as dark areas in the emitting region upon visual examination.

The formation of DLDs and DSDs is the dominant physical degradation mechanism in LEDs. DSDs, which are considered to be caused more by material precipitates than lattice defects, develop much more slowly than DLDs under continuous wave operation. Also, outside of the active region in high radiance devices the current density is very low, therefore, in these regions the rate of formation of DLDs is considerably reduced.

Slow degradation will occur in LEDs even if there are no DLDs or DSDs. This degradation is independent of current density and dependent upon temperature. In terms of optical output power, P(t), one expression which is used for degradation rate is given in equation 3-4.

$$P(t) = P_o e^{-\beta t} \qquad \text{Equation 3-4}^{5}$$

where $P(t)$ is the output power as a function of time

P_o is the initial output power

β is the degradation rate given by

$$\beta = \beta_o e^{-E_a/kT}$$

where β_o is a proportionality constant

E_a is the activation energy in eV

k is Boltzmann's constant in Joules/Kelvin

T is the temperature in Kelvins

These two expressions show that the output power will change with temperature and time. As the temperature increases, β will increase. This will reduce P(t). Likewise, as time progresses, P(t) will decrease. Therefore, equation 3-4 can be used to predict the optical output power at time "t" and temperature "T".

Slow degradation is considered to be a result of 1) diffusion of impurities into the active region from the surrounding material and, 2) the in-migration of metal atoms from the contact materials once contact deterioration has started. Passivation can reduce the in-migration of metal atoms.

2. LASER DEGRADATION

The degradation of semiconductor lasers is a function of numerous parameters. The most important are temperature, humidity, current density, optical power density, processing techniques, materials, mechanical stress, thermal resistance and leakage current. Associated with these parameters are degradation mechanisms having varying degrees of impact on laser performance. The most intensely studied and best understood mechanisms are dark line defects, metallurgical reactions, facet damage and slow bulk degradation.

Experiments performed in various laboratories[10,12] have shown that lasers degrade, as indicated by an increase in threshold current, I_{th}, in three distinct modes. These modes are identified in Fig. 3-20 as the incubation mode, saturable current mode and wear out mode.

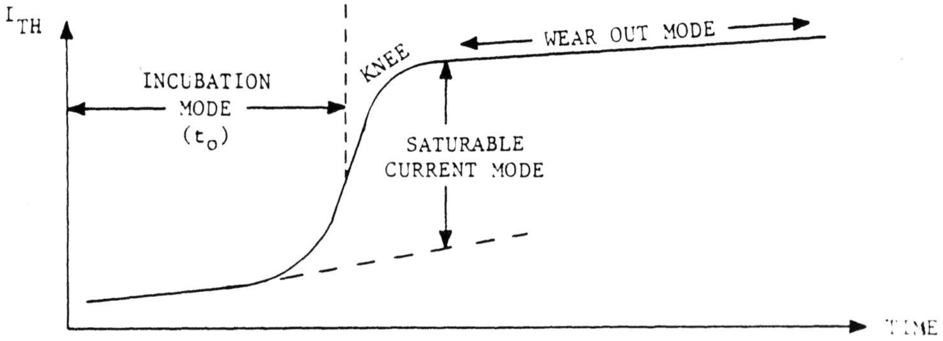

Figure 3-20. Threshold current degradation modes.

Researchers at AT&T Bell Laboratories[10] have empirically described the saturable current degradation and wear out degradation for 1.3 μm BH InGaAsP lasers with equations 3-5 and 3-6, respectively.

$$I_s(t) \propto \frac{S}{1 + e^{\frac{t_o - t}{\tau}}} \qquad \text{Equation 3-5}$$

where
$I_s(t)$ is the saturable current degradation
S is the relative strength of the saturable current
t_o is the incubation period
τ effects the rate of change of the saturable current at t_o

$$I_w(t) \propto 1 + Rt \qquad \text{Equation 3-6}$$

where $I_w(t)$ is the wear out degradation

$$R = R_o e^{-\frac{E_a}{kT}}$$

R_o is a constant

E_a is the activation energy in eV

k is Boltzmann's constant

T is temperature in Kelvins

Of prime importance are the mechanisms that contribute to the saturable current, the formation of the knee, wear out, or long term degradation, and catastrophic degradation. Each of these will be discussed in turn.

a. Saturable current mode -- Investigators at AT&T Bell Laboratories[9] and Hitachi, Ltd.[12] have attributed the saturable current mode to be a result of an increase in leakage or shunt current around the active region.

The amount of leakage current flowing in a laser is primarily dependent upon temperature and terminal current. Even at low lasing threshold currents and room temperatures, some leakage current flows. As the temperature and terminal current are increased, leakage current also increases and in continuous wave operation this can reduce device current flow below the threshold level. At this point (typically 100° C and 250 mA)[12] there is essentially no optical power output. The leakage current can be reduced through passivation of the longitudinal heterojunctions in BH lasers.[2]

One theory attributes the increase in leakage current to the formation of dark spot defects which are formed as a result of the in-migration of contact material into the vicinity of the active region.[9] The current density is larger at these nonradiative DSD centers. The leakage current flowing through these DSDs does not contribute to the device's optical output power and can be considered to be diversionary paths from the active region. This causes a decrease in the optical power output which is compensated for by an increase in the threshold current.

b. Formation of the knee -- The number of DSDs remains fixed after a period of time which coincides with the knee[9] of Fig. 3-20. Since each defect center can only pass a finite amount of current, once the formation of DSDs stabilizes, the threshold current will no longer increase due to this mechanism.

The time required to reach the point at which the threshold current is stabilized will vary significantly with the leakage current. As the terminal current and temperature are increased, more leakage current will flow. At 250 mA and 100° C the threshold current can be stabilized in less than 20 hours[10,12] whereas over 1000 hours[11,12] are required at lasing currents and 60°C.

c. Wear out degradation mode -- This mode of degradation is generally thought to be a result of facet oxidation, contact degradation or crystal grown-in defects.

(1) Facet oxidation - This is a mild form of degradation which manifests itself as stained facets/mirrors due to photo-oxidation. The facets are extremely vulnerable to oxidation when the laser is operated in air without prior passivation. This process is accelerated by the presence of moisture or oxygen in the operating environment.

Facet erosion has deleterious effects on laser performance by reducing the facet reflectivity which in turn increases the nonradiative recombination rate at the facets. This causes an internal temperature rise and a decrease in internal quantum efficiency. This chain of events slowly results in an increased threshold current, i.e., degradation over time.[13]

Facet erosion can be prevented by the application of a coating to the facet which greatly reduces any chemical reactions. This coating should have a thickness of one-half wavelength so as not to alter the reflectivity which, if reduced, will impact the threshold current. Three materials which have been proven to be good coatings are aluminum oxide (Al_2O_3), silicon nitride (Si_3N_4) and silicon dioxide (SiO_2); with Al_2O_3 giving the best results.

(2) Contact degradation -- When the thermal resistance of the contact between the heat sink and semiconductor device increases, the junction temperature rises which is accompanied by an increase in threshold current. The increase in thermal resistance is dependent upon the type of solder used, the current density and the temperature. The increase in junction temperature above that of the heat sink is expressed by equation 3-7.

$$\Delta T = (I^2R_S + IV_i)(1 - \eta_p)D\phi \qquad \text{Equation 3-7}^{13}$$

where

- ΔT = Increase in junction temperature over heat sink temperature
- I = device current
- R_s = device electrical resistance
- V_i = junction voltage
- ϕ = thermal resistance
- D = duty cycle
- η_p = power conversion efficiency

The threshold current density dependence on temperature is expressed by equation 3-8.

$$J_{TH}(T) \simeq J_{TH}(0)e^{\frac{\Delta T}{T_o}}$$

Equation 3-8[13]

where J_{TH} is the threshold current density
T_o is the characteristic temperature
ΔT is the change in temperature in Kelvins

From equations 3-7 and 3-8, it can be seen that the higher the duty cycle, D, and the lower the value of T_o, the greater will be the dependence of threshold current on the quality of the contact.

The thermal resistance is dependent upon the temperature, contact current density and the type of solder used. Indium (In) solder used in conjunction with gold (Au) plated contacts results in the formation of intermetallic compounds from these elements which have high thermal resistivities. Also, the migration of In solder into and around the active region results in voids having very high thermal resistances. The reduction of Au content has been demonstrated to successfully reduce degradation due to thermal resistance whereas the use of Palladium has been used to inhibit the formation of voids where Au and In are used. A recent development used to eliminate contact degradation in buried heterojunction lasers is to construct the device p-side up[15] as shown in Fig. 3-21.

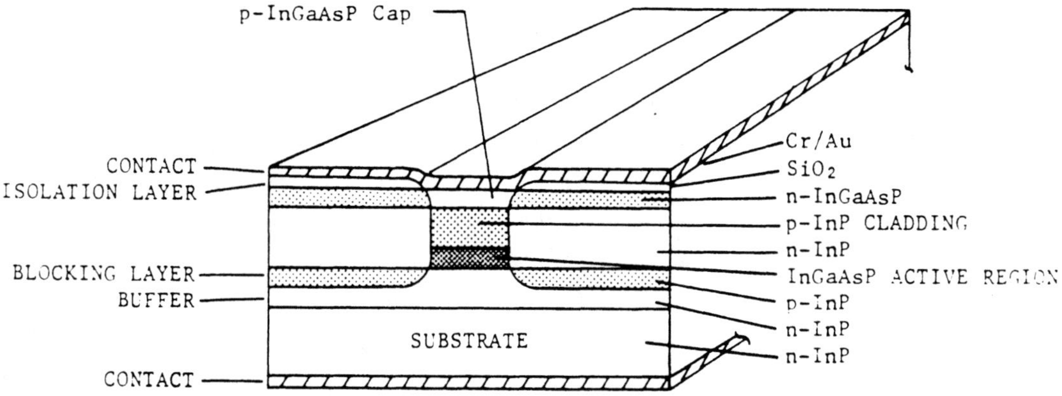

Figure 3-21. P-side up construction of a BH laser.

(3) Grown-in defects -- This is the major cause of long term degradation and is a result of lattice defects introduced into the laser material during the fabrication process. These defective regions are nonradiative recombination centers which do not contribute to the optical power output. Instead, they reduce the internal quantum efficiency and act as nucleation sites from which larger nonradiative 'networks' grow. These networks appear as dark regions and have thus been given the name dark line defects. Once DLDs have formed, the corresponding degradation rate is very dependent upon the current density; the higher the density, the faster the degradation. DLDs are more of a problem in AlGaAs/GaAs lasers than in InGaAsP/InP lasers.[12]

Without the presence of lattice defects, degradation over long periods of time will still occur. This process shows a strong temperature dependence. These devices exhibit a localized reduction in radiative efficiency of the device material. As a result, degradation increases due to the need for a higher threshold current to achieve a given optical output level. The localized areas tend to start at the facets and move in, towards the active region. This has occurred in lasers with Al_2O_3 facet coatings and there has been no indication of mirror damage.[13]

Several precautionary steps can be taken to help improve the degradation rate as caused by internal defects. The most obvious is to use materials with very low lattice defects. Defects in the edges should not be in the current path; only cleanly cleaved edges should be accepted in order to prevent the propagation of the defective area towards the active region. High Zn diffusion concentrations into the active region should be avoided since the Zn can create defects which act as nucleation sites for additional degradation.[13]

d. Catastrophic degradation - Catastrophic or rapid degradation can usually be attributed to facet damage or bonding-solder deterioration resulting in whisker growth.

(1) Facet damage - When optical power densities as high as several milliwatts per micrometer of emitting facet width are reached, physical facet damage is often caused if this conditions prevails. The damage significantly reduces the facet reflectivity causing an abrupt increase in threshold current and a corresponding decrease in quantum efficiency. This is accompanied by an abrupt increase in temperature. AlGaAs/GaAs lasers experience facet damage more frequently than InGaAsP/InP.[12]

The threshold at which catastrophic facet damage occurs can be increased by applying an antireflective coating to the facets, thereby reducing facet absorption. This will decrease the internal power level for a specific optical output power level. Of course this will also cause an increase in threshold current which will generate more internal heat. Therefore, this technique is best suited for pulsed operation of the laser.

(2) Whisker formation -- Rapid degradation due to whisker growth has been associated with the deterioration of solder, especially tin (Sn) rich solder. When whiskers develop, they eventually lead to internal electrical shorts. Tests have shown that ambient temperatures of 50°C will erhance the formation of whisker growth. If lead (Pb) is added to the solder, this degradation mechanism is greatly reduced. Mizuishi et al has demonstrated this by storing devices with two types of solder for a fifty week period at 50°C.[15] The two solders used were Au-Sn (Sn:90 wt %) and Pb-Sn (Pb:40 wt %). The devices using Au-Sn solder developed whiskers within 8 - 20 weeks while the devices using Pb-Sn solder did not develop any whiskers. The amount of Sn was decreased to Au-Sn (Sn:20 wt %) and devices where again tested for whisker formation. No whiskers developed after fifty weeks but the average shearing force required to remove the chips from their submounts was found to be about one half that of the former Sn-rich solder. This means more delicate handling of the devices will be required during fabrication and the vulnerability to damage during mechanical shock and vibration is significantly increased after packaging.

VI. EMITTER TESTING TECHNIQUES

Many test techniques have been developed in an effort to determine the reliability and lifetime of semiconductor emitters. A large number of these techniques have been successfully used by the semiconductor industry for years. However, emitters, especially lasers, have unique properties which demand that modifications or special consideration be given to the way in which tests are performed. Some of the more common tests performed will be presented and any precautionary measures that need to be taken into consideration will be identified.

Tests which are commonly performed on lasers and LEDs are given in Table 3-2.

TABLE 3-2. Emitter Tests.

Test	MIL-STD-750 Test Method	MIL-STD 883 Test Method
Operating Life	1027	N/S*
Temperature Cycling	1051	1010
High Temperature Storage	1032	N/S
Thermal Shock	1056	1011
Humidity Storage	1021	N/S
Hermetic Seal	1071	N/S
Constant Acceleration	2006	2001
Variable Vibration	2056	2007
Mechanical Shock	2016	2002
Salt Atmosphere	1041	1009
Seal, Fine and Gross Leak	N/S	1014
Burn-in	N/S	1015
Moisture Resistance	N/S	1004
Visual Examination	N/S	1010

* None Specified.

The following discussion addresses those tests which are pertinent to laser evaluation.

1. CHARACTERIZATION PRIOR TO TESTING

Characterization of the laser is done prior to stress testing. When the derivative of the optical power output with respect to the current is plotted versus the current (dL/dI-I) before stress testing, kinks in the curves can be identified. A typical family of curves which displays kinks is shown in Fig. 3-22.

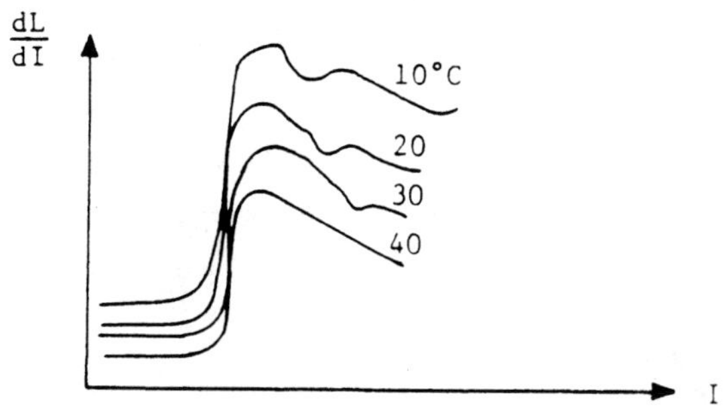

Figure 3-22. Typical dL/dI-I curves.

The occurrence of the kinks have been associated with the following anomalies:[15]

1) Non-linearities in the optical output power
2) Incoherency in the longitudinal modes
3) Longitudinal mode spectrum splitting and,
4) Longitudinal mode spectrum broadening.

Experiments performed on buried heterojunction InGaAsP/InP lasers have shown that only lasers with stripes wider than 2 μm displayed these kinks and undesirable characteristics.[10,15] However, after 1500 hours of aging at 60°C with a constant 5 mW/facet optical output, both the kinks and the associated anomalies disappeared.

2. UNBIASED HUMIDITY AND TEMPERATURE TESTS

Humidity tests have caused the formation of an oxide layer on the facets of lasers with no protective coating applied. This oxide layer will develop on the facet whether the device is reliable or not and, therefore can not be used to predict performance. However, once this coating forms, it will interfere with and reduce the optical output power. For constant output power operation, the device will be driven at a higher current level to adjust for the reduced output. This can lead to catastrophic facet damage.

High temperatures do not cause degradation of lasers under unbiased conditions up to the solder melting temperature. Shelf life has not been found to be a problem under high temperature and unbiased conditions.

3. HIGH TEMPERATURE BURN-IN

Initial high temperature burn-in is performed to eliminate those lasers prone to infant mortality. Devices are operated at normal optical output levels for about 100 hours at high temperatures. However, as is indicated in Fig. 3-20, lasers generally degrade with an initial incubation period and a saturable current mode. Just screening by high temperature burn-in for a relatively short period of time is not sufficient to ensure that the device is operating in the stabilized wear out mode. It has been demonstrated[11,12] that aging under high temperature conditions and normal bias must be conducted in excess of 1000 hours before the devices exhibit wear out degradation at a constant rate.

4. ACTIVE STRESSING

High temperature/high current testing (250mA and 100°C) will stress the shunt paths of lasers which have been determined to be responsible for the rapid degradation seen during the saturable current mode as discussed in paragraph V.2.a. This testing is performed after the initial high temperature burn-in tests are performed.

Tests conducted at AT&T Bell Laboratories[9] and Hitachi, Ltd.[12] have demonstrated that active stress testing at elevated temperatures and current levels will reduce the time it takes to stabilize the degradation rate of lasers. This time was found in both cases to be less than 20 hours as compared to more than 1000 hours for high temperature only testing. Also, AT&T reported that the degradation rates of those devices tested in this way were lower than those not subjected to this test.

5. WEAR OUT DEGRADATION MODE BURN-IN

This burn-in testing is a further refinement to screening out bad devices. It is performed to eliminate those devices which may fail early in the wear out period of operation. This is performed at elevated temperatures (≅ 60°C) for approximately 1000 hours at normal bias levels. After 1000 hours of testing, the degradation rate was found to be insignificant such that testing could be discontinued.

6. TESTING RESULTS PRECAUTIONS

By examining Fig. 3-20, it is apparent that the degradation rate of lasers must have stabilized to the wear out operating mode before any accurate lifetime predictions can be made. Otherwise, if lifetime predictions are based on the incubation portion of the curve, it is apparent that the prediction may change signi icantly after a short period of time. Once this occurs, the system may appear to be experiencing infant mortality as a result of the increased threshold current. On the other hand, if the lifetime predictions are based on data taken from the saturable current mode, then lasers which may actually be very reliable could be rejected. Either way, the outcome is undesirable.

A notable fact regarding the saturable current mode and the stabilized wear out mode is that the activation energy has been consistently found to be high for the former and low for the latter without regard to laser construction.[9,12,13]

The results of the tests performed at AT&T showed that a higher current was needed to maintain a constant optical output when stressed at the elevated temperatures and currents. However, the stressing did not appear to cause any performance deterioration of the devices.

7. DEGRADATION ACCELERANTS

In summary, certain laser parameters have been found to be closely associated with particular types of degradation. Table 3-3 lists these parameters and the degradation to be expected when performing screening tests.

TABLE 3-3. Parameters Causing Laser Degradation.

DEGRADATION	LASER PARAMETER
Facet Damage	Optical Power Density Pulse Width
Facet Erosion	Optical Power Density Ambient Conditions
Grown-in Defects	Current Density Temperature
Contact Deterioration	Current density Temperature

The degradation rate caused by grown-in defects and contact deterioration is similar in LED operation as it is in lasers.

VII. FAILURE PREDICTIONS

1. FAILURE MODELS

Laser failure rate equations have been developed by NTT Corporation[11] researchers for both wear out and catastrophic failures of buried hetero-junction InGaAsP/InP devices. Equations 3-9 and 3-10 give these expressions.

$$\lambda_w(t) = \frac{\sqrt{2}\, e^{\left(-\frac{\left[\ln \frac{t}{t_m}\right]^2}{2\sigma^2}\right)}}{\sqrt{\pi}\, t\, \sigma\, \mathrm{erfc}\left[\frac{\ln \frac{t}{t_m}}{\sqrt{2}\,\sigma}\right]} \qquad \text{Equation 3-9}$$

where λ_w = wear out failure rate

t = service time

t_m = median lifetime

σ = standard deviation

erfc = complementary error function

$$\lambda_c(t) = \frac{n}{N \cdot t} \qquad \text{Equation 3-10}$$

where λ_c = catastrophic failure rate

n = number of failed devices

N = sample size

t = testing time

The equation for catastrophic failures assumes failures to be constant with time.

Equations 3-9 and 3-10 do predict failure rates, however, they are only applicable to long wavelength InGaAsP/InP buried heterojunction lasers. This does not detract from their validity, it only serves to caution the reader that predictions developed from data taken on one type of laser may not be suitable for developing predictions for another type of laser.

2. EXTRAPOLATED LIFETIME PREDICTIONS

The literature is replete with lifetime predictions for various photonic emitters. This obvious interest is a very good sign for the fiber optics industry, however, it is unfortunate that the results vary by orders of magnitude. This certainly gives rise to confusion and a valid reason for concern about the integrity of the predictions. Unfortunately, until testing techniques are more standardized such that results can be more closely correlated, these troubles will continue. Therefore, the following data is presented to give the reader a general idea about current emitter lifetimes. Tables 3-4 and 3-5 give lifetimes* for LEDs and lasers, respectively. These figures have been calculated by extrapolating actual accelerated lifetime test results based on a model which best fits the test data.

* Lifetimes are given in hours of operation. Table I-2 of Appendix I provides equivalent values in years to commonly encountered lifetime figures.

TABLE 3-4. LED Lifetime Predictions.

Reference	Wavelength (μm)	Material	Construction	Output Power (mW)	Forward Current (mA)	Temperature (°C)	E_a (eV)	Mean Time To Failure (10^6 hrs)	Median Value* (10^6 hrs)	Failure Criteria	Date of Source	Comments
5	0.8–0.9	AlGaAs/GaAs	Surface	—**	—	25***	0.56–0.6	1.0–10	—	—	1982	
21	0.85	AlGaAs/GaAs	—	0.055–0.14	100	50	—	1.0	—	—	7/83	
19	0.82	AlGaAs	High Radiance Planar DH chip	—	—	86 junction 65 case	0.43	—	0.0265	50% drop in output power	ca. 1983	Output power characterized using: $P_o(t) = P_o(0)\exp[-(t/\tau)^{1/2}]$ where $P_o(0)$ = initial output power τ = degradation time constant t = total operating time Temperature and current dependence, τ, is given by: $\tau = A(I_f)^x \exp(E_a/kT)$ where I_f = forward current A = constant x = current dependence factor
21	1.25–1.33	InGaAsP/InP	—	0.02	100	25	--	1.0	—	—	7/83	
19	--	AlGaAs/Si	Full Surface LPE chip	—	100	70 junction	--	--	0.24	50% drop in output power	ca. 1983	Output power characterized using: $P_o(t) = P_o(0)\exp[-(t/\tau)^{1/2}]$ where $P_o(0)$ = initial output power τ = degradation time constant t = total operating time Temperature and current dependence, τ, is given by: $\tau = A(I_f)^x \exp(E_a/kT)$ where I_f = forward curent A = constant x = current dependence factor

* Time at which one half of the devices can be expected to fail.
** Data not available.
*** The source data indicated room temperature which has been interpreted to be 25°C.

TABLE 3-5. Laser Lifetime Predictions.

Reference	Wavelength (μm)	Material	Construction	Output Power (mW)	Operating Current* (mA)	Temperature (°C)	E_a (eV)	Mean Time To Failure (10^6 hrs) Confidence (%)	Median Value** (10^6 hrs) Confidence (%)	Failure Criteria	Date of Source	Comments
23	0.83/0.87	AlGaAs/GaAs	CDH (ridge-guide) LPE	3-4	60-100 threshold	25***	0.7	>1.0 std. deviation σ = 1.38	—****	Unable to deliver 3mW output power at 70°C	8/83	Prescreened at room temperature/3mW for 100 hours. Mounted with In solder on Cu heat sink. Passivated emitting facet with half wavelength Al_2O_3. By definition: $MTTF = \tau_m \exp(\sigma^2/2)$ where τ_m = median life; $\sigma = \ln(\tau_m)/\ln(\tau_0)$ where τ_0 is point on log-normal plot that 15.8% of devices fail.
13	0.82-0.85	AlGaAs/GaAs	ODS-CDH chip	≈10 continuous wave	50-150 threshold	22	≈0.7	≈0.08	—	Unable to deliver 5mW output power	2/80	
14	--	AlGaAs/GaAs	DH chip LPE	5	—	70	--	0.13	—	50% increase in drive current	3/83	
21	0.85	AlGaAs/GaAs	—	1.6	90 threshold	50	--	0.4	—	—	7/83	
10	1.3	InGaAsP/InP	BH chip	5	—	10	1.0	—	>20 >5.7 at 98%	100% increase in threshold current	3/85	Prescreened at 50°C/5mW for 100 hours. By definition: Time to Failure = $10^5/[R_o \exp(-E_a/kT)]$ where R_o = constant
15	1.3	InGaAsP/InP	BH p-side up chip	5	30 at 25°C threshold	20	0.6	—	> 0.5	50% increase in drive current	8/83	No mirror coatings. Tested in dry nitrogen environment. Mounted on SiC ceramic using PbSn (Pb:40) eutectic alloy. Pre-screened at 60°C/5mW for 100 hours.
25	1.3	InGaAsP/InP	BH chip	5	10	25***	0.3	0.3	--	100% increase in threshold current	1984	
25	1.3	InGaAsP/InP	BH chip	5	20-50 threshold	25***	0.9	> 10	--	100% increase in threshold current	1985	No facet coating.
25	1.3	InGaAsP/InP	BH chip	5	--	25***	--	10	--	100% increase in threshold current	1984	

See notes at end of table.

TABLE 3-5. Laser Lifetime Predictions (con't).

Reference	Wavelength (μm)	Material	Construction	Output Power (mW)	Operating Current* (mA)	Temperature (°C)	E_a (eV)	Mean Time To Failure (10^6 hrs) Confidence (%)	Median Value** (10^6 hrs) Confidence (%)	Failure Criteria	Date of Source	Comments
12	1.3	InGaAsP/InP	BH chip	5	15-25 threshold Varied to maintain output power	50	0.32	--****	1.0	50% increase in drive current	12/86	Mounted junction-up on SiC using PbSn eutectic solder. No protective coating was used on laser mirrors. Dry nitrogen aging.
25	1.3	InGaAsP/InP	BH chip	5	10	50	0.3	0.08	—	100% increase in threshold current	1984	
25	1.3	InGaAsP/InP	BH chip	5	20-50 threshold	50	0.9	—	0.14 σ = 1.6	100% increase in threshold current	1985	No facet coating.
11	1.3	InGaAsP/InP	DC-PBH chip	5	50	50	0.35	—	0.27 σ = 0.65	150% increase in drive current	12/85	Mounted junction-up on Si heat sink with Au-rich AuSn solder.
11	1.3	InGaAsP/InP	DC-PBH chip	—	70	50	0.35	--	0.14 σ = 0.99	150% increase in drive current	12/85	Mounted junction-up on Si heat sink with Au-rich AuSn solder.
25	1.3	InGaAsP/InP	ODS chip	4	85 threshold	50	--	0.04	—	100% increase in threshold current	1984	
25	1.3	InGaAsP/InP	NO chip	--	150-200 threshold	25***	0.78	10	--	100% increase in threshold current	1984	
25	1.3	InGaAsP/InP	SAS chip	3/5	78-110	50	--	0.3	--	100% increase in threshold current	1984	Used Weibull distribution.
25	1.3	InGaAsP/InP	BC chip	3	10-20	50	--	2.4	--	100% increase in threshold current	1984	
25	1.3	InGaAsP/InP	BC chip	5	10-20	50	--	17	--	100% increase in threshold current	1984	
21	1.33	InGaAsP/InP	--	1.6	70 threshold	25	--	0.4	--	--	7/83	
26	--	InGaAsP/InP	chip	5	--	50	--	> 0.1	--	--	6/85	

See notes at end of table.

TABLE 3-5. Laser Lifetime Predictions (con't).

Reference	Wavelength (μm)	Material	Construction	Output Power (mW)	Operating Current* (mA)	Temperature (°C)	E_a (eV)	Mean Time To Failure (10^6 hrs) Confidence (%)	Median Value** (10^6 hrs) Confidence (%)	Failure Criteria	Date of Source	Comments
22	1.3	--****	chip	1.5	85	20	0.43	0.5	—	—	11/86	
24	1.3	—	chip	5	—	20	0.7	>1.0 at 60%	—	100% increase in operating current	9/86	
22	1.3	—	chip	1.0	80	22	0.7	4.2	—	—	11/86	
22	1.3	—	chip	1.0	75	25***	0.8	1.0	—	—	11/86	
22	1.3	—	chip	1.0	105	25***	0.43	0.22	—	—	11/86	
22	1.3	—	chip	0.7	65	25	—	1.0	—	—	11/86	
22	1.3	—	chip	1.0	65	25	0.7	1.128	—	—	11/86	
22	1.3	—	chip	1.0	75	25	1.1	50	—	—	11/86	
22	1.3	—	chip	1.0	115	25	0.6	0.503	—	—	11/86	
22	1.3	—	chip	1.5	70	25	0.8	1.0 at 40mA	—	—	11/86	
22	1.3	—	chip	3.0	75	25	1.1	50	—	—	11/86	
22	1.3	—	chip	1.0	65	50	—	>0.1 at 5mW	—	—	11/86	

See notes at end of table.

TABLE 3-5. Laser Lifetime Predictions (con't).

Reference	Wavelength (μm)	Material	Construction	Output Power (mW)	Operating Current* (mA)	Temperature (°C)	E_a (eV)	Mean Time To Failure (10^6 hrs) Confidence (%)	Median Value** (10^6 hrs) Confidence (%)	Failure Criteria	Date of Source	Comments
24	1.3	--****	Module	--	—	20	--	≥ 0.72	—	100% increase in operating current	9/86	
22	1.3	InGaAsP/InP	Pigtail, TEC, Monitor Detector LPE Module	1.0	90	50	—	> 0.219 at 60%	—	50% increase in drive current	11/86	Ti:Pt:Au Schottky p-side contact. AuSn:Au alloyed n-side contact.
22	1.3	—	Pigtail, TEC, Monitor Detector LPE Module	—	—	—	0.43	0.35	—	—	11/86	
22	1.3	—	Pigtail, TEC, Monitor Detector LPE Module	--	—	--	0.6	0.173 at 90% 27.7 at 60%	--	--	11/86	

* Includes threshold and modulation current (difference between threshold and current required to deliver specified power output) unless noted.
** Time at which one half of the devices can be expected to fail.
*** The source data indicated room temperature which has been interpreted to be 25°C.
**** Data not available.

Generally speaking, AlGaAs/GaAs surface emitting LEDs have extrapolated room temperature lifetimes which are at least an order of magnitude greater than semiconductor lasers. Also, room temperature lifetimes of InGaAsP family LEDs increase exponentially with decreasing band gap. Therefore, long wavelength LED devices should outlast short wavelength devices by orders of magnitude. This is indicated in Table 3-4.

Fig. 3-23 illustrates the typical Arrhenius relationship between emitter lifetime and case temperature.

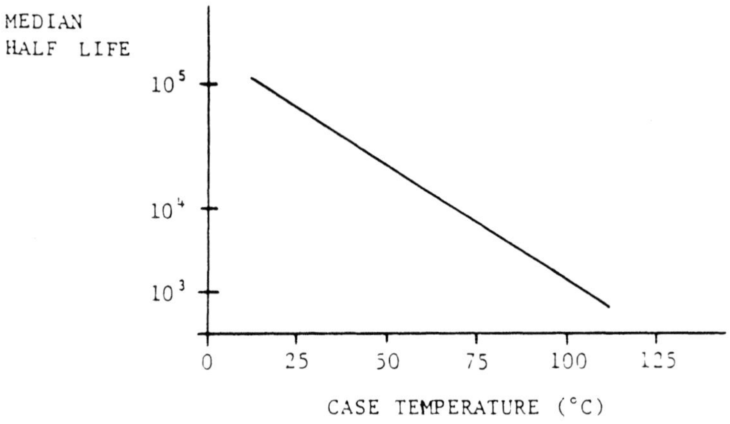

Figure 3-23. Log-normal plot showing the general relationship between emitter lifetime and case temperature.

The rate of degradation, that is the slope of the curve, is directly dependent upon the activation energy, E_a, for a given device, therefore, it is extremely important that the value of this parameter be accurate. In practice this has proven to be a difficult task. Evidence of this can be seen in the wide range of values used throughout the literature, typically 0.3 to 1.0 eV.

VIII. TRANSMITTERS

Only in the laboratory environment are emitters used in an unpackaged state. Otherwise they are housed in discrete packages or incorporated with additional circuitry in one package to form a transmitter. This circuitry and package serve the functions of providing a safe environment for the emitter, provide the means to maintain a stable light output which continues to satisfy the requirements of the particular application, provide modulation to the emitter, supply bias voltages to the emitter and provide the means to couple the optical power from the emitter to the output fiber or connector. Many types of coding schemes are currently in use along with a tremendous variety of driving circuits. This is not the place to attempt an exhaustive review of these transmitter features. Instead, the basic elements of a transmitter, construction design and testing techniques will be discussed along with the areas which have been found to be a concern in terms of reliability.

1. BASIC COMPONENTS

The major components of an optical transmitter, excluding the electronic circuitry required for signal processing, are the emitter, coupling system, monitor photodetector and integral cooling system. The latter two elements can be considered optional depending upon the quality of performance required, however, their inclusion is highly recommended since they function to ensure stable operation over a wider range of conditions.

a. Emitter - The characteristics of the emitter have been previously discussed.

b. Coupling system - Three coupling systems are most commonly used by manufacturers. The first design uses discrete optical lenses to collimate and focus the emitter output onto the optical fiber. The second design has no discrete lenses but uses the optical properties of the fiber end after it has been heated or chemically etched into a conical shape. The third design butt couples the cleaved and polished fiber to the emitter output.

The purpose of the coupling systems is to provide maximum optical coupling efficiency between the emitter output power and the optical fiber and to ensure that once this proper alignment is reached, it is maintained throughout the life of the device.

The use of a discrete lens or several lenses provides a higher coupling efficiency than that of the butt couple method. Typically, a collimating lens is used directly in front of the emitter output. Following this is a graded index (GRIN) lens which is constructed and positioned to focus the collimated light onto the fiber core. This design suffers from significant reflection back into the emitter when the front face of the GRIN is situated at the focal point of the collimating lens. This decreases emitter stability and coupling efficiency. This problem is very effectively reduced if the GRIN is placed just short of this focal length. The resulting reflections are then divergent and will not reenter the emitter. In addition, moving the GRIN lens closer to the collimating lens reduces the spherical aberration of the lens. Further reductions in reflections can be obtained by applying an antireflection coating to the lenses. Although this coupling technique provides good coupling efficiency, it is more difficult to manufacture because of the additional elements. The fiber, GRIN, collimating lens and laser must all be aligned and secured in that aligned position.

Tapering of the fiber tip and establishing a conical shape significantly reduces the reflections off the fiber end. With no tapering, these reflections often reenter the emitter resulting in instability exhibited as mode spectrum shift and noise which degrades system performance. The tapering improves coupling efficiency by decreasing the number of reflecting surfaces and by increasing the spot size of the fiber. This design can provide a theoretical coupling efficiency of 75%[16] but typically provides 25%[17], whereas the butt coupled fiber typically yields a 10%-12%[17] coupling efficiency. The increased spot size also reduces the sensitivity of core-to-cladding eccentricity at the fiber endface. However, very tight physical constraints are traded off for the increased coupling efficiency. To achieve the theoretical maximum coupling efficiency, it is necessary to maintain alignment precision to tenths of micrometers. For single mode fiber, transverse misalignment on the order of 0.25 μm will cause a 10% reduction in coupled power and a 0.75 μm misalignment will result in a 50% reduction in coupled power. This is in contrast to the losses incurred when using butt coupling which requires a 1 μm transverse misalignment before there will be a 10% coupled power reduction. When compared to the butt coupling technique, the tapered fiber is relatively insensitive to longitudinal displacement. The butt couple suffers a 15% power change with a 0.33 μm longitudinal shift due to half-wave reflections. By comparison, the taper fiber can tolerate a 1 μm displacement with only a 5% change in power. Adding an antireflection coating to the butt fiber endface can significantly reduce the losses. As can be seen, the different sensitivities to misalignment vary significantly with the coupling technique used. The misalignments may be incurred as a result of shock and vibration or temperature changes which results in thermal expansion and contraction of the various elements within the transmitter. The transmitter package must be mounted in such a way as to translate the maximum mechanical movement to the most tolerant axis of the coupling scheme used.

The taper and lens can be formed by one of several techniques. One method used is to etch the fiber end with hydrofluoric acid using computer control. Another technique follows the process of tapering the fiber end using an arc then cleaving the end of the taper. Dipping this end into a low melting glass having the proper refractive index for the given application will result in a fiber with a separate lens deposited onto the endface. Another method uses fiber heating, then drawing to a taper and fire-polishing the fiber to the desired geometry. It is difficult to maintain the lens concentricity and radius using this last method.

Major advantages with this technique are the reduced number of reflecting surfaces, fewer components to align and assemble and better emitter stability

The butt coupled technique has the lowest coupling efficiency. This is a result of the lack of optical beam size control and the reflections off the fiber endface back into the emitter. Typical coupling efficiencies achieved are between 10% and 12%.

c. Coolers -- The extracting of thermal energy from the active layer junction is an extremely critical function when operating a laser and many types of LEDs. As has been pointed out, when the active device temperature increases, the power output decreases. If the device is being operated in a constant output power mode, the driving current will increase to accommodate the reduction in output power. If the device continues to heat up, th driving current will rise, thus decreasing the lifetime of the device. Transmitters are fitted with an integral or external cooler to provide the essential function of preventing a continual build-up of heat.

The most common method used to cool fiber optic transmitters is through Peltier cooling. This is commonly referred to as thermoelectric (TE) cooling. These devices are constructed using oppositely doped bismuth-telluride junctions which are connected in series electrically. When a current is passed through them, heat transfer will take place between the junctions and the surrounding area. The amount of heat transferred is proportional to the current flowing through the junctions. When a TE cooler is used in a transmitter, heat is extracted from the temperature sensitive emitter and liberated to the heat sink to which the TE cooler is attached. The maximum temperature differential between the device being cooled and the TE cooler obtainable for a single stage TE cooler is 60°C. This is a limitation due to the properties of the materials used. Cascading stages can provide a differential up to 125°C.[18]

The size and shape of the heat sink is critical in the performance of a TE cooler. The larger the heat sink is, the greater will be the temperature differential and consequently the cooler the heat sink and device can become. However, a large heat sink is bulky and often undesirable if not completely impractical.

d. Monitor photodetectors -- The monitor photodetector is typically situated at the backface (rear facet) of the emitter. Only a small percentage of the light leaving the front facet exists via the backface. The monitor photodetector is used to control the power output of the emitter by providing feedback to the emitter driving circuitry based on the amount of light detected. This is needed to make adjustments for increases in threshold current with aging and to counter the results of temperature fluctuations. Therefore, it is essential that the backface signal not fluctuate with respect to the front facet signal as temperature changes. Maintaining this signal ratio with changes in temperature is termed tracking and the ability to properly track must be considered when selecting a hardy transmitter.

Several techniques are used to couple light from the emitter to the photodetector. In all methods, reflections back into the emitter need to be minimized to prevent output power fluctuations and instability. One method has the photodetector situated in the backface optical output path with no additional elements. The photodetector is angled obliquely to percent reflections into the emitter. Another technique uses a GRIN lens to couple the backface light onto the photodetector. The lens is used as a focusing instrument rather than a collimator and is, therefore, longer than one quarter wavelength. To prevent unwanted reflections, antireflection coatings are applied to the front and back sides of the GRIN and to the window of the photodetector. Also, the coupling elements are positioned such that light imaged onto the photodetector is deliberately out of focus. This reduces the emitter's sensitivity to reflections by preventing them from being in focus upon returning to the emitter backface.

IX. TRANSMITTER CONSTRUCTION

Three of the primary achievements that a transmitter construction strives to accomplish is the matching of materials with respect to coefficients of thermal expansion, maintaining alignment and hermeticity at the fiber entrance.

1. THERMAL CHARACTERISTIC MATCHING

To design a transmitter with good thermal characteristics, there must be good thermal stability between the individual elements. Therefore, the coefficient of thermal expansion of the emitter mount material must be low and must match the characteristics of the monitor photodetector mount and also that of the fiber carrier. In addition, this material must have very high thermal conductivity in order to conduct the heat away from the heat sensitive regions.

The goal of material matching is to achieve position stability with temperature. Variations in vertical position are very critical to control, especially when a lensed coupling system is used. The primary method used is to match material thermal properties. Variations in horizontal position are controlled by the use of symmetry in the structure of the individual elements. This will result in one horizontal force countering another. Variations in the longitudinal direction can not be accommodated by using symmetry due to the layout of the transmitter elements. However, to prevent the power fluctuation penalties incurred as a result of longitudinal motion, the use of very low expansion materials must be used. Many different materials are used to achieve the above characteristics, depending upon the end use of the package.

2. MAINTAINING ALIGNMENT

Securing the optical fiber, emitter, coupling mechanisms and monitor photodetector such that optical coupling is optimized is a very demanding task. The photodetector is the easiest to align since its optically sensitive area is extremely large compared to the submicron tolerances the other elements must be held to. Once all of the elements have been aligned, securing them is usually performed in one of three ways; epoxy, solder or welding.

Epoxy is widely used by many manufacturers. However, its use in critical applications or harsh environments is not recommended. There are two reasons for this. The first is that epoxy softens at high temperatures. This allows relative movement between the individual elements of the transmitter which degrades the coupling efficiency. This temperature is not extremely high. Commonly encountered storage temperatures, such as 70°C, are adequate to soften the epoxy. The second reason is that epoxy shrinks when it cures. This also results in the optical alignment being degraded. Research is being performed in these areas to eliminate these problems.

Solder has the same problem with shrinkage as does epoxy. Often low temperature melting point solders like Indium are used to facilitate active alignment of the elements. However, these solders are subject to creep which exerts forces on elements causing them to move. Active alignment is a very good technique but using high temperature melting point solders makes this technique extremely difficult to perform. However, these solders can be used without any problem in high temperature environments.

Welding is the most recently introduced method for securing the transmitter elements. The welds are made using a commercial, pulsed laser welder, typically a neodymium:yttrium aluminum garnet (Nd:YAG) laser. These are metal-to-metal spot welds performed at extremely high temperatures. These welds are not effected by high temperatures and the process does not appreciably increase the temperature of the transmitter assembly. However, this technique is very capital intensive.

When welding, the process of securing a fiber is performed by first metallizing (coating with metal) the fiber and sliding a metal sleeve over the fiber which functions only to hold the fiber to the mount. This sleeve does not protrude from the transmitter housing. The metallized fiber is then soldered to the inside of the sleeve. Next, a flanged, U-shaped clip is placed over the sleeve which is attached to the mounting assembly. The fiber is then properly aligned and the sleeve is welded to the clip which is then welded to the mounting assembly.

3. HERMETIC SEALING

Hermetic sealing of a transmitter package at the fiber entrance involves two areas; the treatment of the fiber and the enclosure surrounding the fiber at the fiber-housing interface.

To affect a true hermetic seal, epoxies can not be used as a sealant. Typically, a segment of the fiber length is sputter coated with metal, which is termed metallizing. Then the fiber is placed through the wall of the housing and soldered using eutectic solder. The components which make up any eutectic solder are in a ratio that gives it the lowest possible melting p int. Flux is not used in the soldering operation to eliminate the possibility of a long term chemical interaction which could degrade the optical fiber or housing. After the solder is complete, the remaining elements of the transmitter are assembled.

Another method used to hermetically seal the fiber-housing interface uses a glass ring formed from low melting point glass. This ring is slid onto the fiber and aligned with the inner wall of the housing ferrule. High frequency heating is then used to melt the glass, thus forming a hermetic seal between the fiber and ferrule. Now, the ferrule is hermetically sealed to the transmitter housing using conventional soldering techniques. This method has the advantage that the fiber does not require any prior preparation.

X. TESTING

The most common mechanical and environmental tests performed on packaged transmitters are listed in Table 3-6.

TABLE 3-6. Environmental and Mechanical Tests Performed on Packaged Transmitters

Test	MIL-STD-883 Method
Temperature Cycling	1010
Temperature Shock	1011
Vibration	2007
Thermal Shock	1011
Shock Test	2002
Fiber Pull	*
Constant Acceleration	2001

* Varies with the intended application.

These tests are performed at various degrees of severity depending upon the environmental and mechanical harshness in which the device is intended to operate.

XI. LIFETIME PREDICTIONS

Very little lifetime or failure data is available on transmitters. However, the two most prevalent forms of failure are:

1. At high temperatures, the electromigration of solder, especially Indium solder, used in mounting the emitter.

2. Mechanical strain at bonded surfaces between different materials resulting from coefficient of thermal expansion mismatch.

XII. SUMMARY

Emitters are the heart of the transmitters used in fiber optic systems. These devices have very high demands placed on them which must be satisfied for the life of the system which is usually 20-25 years. This means that the devices must be predicted to operate reliably for a significantly longer period of time. As Tables 3-4 and 3-5 show, emitters are being developed that meet this challenge.

There are numerous mechanisms which have been attributed to emitter degradation and failure. Those most widely researched include: dark line defects, dark spot defects, in-migration of impurities into the active region, deterioration of solder, leakage currents, facet deterioration and an increase in thermal resistance due to contact deterioration. Selecting high quality materials, fabricating the devices in such a manner to minimize contamination and subjecting the devices to adequate screening procedures will probabilistically insure long operating lives.

Researchers have shown that it is of the utmost importance to stress test lasers until they are operating in the long term wear out mode. Otherwise, a true evaluation of their predicted performance is unlikely. This will impact system design by resulting in an overly optimistic or highly pessimistic predicted lifetime.

From life tests performed on state-of-the-art devices, LEDs will significantly outlive semiconductor lasers. In addition, lasers frequently require relatively stringent environmental controls, especially with regard to temperature. However, lasers can also provide high power single mode outputs which allow a longer system repeater separation.

Emitter reliability research is a very active science with new device construction and materials being developed to increase device performance and longevity. Such little failure rate data is available from deployed operating systems that a truly valid assessment of their reliability is difficult to make.

Chapter 4. Photodetectors and Receivers

An optical receiver performs the operation of converting optical signals into electrical signals through photodetection. This electrical signal is then amplified to a useful level for subsequent decoding and processing. The primary optical component in a receiver is the photodetector, on which this discussion will concentrate. However, the type of amplifier selected will have an impact on important performance parameters which will be brought to the attention of the reader.

First, a brief discussion on photodetector operation and construction is presented.

I. BASIC THEORY OF OPERATION

Semiconductor devices (emitters, detectors, transistors, etc.), are made from materials doped n- or p-type. When these two types of materials are used in one structure, the resulting p-n junction has electrical qualities which form the basis of many optical and electrical devices.

The majority carriers of a p-n junction are electrons in the n side and holes in the p side. Until equilibrium is reached, current will be generated by majority electrons diffusing across the p-n junction to the p side and majority holes diffusing to the n side. This causes a potential difference to be formed across the junction which prevents any further net diffusion of charges across the junction in the equilibrium state. As a result, the area around the junction has no more mobile carriers and is referred to as the depletion region and its width is designated W_d. The depletion region of a p-n junction is shown in Fig. 4-1.

When an external voltage is used to reverse bias the junction (positive terminal to n side and negative terminal to p side), the width of the depletion region is increased. This further prevents any diffusion of majority carriers across the junction. However, the minority carriers (electrons on the p side and holes on the n side) can drift with the electric

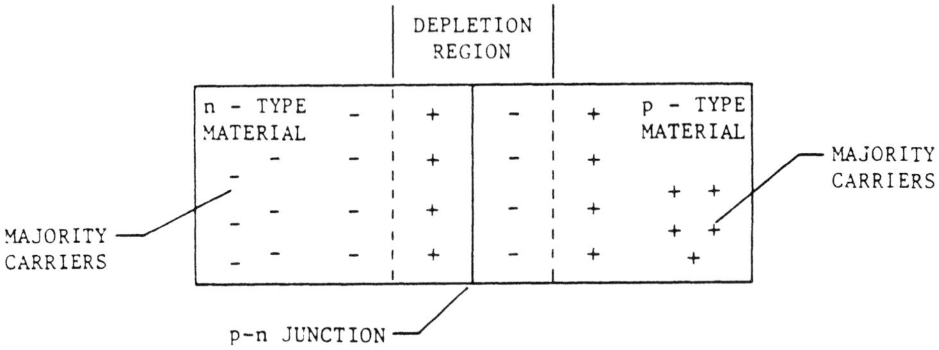

Figure 4-1. Depletion region of a p-n junction at equilibrium.

field created by the application of the applied voltage resulting in a drift current. This current is typically too small to be of value at low temperatures and low voltages. However, upon the application of light to the photodetector p-n junction, many electron-hole pairs are created in the depletion region. Under the influence of the electric field in this region, they do not recombine but are quickly separated and contribute to the drift current. This activity is shown in Fig. 4-2 along with the associated electric field distribution.

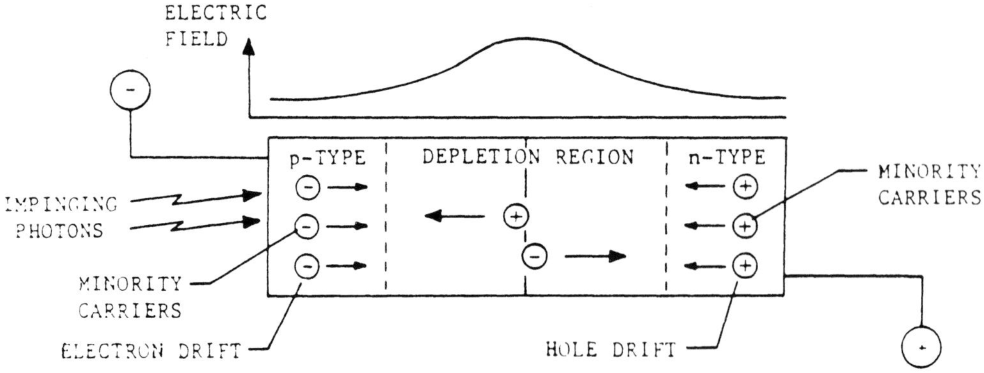

Figure 4-2. Reverse biased p-n junction with incident light.

In addition to the drift current developed in the depletion region, current is generated in a space just outside of the depletion region. This region has a width equal to the diffusion length of the minority carrier for that type material. The diffusion length is the average distance that a carrier can travel before recombining. Therefore, carriers within this length do not recombine but are prone to diffuse into the depletion region where they are influenced by the electric field and thereby contribute to current flow.

It is the combination of drift current and diffusion current produced within the depletion region and within the diffusion length that make up the photocurrent of a photodetector.

The number of electron-hole pairs generated per impinging photon describes the efficiency of a photodetector. The more pairs generated per photon, the higher the photocurrent. This is referred to as the quantum efficiency and is given by equation 4-1.

$$\eta q = 1 - \frac{e^{-\alpha W_d}}{1 - \alpha L_p} \qquad \text{Equation 4-1}$$

where, ηq = quantum efficiency

α = interband absorption coefficient

W_d = depletion width

L_p = hole diffusion length

The absorption coefficient in equation 4-1 determines how far impinging photons can penetrate the semiconductor material before being absorbed. For photons to contribute to the photocurrent, it is necessary that they reach the region $W_d + L_p$ or $W_d + L_n$ which is shown in Fig. 4-3. If the absorption coefficient is not large enough, then many photons will not be absorbed in the depletion region, but will pass through resulting in a low quantum efficiency. Likewise, if the absorption coefficient is too large, photons will be absorbed before they even reach the region where they can diffuse into the depletion region.

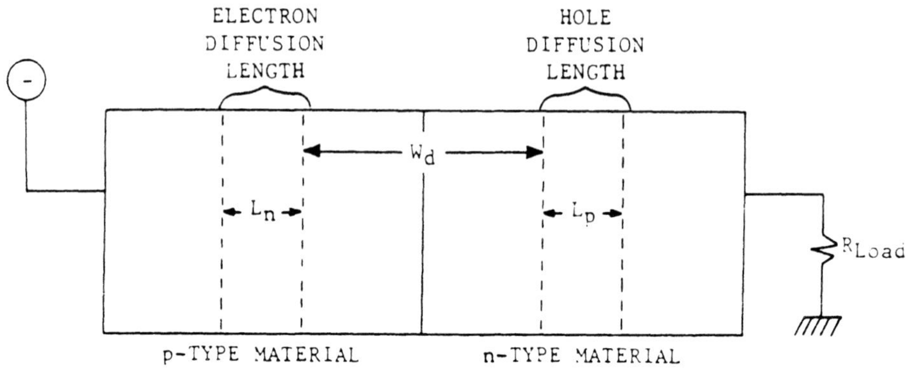

Figure 4-3. Depletion width and diffusion lengths.

To improve the quantum efficiency, the depletion width can be widened to increase the probability that photons will be absorbed. Therefore, for a given amount of incident optical power at a given wavelength, a higher photocurrent will be generated relative to that generated by a narrower depletion width. Since it is more intuitive to relate to a photodetector responding to a given optical power level than to electron-hole pairs generated per photon, the former relationship is used to describe the performance of a photodetector and is termed the responsivity. Responsivity is a measure of the electric current produced in response to an incident optical power level.

It is important to note that the depletion width cannot be made arbitrarily wide. The depletion width defines the distance that the carriers must travel to cross the p-n junction and, therefore, directly impacts the response time. As a result, a trade-off must be made between the quantum efficiency (and therefore responsivity) and response time.

II. PHOTODETECTOR CONSTRUCTION

1. DEPLETION-LAYER P-N PHOTODETECTOR

The most basic detector is the depletion-layer p-n photodetector which is a reverse-biased semiconductor diode. When reverse-biased, the diode is being operated in the photodetector mode. When operated with no external bias, the device is being operated in the photovoltaic mode where it acts as its own electrical source. For fiber optic applications, the photodetector mode is used almost exclusively.

The construction of a practical mesa depletion-layer p-n photodetector is shown in Fig. 4-4. For GaAs based devices, the p region is made much thinner than the n region and is also much more heavily doped than the n region. This keeps the series resistance very low and the p region effectively becomes a contact layer.

The mesa construction has one significant disadvantage in that the p-n junction perimeter is not protected since this design does not lend itself to passivation. This can result in contaminants on the wall of the mesa which will lead to leakage currents. Also, the in-migration of contaminants into the junction can increase the occurrence of generation-recombination current.

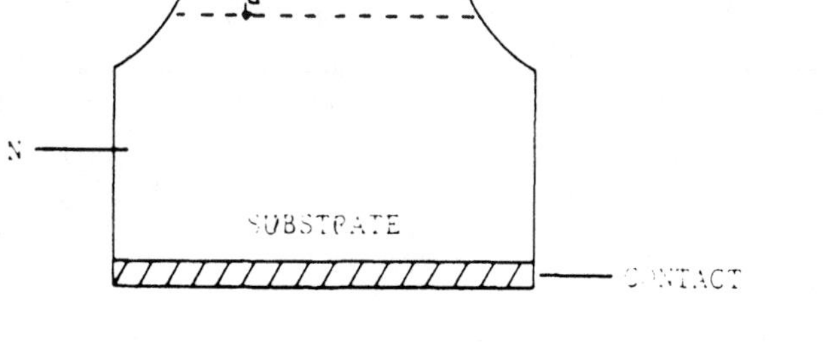

Figure 4-4. Basic mesa depletion-layer p-n photodetector construction.

2. AVALANCHE PHOTODETECTOR

Unlike the p-n photodetector which has a maximum current gain of unity, the Avalanche Photodetector (APD) has been reported to have a gain as high as 1000[1], but a gain of several hundred is more typical. This is achieved through the process of avalanche multiplication. The APD is constructed such that carriers must traverse a region where there is a very high electric field. The intensity of this field is such that a hole or electron can be imparted with sufficient energy that it knocks free bound electrons in the valence band when they collide. This chain of events is known as impact ionization. The electrons that have been knocked free interact with the electric field and gain sufficient energy to ionize other bound electrons. This process snowballs and is called the avalanche effect. Eventually these electrons reach the conduction band where they contribute to current flow. The construction of a typical APD and the associated electric field distribution is shown in Fig. 4-5.

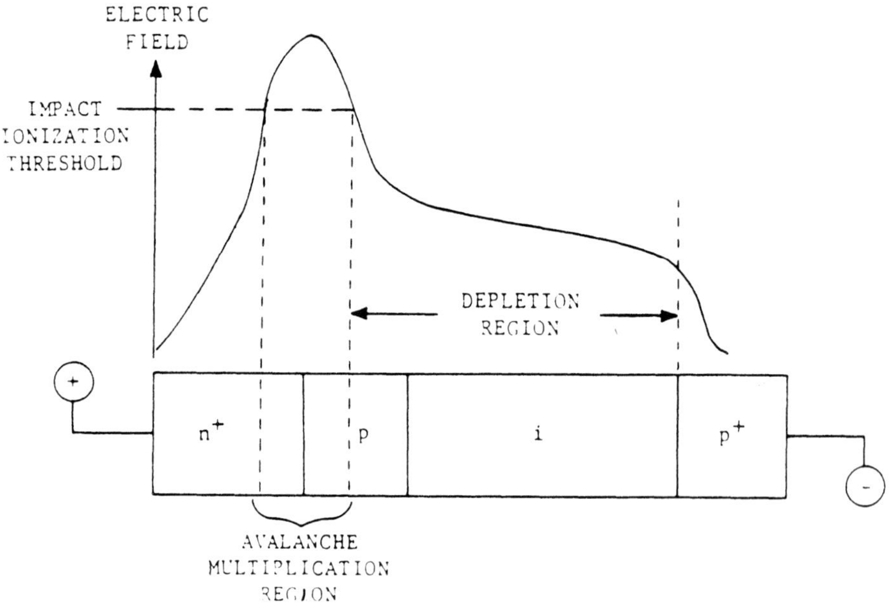

Figure 4-5. APD structure and electric field distribution.

As Fig. 4-5 shows, the APD is typically made up of four layers. Photons enter the device through the p+ layer and are absorbed by the high resistivity intrinsic p-type layer (usually called the pi layer) where electron-hole pairs are created. The relatively weak electric field in this region then separates the carriers causing the electrons or holes to drift into the high electric field region where avalanche multiplication occurs resulting in a current gain. This gain can be used advantageously to achieve a wider spacing between repeaters. However, as the gain increases, the bandwidth decreases since the avalanche effect takes time to build up.

For a given level of incident optical power, the APD will produce a photocurrent which is larger than that produced by a depletion layer p-n photodetector by the multiplication factor achieved through the avalanche affect. Therefore, the APD responsivity, and therefore sensitivity, will also be higher by this factor.

For the avalanche effect to take place, a very high electric field must be present which means the application of a high voltage. Dependent upon the materials used, this can vary between 100-400 Vdc. Not until the avalanching voltage level is reached will the APD provide a significant output current for a given illumination level. This voltage is called the avalanche breakdown

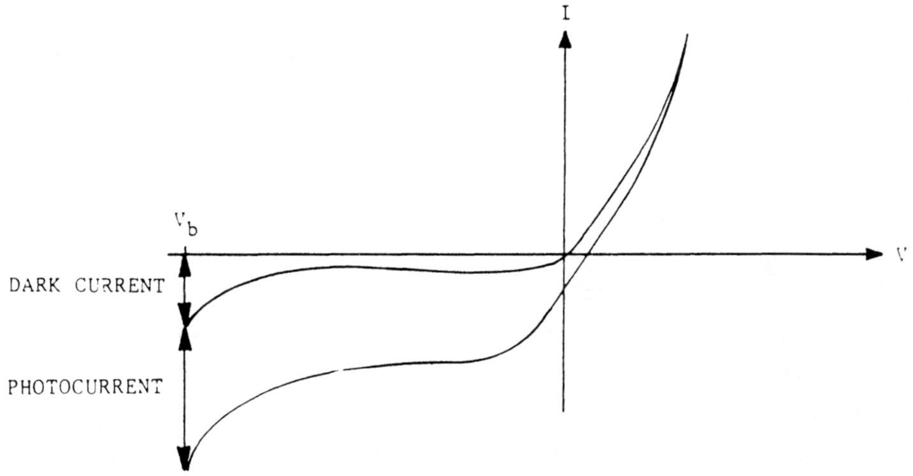

Figure 4-6. Voltage versus current response curves for an APD.

voltage, V_b. A typical plot of voltage versus current for an APD is shown in Fig. 4-6. There are two curves shown in this figure. The lower curve in Fig. 4-6 shows the response of the APD when illuminated. Under low bias voltages, the APD exhibits a constant output current, however, when the breakdown voltage is reached, the photocurrent increases sharply. Under breakdown voltage conditions, the gain of the APD is very temperature sensitive, especially at very high bias voltages. This makes it necessary to control the ambient temperature, use a thermoelectric cooler to control the photodetector temperature, or to automatically compensate for changes in the gain by adjusting the bias. Ambient temperature control is costly and the latter two methods require additional electronic circuitry which increases the chance of failure.

The upper curve in Fig. 4-6 shows the current which flows in the bias circuitry when the diode is in a darkened environment under bias voltage conditions. This current is called the dark current, I_d, and has three components; bulk, surface leakage and tunnel current. The bulk current is due to the thermal generation of carriers in the p-n junction. The surface leakage current is influenced by the bias voltage and defects in, cleanliness of, and size of the illuminated surface area. A guard ring is often used to shunt the leakage current away from the load circuitry. This ring surrounds the periphery of the illuminated surface area. Tunneling current results when electrons transition from the valence band to the conduction band either directly or via mid-band traps due to defects in the crystal lattice. Tunneling in an APD will generally only occur in the high electric field region and, therefore, only under high reverse bias voltages. This bias causes the energy bands to move closer and closer together as the reverse bias is increased which reduces the energy needed to transition the gap. The resulting tunneling current is not dependent upon impinging photons and once the necessary tunneling breakdown voltage is reached, it will exist at a relatively constant level for a given bias. This current is, therefore, a constant source of noise. The tunneling current can be reduced by constructing the photodetector such that the high electric field occurs in a wide band gap region or by ensuring that the avalanche breakdown voltage is lower than the tunneling breakdown voltage. Of the three components which comprise the dark current, only the bulk current is multiplied by the APD avalanche effect.

The magnitude of the dark current is material-dependent with germanium-based photodetectors having a relatively high dark current and silicon-based devices having a relatively low dark current.[2]

The dark current is a source of noise for the APD in addition to quantum noise, thermal noise and noise due to the random nature of the avalanche effect.

The quantum noise results from the impinging photons and will always exist, thus, theoretically establishing the lower limit on photodetector sensitivity. This noise source is also multiplied by the avalanche effect.

Thermal noise is caused by the APD load resistor current and can be decreased by using a larger resistance value, however, the response time and bandwidth of the photodetector will be degraded as this value is increased.

The final noise source arises from the random nature of the multiplication process which is a function of the materials used and increases with gain. This random process can be considered in terms of a gain density distribution in that individual electrons will ionize a different number of additional electrons through collisions with the crystal lattice. The more electrons which are ionized, the greater the gain will be. However, a high gain value will prevail over lower gain values. These lower gain values constitute noise in the form of fluctuations which is referred to as excess noise. This term represents that noise which is in excess of the noise level that would exist if all electron-hole pairs ionized the same number of electrons.

The combined noise level from these various sources defines the minimum signal level that must be received before intelligent information can be extracted from the photodetector output. It is this signal level that defines the sensitivity of a photodetector.

3. PIN PHOTODETECTOR

The Positive-Intrinsic-Negative (PIN) photodetector is the most commonly used photodetector. Just as in the APD, the PIN operates on the principle of electron-hole pairs being generated in the depletion region by the absorption of incident photons. The subsequent separation of the carriers by the bias voltage electric field results in the device photocurrent. However, one significant difference between the APD and PIN photodiode is that the PIN operates under a reverse bias of 5-12 Vdc compared to several hundred for the APD. Also, the PIN does not operate using the avalanche effect which prevents the device from achieving a current gain greater than unity. The impact of this is reduced noise and also a reduction in responsivity by an amount equal to the gain of an APD.

One common construction is shown in Fig. 4-7 where it can be seen that it gets its name from its construction. Very heavily doped p type and n type layers are separated by an intrinsic or essentially pure layer of silicon. The p and n type layers have a very low resistivity while the intrinsic material has an extremely high resistivity.

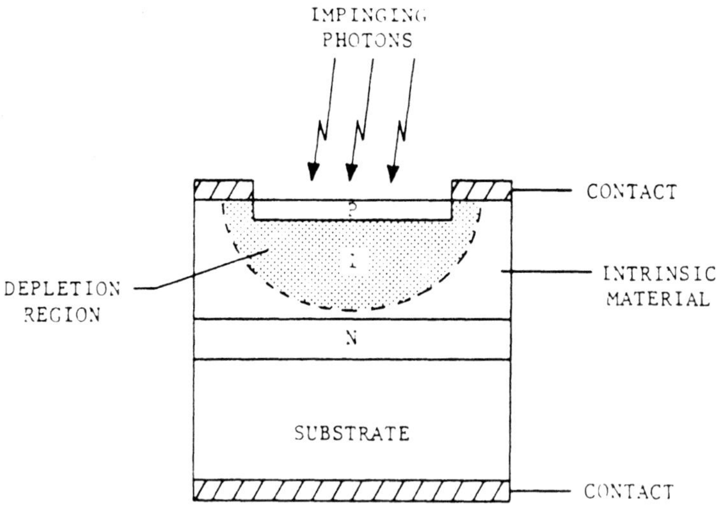

Figure 4-7. PIN photodetector construction.

A more recently developed construction, the back-illuminated PIN, is shown in Fig. 4-8. Immediately different is the illumination of the device through the substrate side. The incident surface is coated with a quarter wavelength thick film which eliminates reflection. Back-illumination increases quantum efficiency by reducing carrier surface recombination which occurs with conventional photodetectors. The increased efficiency is a result of the depletion region extending to the substrate which causes all photocarriers to be generated away from any surface. Therefore, the number of carriers reaching the depletion region increases. Also, this photodetector can be made smaller in size which gives it a lower capacitance and therefore an improved bandwidth.

The diffusion mask is used to define the P-region and the passivation layer defines the contact area. The passivation layer serves the primary function of sealing the surface from contamination and to reduce the recombination of holes and electrons at the surface.

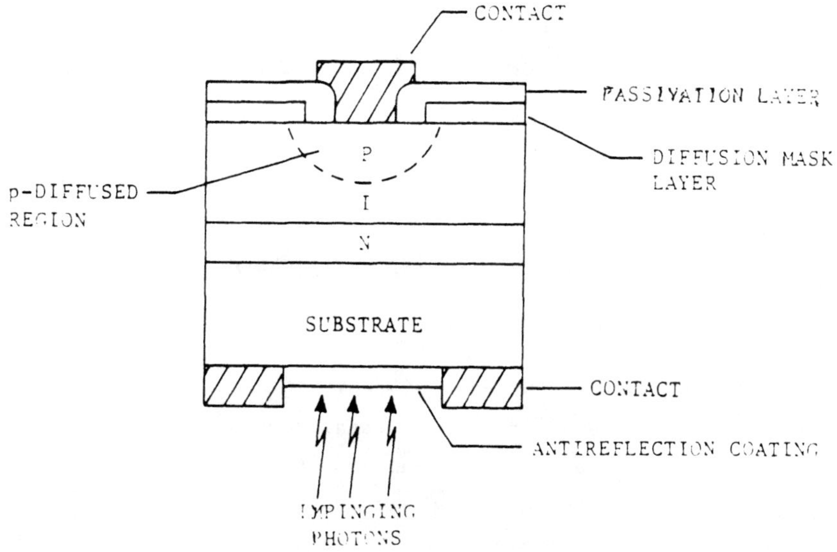

Figure 4-8. Back-illuminated planar PIN photodetector.

The depletion region width of PINs is dependent upon the applied voltage and the resistivity of the intrinsic material as shown in Fig. 4-9. As already stated, this will determine the response time and the quantum efficiency (and thus responsivity) of the photodetector. The same trade-off must be made between these two parameters as is made with the APD.

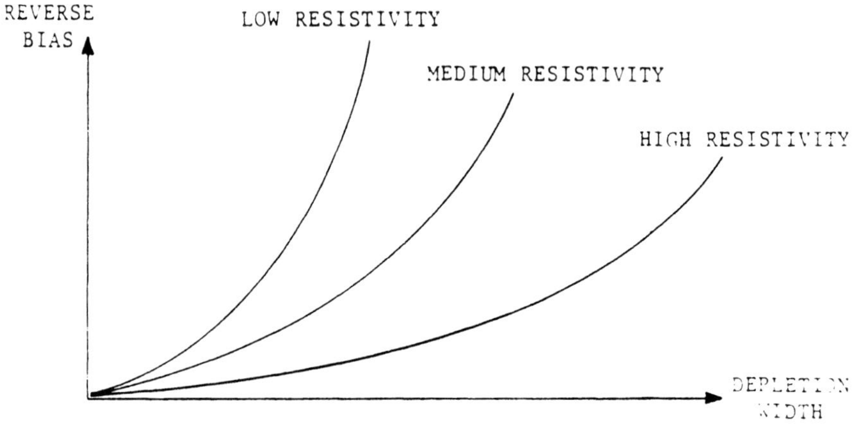

Figure 4-9. PIN depletion width as a function of reverse bias.

The noise sources for PINs are the same as those for the APD except there is no avalanche effect and therefore no avalanche noise nor are the bulk current and quantum noises increased by avalanche gain.

The PIN dark current is not as significant of a noise factor as it is in APDs. This is especially true at short wavelengths. To keep the dark current very low, a layer of InGaAs or InP is often grown onto the active region to reduce surface dark currents. The band gap of this layer is wider than the active layer material which prevents electrons from transitioning this band gap, thereby reducing surface currents. The thermal noise generated by the PIN load resistor current is this devices' predominant source of noise.

4. APD/PIN PERFORMANCE COMPARISON

To achieve the best compromise between bandwidth and sensitivity, the PIN is the proper selection for very low and very high bit rates. For intermediate bit rates, 100Mb/s to 4Gb/s, the APD is the best photodetector selection.[11] In analog operation, the APD provides the largest signal to noise ratio (SNR) when the received optical signal level is low. For large signal levels, the PIN photodetector provides the larger SNR.[1]

In terms of repeater spacing, the high gain and superior sensitivity of the APD over the PIN will generally allow fewer repeaters to be used in a system while achieving the same level of performance. These comparisons are rules of thumb to keep in mind when planning a system. Many factors must be taken into consideration before a final photodetector selection is made.

III. MATERIALS SELECTION

The importance of the materials used in the fabrication of photodetectors is not to be ignored. In general, the following statements can be made:

- InGaAs PIN photodetectors have a lower surface leakage current than Ge PINs.

- InGaAs APD leakage current is dominated by generation-recombination at low voltages and tunneling at higher voltages.

- InGaAs devices are sensitive to moisture and should be passivated or hermetically packaged.

- Ge, InGaAs and InGaAsP devices operate from 1.3 μm to 1.55 μm where chromatic dispersion and attenuation are lowest, respectively.

- Si devices operate from 0.8 μm and 1.04 μm.

- GaAs devices operate from 0.8 μm to 0.9 μm.

In addition to these general comments, the performance of photodetectors is significantly affected by the materials used as is discussed in the following sections.

1. IMPACT OF MATERIALS ON APD PERFORMANCE

As has already been discussed, APD gain is achieved through the avalanche effect due to impact ionization. An important concept with regard to ionization is the ionization rate. This is the average number of electron-hole pairs which are generated by a carrier per unit distance traveled. This number is not equal for electrons and holes of the same material. For the commonly used elements and compounds, Si has the largest difference between electron and hole ionization rates followed by Ge then InGaAs. The ratio of these two rates is denoted by k and the smaller the value of k, the less noise, better sensitivity and higher gain-bandwidth product a photodetector will exhibit. This can be explained by referring to Fig. 4-10 which shows the basic structure of an APD along with the electric field distribution and a simplification of the avalanche process.

For simplicity, only one initial electron-hole pair will be considered as shown immediately below the intrinsic region. Upon separation of this electron-hole pair by the electric field, the initial electron will be drawn to the high electric field region where impact ionization occurs thus generating the additional primary electron-hole pairs. The initial hole will be drawn in the opposite direction to that of the electron and if the electric field it encounters is strong enough, it will produce secondary electron-hole pairs through impact ionization. The electrons produced from these secondary electron-hole pairs will be drawn to the high electric field region where they will initiate impact ionization. The long transit time it takes these electrons to travel from the p+ layer to the n+ layer is excessive and degrades the photodetector gain-bandwidth product. This entire chain of events will continue as long as the proper conditions prevail.

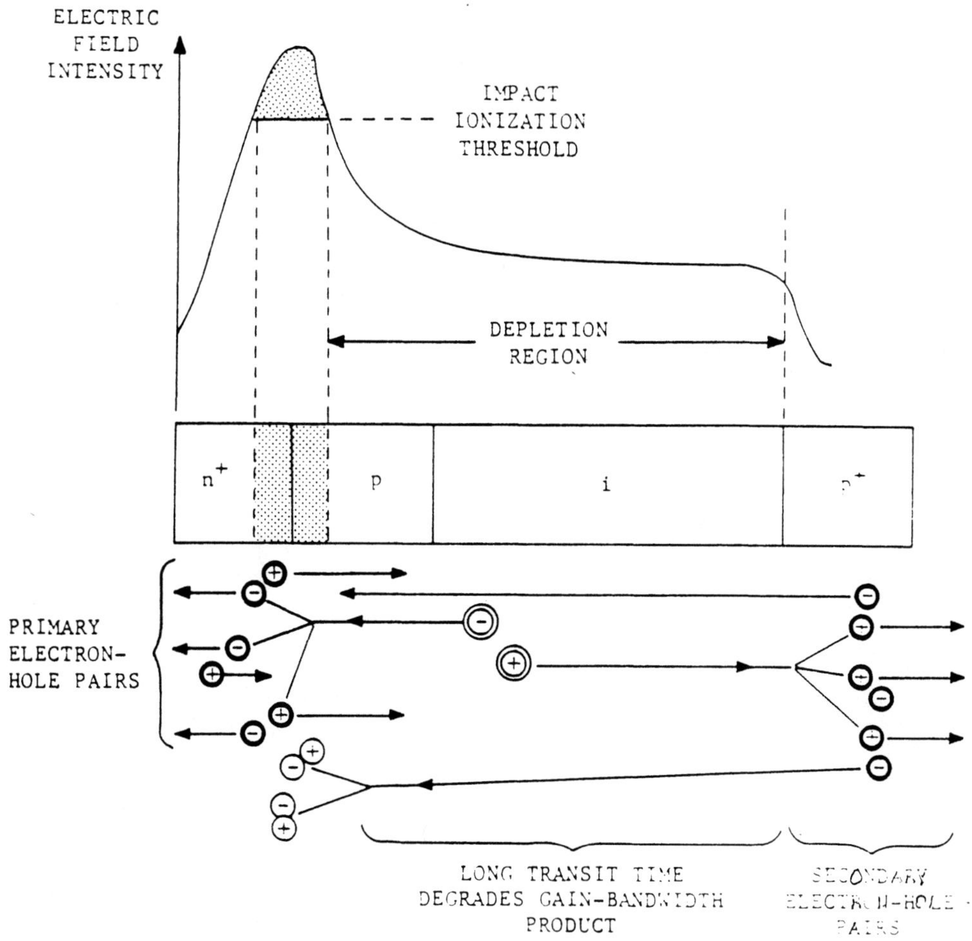

Figure 4-10. Effect of high k value on APD avalanche process.

By design, the electric field in the n+ layer is much higher than that in the p+ layer. This translates into the holes in the p+ layer being accelerated much less than the electrons in the n+ layer, therefore, the initial holes will be much less energetic resulting in fewer electron-hole pairs being knocked loose. This means those holes contribute a much lower gain. The importance of this becomes evident since this variation in gain is exactly the situation that produces excess noise described in Section II.2.

To reduce the excess noise, it is necessary to have as few hole-initiated ionizations as possible which can be accomplished by using materials that have a low hole ionization rate. However, the electron ionization rate must be kept high to achieve the gain characteristic of an APD. By the definition of k given on page 244, these two conditions yield a low value for k. Since sensitivity is degraded by an increase in noise, having a low value of k is very desirable. (A low value of k can be equally achieved by using a material with a high hole ionization rate and low electron ionization rate.) Eq. 4-2 identifies those parameters which influence the sensitivity of a receiver which uses an APD.

$$P_{APD} = \frac{hc\beta K}{2\lambda\eta}\left\{K\Delta f F_p + \frac{2}{M_p}\left(Z + \frac{\Delta f}{q\beta}\left[I_{dm}M_d{}^2F_d\right]\right)^{\frac{1}{2}}\right\} \qquad \text{Equation 4-2}^{9}$$

where P_{APD} = APD based receiver sensitivity (A/W)
h = Planck's constant
c = Speed of light
β = bit rate of pulses
K = BER determined factor
λ = operating wavelength
η = quantum efficiency
Δf = noise equivalent bandwidth
F_p = photocurrent excess noise factor
M_p = photocurrent gain
Z = term containing photocurrent quantum (shot) noise of thermal noise of load R
q = electron charge
I_{dm} = leakage current quantum (shot) noise multiplied
M_d = leakage current quantum (shot) noise gain
F_d = leakage current excess noise factor

Since F_p and F_d are directly dependent upon the ionization rates of the carriers, having a low value of k will keep these terms low and also keep the sensitivity low. If I_{dm} is small, then increasing the photocurrent gain, M_p, will reduce the effects that the thermal terms in Z have on the sensitivity by reducing the coefficient $2/M_p$. However, as the gain continues to increase, I_{dm} becomes the dominate factor since it is multiplied by the leakage current gain term, M_d^2, which also increases. Therefore, there is an optimum value of photocurrent gain, M_p, that will yield the best (lowest) sensitivity, P_{APD}. The optimum value is a function of the terms Z and I_{dm} in equation 4-2.

If t e ionization ratio could be reduced to zero, the response time of the APD would only be limited by the time required to transition from the intrinsic layer to the n^+ layer. This would result in a device which could respond to extremely high bit rates.

On the other hand, if the ionization rate equaled unity, secondary carriers would be continually generated and they would traverse the intrinsic layer. This would result in a very high gain but the response time of the APD would be dramatically reduced. Therefore, the gain-bandwidth product is greatest when the ionization ratio is minimized.

2. IMPACT OF MATERIALS ON PIN PERFORMANCE

Since the PIN photodetector does not exhibit gain, ionization rates and excess noise are of no concern. This simplifies the expression for the sensitivity of a receiver which uses a PIN as shown in Eq. 4-3.

$$P_{PIN} = \frac{hc\beta K}{2\lambda\eta}\left(K\Delta f + 2\sqrt{Z}\right) \qquad \text{Equation 4-3}^{9}$$

where P_{PIN} = PIN based receiver sensitivity (A/W)
h = Planck's constant
c = speed of light
β = bit rate of pulses
K = bit error rate determined factor
λ = operating wavelength
η = quantum efficiency
f = noise equivalent bandwidth
Z = term containing photocurrent quantum noise and thermal noise from load resistor

It is evident that the sensitivity is improved by keeping the thermal noise low which can be accomplished by using a large-valued bias resistor. High quantum efficiency will also enhance the sensitivity which is accomplished by using only defect-free materials.

IV. RECEIVER SELECTION

Generally, there are two types of receivers that are used with photodetectors, those using high impedance amplifiers and those using transimpedance amplifiers. These two types will be discussed in general terms since an endless variety of specific designs are available.

1. HIGH IMPEDANCE AMPLIFIER

This class of amplifier has less noise than the transimpedance amplifiers. There are two limitations associated with this class of receiver in that it requires equalization filter circuitry when used in broadband applications and it has a limited dynamic range.

a. Bipolar junction transistor amplifying stage -- When a bipolar transistor is used as the first amplifying stage, the total input resistance is effectively equal to the input resistance of the transistor. Both, this input resistance and input capacitance are large enough to impose severe bandwidth limitations on the receiver. If one or both of these values is decreased, the bandwidth can be extended, however, this is not typically done. Instead, an equalization filter is usually added which can be comprised of a resistor and capacitor network.[13]

The sensitivity of this type receiver is primarily dependent upon three parameters: transistor input capacitance, transistor base resistance and transistor current gain. A low input capacitance value will avoid the shunting of the input signal away from the amplifier at high frequencies, thereby ma'ntaining a sufficient signal level for good sensitivity. A low-value base resistance will minimize the thermal noise associated with this resistance. A high current gain will allow the base current to be low which will reduce the shot noise associated with this current. This can be achieved by using a large-value feedback resistor. This will have two additional affects. One is a reduction in thermal noise associated with the feedback resistor which will enhance the sensitivity. The other is a reduction in the dynamic range since the maximum allowable input signal is reduced. Therefore, using a transistor with these relative values for the indicated parameters will maximize the receiver sensitivity by extending the minimum required received signal level to inc'.ude high frequencies and reducing noise sources at the expense of a reduced dynamic range.[13] This receiver type is characterized by a low dynamic range and low noise.

b. Field Effect Transistor (FET) amplifying stage -- When a FET is used as the first amplifying stage, the inherent high input resistance of the FET amplifier reduces the total input resistance of the receiver to effectively that of the photodetector bias resistance. Such a large value for the amplifier input resistance results in a very small thermal noise component associated with this resistance.[1] The primary noise source in this type amplifier is the thermal noise of the FET conducting channel. Using a very large-value photodetector bias resistor will minimize this noise and achieve very good sensitivity. However, the result is a narrowed bandwidth. This is typically compensated for by the addition of an equalization filter as is done with the bipolar transistor amplifier.[1]

Also, to have good sensitivity at low bit rates, the FET must be free of carrier trapping centers in the FET gate-source junction. If not, the capacitance of this junction will increase at bit rates below several megahertz resulting in a deterioration of the receivers' sensitivity at these frequencies.

The FET amplifier is characterized by low noise and low dynamic range.

2. TRANSIMPEDANCE AMPLIFIER

This class of amplifier uses feedback to achieve high gain and high impedance yielding a low noise and large dynamic range. The sacrifice is a less sensitive receiver.

The bandwidth of this type amplifier is not limited as is the case with the high-impedance design. This is a result of the time constant of the photodetector being kept very small by the small-value input resistance and feedback resistance combination.

The transimpedance amplifier has a lower sensitivity than high impedance amplifiers. The reduction in sensitivity can be as little as 2dB[1] or as high as 10dB.[12] The poor sensitivity is a result of using a low-value feedback resistor to avoid signal integration, however, as a result, the resistor produces a high thermal noise. This noise source is what reduces the sensitivity. If a large-value feedback resistor is used, the associated high parasitic capacitance increases the feedback at high frequencies which effectively reduces the transimpedance of the amplifier. Stated differently, the feedback resistor begins to behave like a feedback capacitor. This results in signal integration which imposes the same limitations on the transimpedance amplifier as that of the high-impedance amplifier.

V. DEGRADATION MECHANISMS

The photodetector is typically used for two purposes; as the photodetection mechanism in a receiver and as an emitter back-face monitor providing feedback to control circuitry to ensure a stable optical output. Used in a receiver, an increase in photodetector junction capacitance or dark current are the parameters that will degrade receiver performance most significantly. Used as a monitor, quantum efficiency is the primary parameter that must be considered. However, the results of accelerated aging tests do not indicate that junction capacitance or quantum efficiency are failure mechanisms, either catastrophically or gradually. Therefore, the most important device parameter to use in indicating the degradation status of a photodetector is dark current. Other failure mechanisms which have been reported are device shorting and device opens.

1. DARK CURRENT

An increase in dark current, I_d, is the most commonly used indicator of photodetector failure with a ten-fold increase from the initial value often used to signify end-of-life. Depending upon the mechanism responsible for the increase in I_d, the degradation may be permanent or temporary. Researchers have not come to full agreement on the causes of either type of degradation, however, those mechanisms which appear to have general support are hole trapping, contact degradation and surface contamination.

a. <u>APD dark current</u> -- When compared to the PIN, the APD has a much higher dark current at 0.8 μm, whereas at 1.3 μm, it has a much lower dark current.

Increases in APD dark current which are recoverable almost to their non-degraded value can be attributed to surface deterioration.[5] This results in hole traps in the surface layer while under reverse breakdown voltage. These traps serve to attract electrons which accumulate over time. Once trapped, the normal electron-hole pairing process is delayed, thereby allowing thermal re-excitation of the electrons to the conduction band where they can contribute to leakage current. Recovery occurs when these devices are stored at elevated ambient temperatures (below 200°C) without bias. The surface deterioration becomes irrevocable when stored in higher ambient temperatures.

Non-recoverable dark current degradation in APDs is due to thermal deterioration of the metal contact.[5] The result is a metallurgical reaction between the contact material and the surface layer. When the resulting precipitates reach the depletion region, generation-recombination centers are created. At these centers, carriers can be generated and subsequently acted upon by the electric field before recombination occurs resulting in a net current flow.[6] This current flow is a source of noise. Permanent degradation with contacts made from AuZn material was reported at temperatures above 180°C.

An additional irrecoverable APD dark current degradation often occurs when a guard ring is not used. Under high APD reverse bias voltages, the perimeter of the p-n junction can break down which accelerates the degradation process relative to APDs which use guard rings.

b. <u>PIN dark current</u> -- Recoverable dark current degradation in PINs is reported to result from the contamination of the passivation (protective) layer.[7] Mobile ions present in the passivation layer are acted upon by the electric field causing them to accumulate at the p-n junction perimeter where they act to enhance dark current. Upon storage in elevated ambient temperatures of 150°C without bias voltage, the dark current decreases to acceptable levels.

2. HUMIDITY

In addition to increases in dark current, humidity testing[3] at relative humidities up to 85% have shown that devices exhibit effective electrical shorts through the process of electro-chemical oxidation. As the humidity level increases, tests have indicated that the lifetime of photodetectors decrease proportionately.

Hermetic sealing of the device will eliminate this problem as long as the critical concentration (parts per million) of moisture for the device material is sealed out.

3. TEMPERATURE CYCLING

Temperature cycling from -55°C to 125°C has been reported[3] to cause opens in lead-bonds of devices using plated contacts. This problem was effectively eliminated by using devices with evaporated contacts.

VI. SCREENING AND TESTING TECHNIQUES

1. SCREENING EARLY FAILURE DEVICES

As with emitters, there are certain screening techniques which have been found to be most effective in eliminating devices which are prone to early failure. One such technique being adapted for photodetectors involves high temperature-high bias voltage testing followed by a standard high temperature burn-in test.[3,8]

Most manufacturers perform high temperature, normal bias burn-in tests to eliminate those devices which are susceptible to infant mortality. However, photodetectors may not follow predictable dark current degradation rates. That is, the dark current in InGaAs/InP PINs and Si-based devices usually increases gradually and then increases rather sharply. However, not all photodetectors degrade this way. Some photodetectors begin degrading rapidly without the initial gradual increase in dark current. This suggests that conventional tests (high temperature, normal bias) must be performed for very long times to ensure all early failures have been discovered. Test results[3] have shown that this type of testing must be carried out for 500-1000 hours to give accurate results. Also, there was no clear distinction between where early failures ended and long-lived devices began, even after approximately 10,000 hours of testing.

AT&T researchers have implemented a high-temperature, high-bias voltage accelerated aging test to evaluate the quality of the InGaAs/InP receiver photodetectors for use in the TAT-8 submarine cable system. Their findings indicate that subjecting the photodetectors to tests for 10 hours at 200°C at two to three times the maximum rated reverse bias voltage will accurately identify good devices.

The dark current is the indicator which is used to determine device status. This test caused those devices prone to infant mortality to rapidly fail thereby allowing these devices to be eliminated from the test. Those devices which showed a significant change in dark current, either as an increase or decrease, are considered to be unstable and, therefore, prone to failure early in the life cycle of the device. These devices are also eliminated.

This test method has been verified by performing subsequent high-temperature burn-in aging for approximately 2000 hours on those devices which passed this screening method with confirming results.

2. MECHANICAL AND ENVIRONMENTAL TESTING

In addition to determining the integrity of photodetectors with respect to their anticipated lifetime, it is necessary to ensure that they will withstand the environment in which they were designed to operate. This is accomplished by subjecting the photodetectors to mechanical and environmental tests. Some of the more common tests serving this purpose are listed in Table 4-1.

TABLE 4-1. Photodetector Mechanical and Environmental Tests.

Test	Standard
Mechanical Shock	MIL-STD-750 Method 20162
Constant Acceleration	MIL-STD-750 Method 2006
Visual Inspection	MIL-STD-20722
Hermetic Seal	MIL-STD-750 Method 10712 Conditions C and H
Temperature Cycle	MIL-STD-750 Method 10512
Thermal Shock	MIL-STD-750 Method 10561
High Temperature Storage	MIL-STD-750 Method 10321
Humidity Storage	MIL-STD-750 Method 10211
Solderability	MIL-STD-750 Method 20311

The tests listed in Table 4-1 are only representative and not exhaustive. Solder heat resistance, bench drop, terminal strength, salt water spray and anti-solvent tests can also be performed on photodetectors, depending upon the intended application.

VII. ACCELERATED LIFETIME PREDICTIONS

Manufacturers do not typically supply predicted lifetimes for their devices. However, lifetime predictions based on accelerated test results are given in Table 4-2 for photodetector chips and modules. There was an insufficient number of data points available to the author to correlate these figures with respect to materials, construction, test techniques, activation energies assigned, and other very important considerations. Therefore, this lifetime data is only presented to give the reader a general idea as to how long these devices may last.

From this information it can be seen that photodetector lifetime predictions vary considerably and no information was uncovered indicating that one photodetector construction is more reliable or has a longer lifetime than another. However, because of the high voltages that are used, the APD appears to be a highly stressed device and good reliability seems predicated upon using only those crystals which are defect-free to avoid high current paths and subsequent localized heating which leads to early failures.

Table 4-3 gives field data on photodetectors and photocouplers with phototransistor outputs. All of these devices were operated in a benign environment at 55% rated power. A generic failure rate was calculated for the photocouplers in Table 4-3 by dividing the accumulated failures by the accumulated device hours. This gives:

$$\text{Failure Rate} = \frac{149}{306.9235 \times 10^6} = 0.485 \times 10^6 \ \frac{\text{Failures}}{\text{Hour}}$$

TABLE 4-2 Photodetector Lifetime Predictions.

Reference	Wavelength (μm)	Material	Construction	Reverse Bias (Vdc)	Temperature (°C)	E_a (eV)	Mean Time To Failure (10^6 hrs)	Failure Criteria	Date of Source	Comments
1	--*	--	APD	--	170	--	0.1	--	1984	Hermetically sealed package.
4	--	--	APD chip	--	--	--	0.01-0.1	--	3/83	
14	--	Ge	APD chip	25	260	1.0	0.004 (median) $\sigma = 1.0$	Dark Current > 2 μA	12/84	Monitoring photodetector. Wear out failure rate $\lambda(t)$ modeled by: $\lambda(t) = \dfrac{\sqrt{2}\exp\left[\dfrac{-1}{2\sigma^2}\left(\ln\left(\dfrac{t}{t_m}\right)\right)^2\right]}{\sqrt{\pi}\,t\sigma\,\mathrm{erfc}\left[\dfrac{\sigma}{\sqrt{2}}\ln\left(\dfrac{t}{t_m}\right)\right]}$
7	--	InGaAs/InP	PIN planar Zn diffused	15	25	0.85	10	Ten times increase in initial value of dark current at -5Vdc	7/86	LPE fabricated. Spin coat passivated with polyimide after removal of SiN_x film. Evaporated Ti/Au p-contact and Au:Ge:Ni alloy n-contact. SiN_x anti-reflection coating applied using plasma CVD method. Mounted and sealed in dry nitrogen ambient in a ceramic package. Prescreened 80°C/-15Vdc for 48 hours. $I(t) = I(o) + C_1 \exp\left(\dfrac{-E_a}{kT}\right) t^m$ where, t_m = median life, $C_1 = 9 \times 10^8 \dfrac{nA}{h}$, $m = 1$
3	1.0-1.6	InGaAs/InP	planar back-illuminated	10	30 / 85	0.55-1.1	≥ 2.0 / >> 0.02 $\sigma = 0.8$	Dark current ≥100nA	3/85	SiN_x passivated receiver photodetector. E-beam evaporated contacts. SiN_x substrate coating for anti-reflection.
4	--	--	PIN chip	--	--	--	> 1.0	--	3/83	

* Data not available.

TABLE 4-2. Photodetector Lifetime Predictions (con't).

Reference	Wavelength (μm)	Material	Construction	Reverse Bias (Vdc)	Temperature (°C)	E_a (eV)	Mean Time To Failure* (10^6 hrs)	Failure Criteria	Date of Source	Comments
16	1.3	Ge	APD Module	25-45	25**	--***	10	--	7/83	
16	0.85	Si	APD Module	205	25**	--	10	--	7/83	
15	1.3	InGaAs	PIN planar back-illuminated Module	--	--	--	408	--	12/84	Passivated. Anti-reflection coated.

* Mean Time To Failure is given unless noted otherwise.
** The source data indicated room temperature which has been interpreted to be 25°C.
*** Data not available.

TABLE 4-3 Operational Lifetime Data On Photocouplers With Phototransistor Outputs And Photodetectors.

Device Type	Material	Duty Cycle (%)	Quality Level	Ambient Temperature (°C)	Maximum Junction Temperature (°C)	Number Fielded	Device Hours (10^6)	Number Failed	Failure Rate (10^{-6})	Power (mW) Rated/Actual	Voltage (Vdc) Rated	Current (mA) Rated	Contact Construction	Environment
Photocoupler	GaAs	30	Plastic	40	100	2464	3.2032	0	--**	250/--	3.0	80	Non-metallurgic	Ground Benign
Photocoupler	--	30	Plastic	40	70	4872	6.3336	0	--	80/44	10	30	--	Ground Benign
Photocoupler	--	30	Plastic	40	100	21018	27.3234	14	0.512	100/55	5.0	25	--	Ground Benign
Photocoupler	--	30	Plastic	40	100	90691	117.8983	51	0.432	45/24.75	5.0	25	--	Ground Benign
Photocoupler	--	30	Plastic	40	70	33	0.0429	0	--	40/22	0.95	60	--	Ground Benign
Photocoupler	GaAs	30	Plastic	40	100	315	0.4095	0	--	260/143	6.0	40	Non-metallurgic	Ground Benign
Photocoupler	GaAs	30	Plastic	40	100	3785	4.9205	9	1.829	40/22	3.0	60	--	Ground Benign
Photocoupler	GaAs	30	Plastic	40	100	232	0.3016	0	--	260/143	3.0	60	Non-metallurgic	Ground Benign
Photocoupler	GaAs	30	Plastic	40	100	22183	28.8379	34	1.179	250/137.5	3.0	100	Non-Metallurgic	Ground Benign
Photocoupler	GaAs	30	Plastic	40	70	90205	117.6526	41	0.348	85/46.75	5.0	20	--	Ground Benign

* In all cases where rated and actual power is given, the devices were operated at 55% rated power.
** Data not available.

TABLE 4-3 Operational Lifetime Data On Photocouplers With Phototransistor Outputs And Photodetectors (con't).

Device Type	Material	Duty Cycle (%)	Quality Level	Ambient Temperature (°C)	Maximum Junction Temperature (°C)	Number Fielded	Device Hours (10^6)	Number Failed	Failure Rate (10^{-6})	Power* (mW) Rated / Actual	Voltage (Vdc) Rated	Current (mA) Rated	Contact Construction	Environment
Photodetector	Si	30	Plastic	40	125	148	0.1924	0	--**	100 / 55	20	--	Metallurgical	Ground Benign
Photodetector	Si	30	Plastic	40	125	5	0.0065	0	--	100 / 55	50.0	--	Metallurgical	Ground Benign

* In all cases where rated and actual power is given, the devices were operated at 55% rated power.
** Data not available.

Ref: 18

VIII. MAINTENANCE

Only when used as discrete components versus within modules will photodetectors require maintenance. This consists of cleaning the window which would occur just prior to or after installation. All discrete photodetectors have a window through which incident light travels. This window may have a filter or it may be lensed. In any case, cleaning of the window should be performed prior to use. Although ethyl alcohol is commonly used, the proper cleaning agent for a given device should be identified in the manufacturer's literature.

Photodetectors windows should not be touched with bare fingers since oils will be transferred to the window which will significantly impact the device transmittance.

Packaged receivers do not typically provide direct access to the photodetector. The receiver is either pigtailed with fiber or the photodetector is recessed within the package, mounted immediately following the connector receptacle. The connector should always be covered with a dust cap to prevent contamination from reaching the photodetector.

IX. LOGISTICS

Even though the construction of photodetectors will continue to change and improve over time, the physics of solid state photocurrent generation will remain unchanged.

These operating principles place physical limits on the materials that can be used. Many of the materials that can meet the demands of these principles have been identified and developed. Some of these materials are already capable of operating near their theoretical limits. This level of maturity is the result of intense research over a short period of time. This incredible development effort was a response to the growing fiber optic market which is still expanding.

Unfortunately, very few parameters have been standardized in this technology. However, the wavelengths of interest are firmly established, 820nm to 1550nm, for communications purposes. This single fact limits those materials which can be used in photodetector fabrication. This may very well be the key to continued availability of spare parts during the 20-25 year life cycle of a given system.

However, whether the current state of the technology remains available at a reasonable cost 25 years from now depends upon new developments. There is no way future developments can be prevented from happening. The only hope is that they will be developed with compatibility to current technology in mind.

X. SUMMARY

Optical detectors are generally of two types; PIN or APD. They serve the purpose of converting optical signals into electrical signals in system receivers and in monitoring the output of lasers. The photodetectors deliver a very low amplitude signal which requires the receiver amplifier to have very low noise if the optical signal is to be accurately reproduced while maintaining an acceptable receiver sensitivity.

The primary failure mechanism exhibited by photodetectors is an increase in dark current. Since changes in dark current may not follow a predictable pattern, manufacturers must make certain that the screening and quality assurance techniques that they employ allow the photodetector to reach a stable operating point before device integrity is assessed.

Chapter 5. Multiplexers, Demultiplexers and Couplers

The control and diversion of optical power using passive techniques is achieved through the use of multiplexers, demultiplexers and couplers which opera.e in the optical domain. Since agreement has not been reached on the terminology used to designate these devices, they are often grouped under the general heading of couplers. This discussion will address the construction and operation of these devices separately, whereas their reliability, testing and performance considerations will be combined.

Optical multiplexers and demultiplexers are used in fiber optic systems to allow the simultaneous transfer of more than one optical signal over the same fiber. This reduces the overall system cost and weight, dramatically increases the information carrying capacity per fiber and eliminates much of the complexity associated with systems that require redundancy. Also, these devices allow the simultaneous transmission, over the same fiber, of signals which use different modulation schemes and they provide the capability for channel expandability after system installation.

Optical couplers provide the function of distributing optical power, whether it be in a combining or dividing manner. These devices reduce system complexity and costs and provide the designer with a significant degree of flexibility.

Many different designs for these devices have been commercially available for several years now. However, only a few of these designs have become commonplace. The development of efficient manufacturing techniques has helped to improve product repeatability and to keep the cost of production down which has also contributed to their widespread use. As the photonic technology matures, we will see the mass production of higher quality products with a minimum amount of complexity.

The reliability and environmental issues impacting these two categories of components will be discussed after a review of their basic design and operation.

I. MULTIPLEXERS AND DEMULTIPLEXERS

The multiplexing and demultiplexing devices to be addressed will perform these functions optically by using the technique of wavelength division multiplexing (WDM). Many devices are called fiber optic multiplexers or demultiplexers when in fact they perform these functions electrically. Typically, these devices perform time division multiplexing electrically then this electrical signal is sent through a line driver which drives an optical emitter. Therefore, electrical signals are received, multiplexed, converted into an optical signal, then this optical signal is transmitted. This is depicted in Fig. 5-1. In contrast, true optical multiplexers receive optical signals, multiplex these optical signals, then transmit this multiplexed signal. All manipulations are performed on optical signals as shown in Fig. 5-2. Only those multiplexers and demultiplexers operating in the latter fashion were considered in this study.

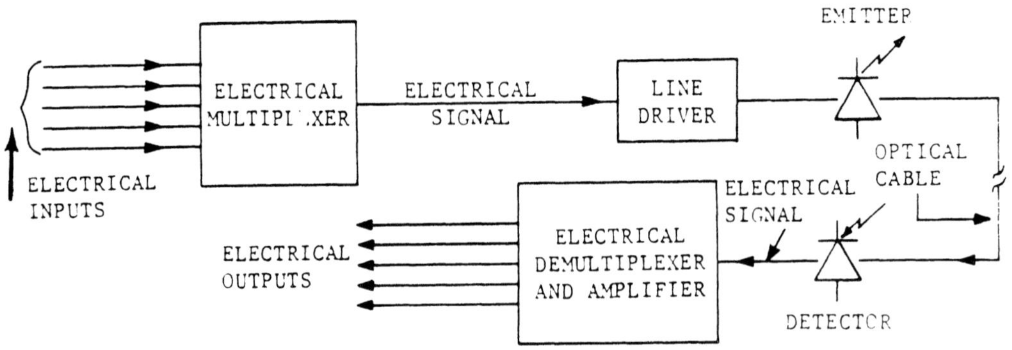

Figure 5-1. Optical multiplexing in the electrical domain.

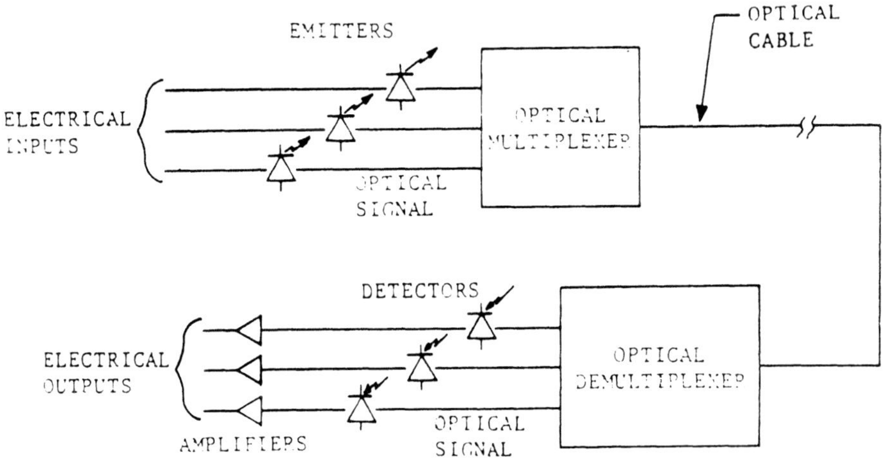

Figure 5-2. Optical multiplexing in the optical domain.

1. WAVELENGTH DISPERSIVE DEVICES

These devices are wavelength selective and make use of elements such as prisms and gratings that disperse incident light into its spectral components. Although they are used as both multiplexers and demultiplexers, they perform better as demultiplexers. This stems from the higher coupling efficiency which is achieved when the multiplexer output fiber pigtail has a core diameter which is smaller than that of the input fiber. However, the output fiber is the transmission fiber and it typically has a rather small core. Therefore, this will result in a low coupling efficiency. On the other hand, the demultiplexer couples light from the small core transmission fiber into a larger core pigtail, therefore, the coupling efficiency will be relatively high.

Wavelength dispersive devices are compact and have relatively uncomplicated designs. They are advantageous in that they require only one dispersive element to easily control up to three channels. However, due to the characteristics of the grating or prism material used, they are unable to achieve a low insertion loss at both the 0.8 and 1.3 μm wavelength regions.[1] These are the wavelengths at which optical fiber exhibits low attenuation. Therefore, a fundamental wavelength incompatibility exists between this type of device and optical fibers.

a. Diffraction grating characteristics -- A mirrored diffraction grating will reflect incident light an amount which is a function of the spectral content of that light. That is, the shorter wavelength components will be diffracted more than the longer wavelength components. This property allows individual wavelengths to be directed to precise locations such as input and output fiber cores.

The important parameters that impact grating performance are designated in Fig. 5-3, and the relationship between these parameters is given by equation 5-1[2] where a paraxial approximation has been assumed for the angle of incidence. That is, incident light is assumed to be close to and parallel to the optical axis.

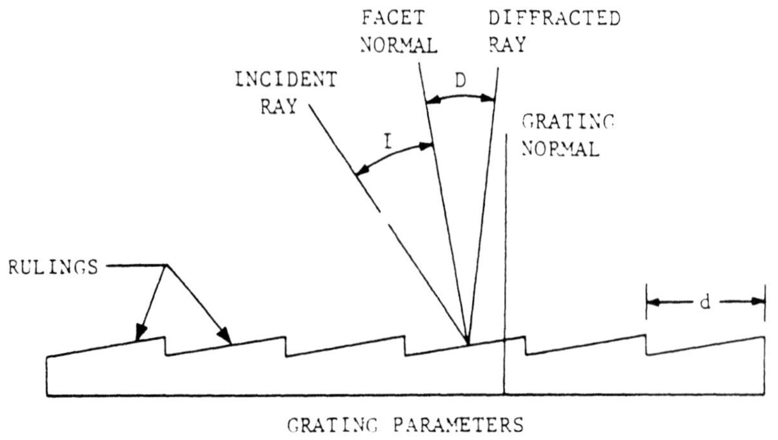

Figure 5-3. Fundamental diffraction grating parameters.

$$\lambda = d\left(\text{Sin}(I) + \text{Sin}(D)\right) \qquad \text{Equation 5-1}$$

where λ = wavelength

d = grating spacing

I = angle of incidence

D = angle at which light is diffracted from the normal

There are two common wavelength dispersive device constructions which use gratings: slant rod and Littrow mount prism grating. Both types operate on the same principle.

(1) Slant rod devices -- The slant rod is constructed by mounting a reflection grating directly onto the oblique end of a graded index (GRIN) glass rod which has a quarter period length as shown in Fig. 5-4.

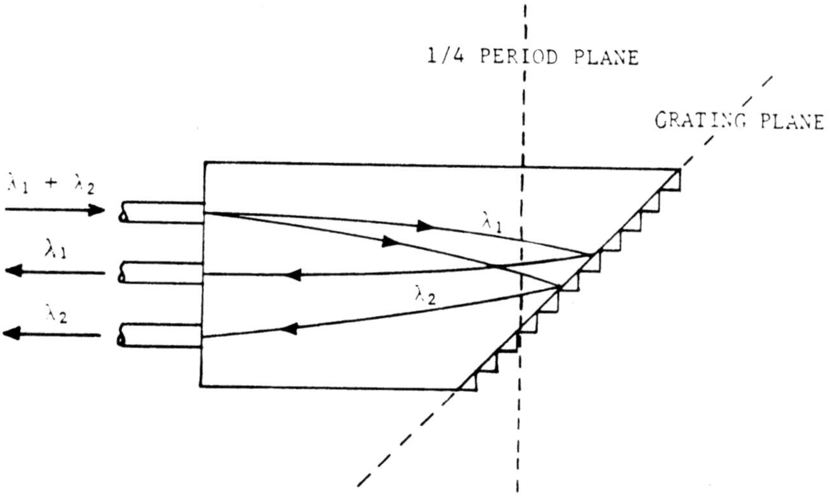

Figure 5-4. GRIN rod functioning as a demultiplexer.

The GRIN rod angularly disperses light as a function of wavelength, like graded index fiber, as shown in Fig. 5-5. The quarter period length is important since this length will allow the GRIN rod to collimate incident light comprised of multiple wavelengths and to focus individual wavelengths onto one point. This results from the controlled variations in the index of refraction of the rod as a function of its radius.

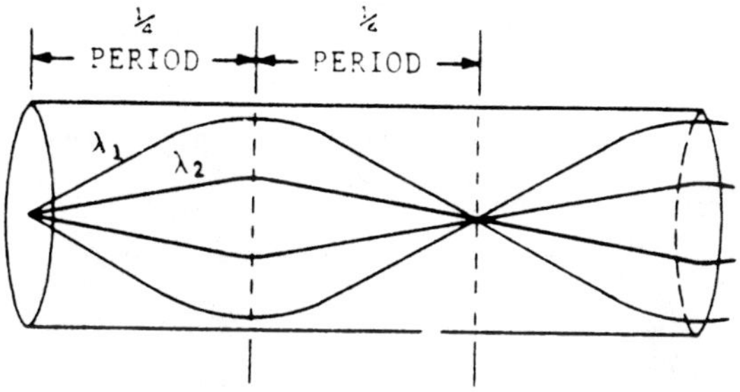

Figure 5-5. Collimating and focusing action of a GRIN rod.

The mirrored grating acts to reflect and diffract the collimated light according to wavelength. Once reflected, the individual spectral components are angularly dispersed by the GRIN rod. The dispersion of this light results in the individual components being directed to different locations on the incident plane where they are coupled to separate optical fibers.

GRIN slant rod devices are very compact and have few elements which makes them less susceptible to environmental disturbances and makes it easier to properly align the input and output fibers. However, alignment becomes much more difficult if more than three channels are to be multiplexed or demultiplexed.[1]

One significant drawback of the GRIN rod construction is that <u>collimated</u> light only reaches a single point on the grating. This is because true collimation only occurs at odd multiples of the GRIN quarter length. As can be seen in Fig. 5-4, the quarter length plane intersects the plane of the grating at just one point. Only the light reflected from the line defined by the intersection of these planes will be in focus when it exits from the GRIN rod. Light reflecting from all other points on the grating will experience chromatic aberration when it reaches the end of the GRIN rod since precise odd multiples of the quarter length have not been traveled. The focal point for these rays will either be in front of or behind the end of the GRIN. The result is an increased spot size at the GRIN which reduces coupling efficiency. This problem is significantly reduced in the prism grating construction.

(2) <u>Prism grating devices</u> -- Fig. 5-6 shows the construction of a prism grating where it is functioning as a demultiplexer. A reflection grating is epoxyed onto the hypotenuse of a right triangle prism and Littrow mounted to the end of a GRIN rod as shown.

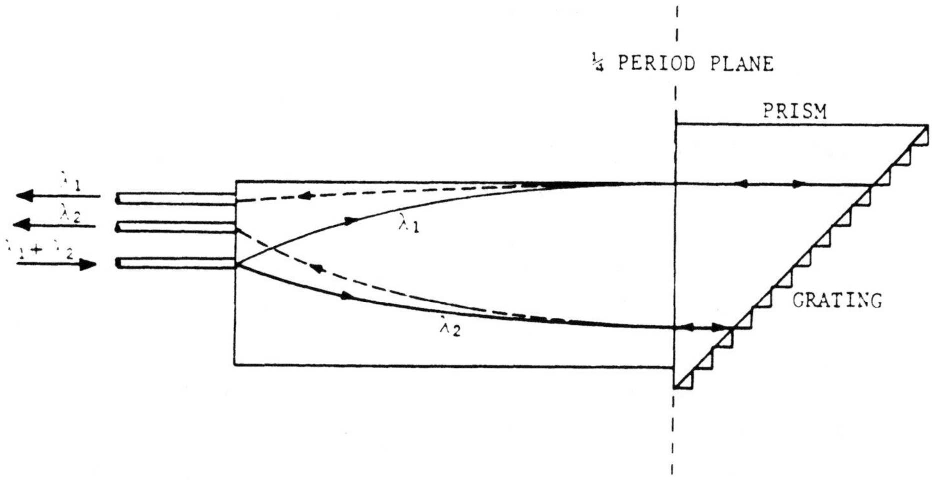

Figure 5-6. Prism grating construction.

The principle of operation of the prism grating is the same as that of the slant rod grating, however there is much less aberration. As Fig. 5-6 shows, the GRIN rod is only one quarter period in length. Therefore, all light entering the GRIN will be collimated when it reaches the prism. The prism will angularly disperse the individual wavelengths much less than that of a GRIN having the same dimensions as the prism, as in the slant rod. This results in the reflected rays being better collimated as they enter the GRIN rod from the prism after reflection. Therefore, they will experience less aberration when exiting the GRIN rod.

Since the angular dispersion characteristics of a prism are relatively low, they are typically only used in conjunction with other optical elements to perform multiplexing or demultiplexing operations.

2. WAVELENGTH SELECTIVE DEVICES

Multiplexing and demultiplexing can be accomplished with two types of filters; edge interference filters and bandpass interference filters. These devices are often categorized as Dielectric Thin-film Filters (DTFs). They can be used equally well as multiplexers and/or demultiplexers by simply reversing the input and output ports. DTFs are fabricated from alternating layers of high and low refractive index dielectric films which forms a Fabry-Perot interferometer when fully assembled.

a. Interference filter characateristics -- Generally, interference filters are designed for applications where the incident light is collimated and normal to the filter surface. If normal incidence is not the case, then the peak transmission wavelength will decrease as a function of the angle of incidence. This characteristic can be used to tune the peak transmission wavelength. However, this is undesirable when not controlled for it increases the attenuation of the filter. This must be taken into consideration for high shock and vibration environments. There will only be minor changes in the passband width and transmission characteristics. To determine the peak wavelength, equation 5-2[3] can be used for edge filters with off-normal angles <15 degrees and for bandpass filters with off-normal angles <30 degrees. Equation 5-2 shows that for higher values of n*, the smaller will be the wavelength shift.

$$\lambda_{\phi} = \frac{\lambda_{o}\sqrt{n^{*2} - \sin^{2}\phi}}{n^{*}} \qquad \text{Equation 5-2}$$

where λ_{ϕ} = central output wavelength at off-normal incidence
λ_{o} = central output wavelength at normal incidence
n^{*} = effective index of refraction of total filter
ϕ = off-normal angle

If these filters are used with uncollimated light, the resulting condition is equivalent to having an off-normal angle for a multitude of rays. As predicted by equation 5-2, this results in individual rays being wavelength shifted by different amounts. This effects the filter output by broadening its bandwidth and reducing the peak transmission.

(1) Edge interference filter devices -- Edge interference filters (also called dichroic filters) operate on the principles of constructive and destructive interference. There are two edge interference filter types; those with a short-wavelength pass band and an accompanying long-wavelength reflection band and those with a long-wavelength pass band and accompanying short-wavelength reflection band. The optical characteristics of these filters are shown in Fig. 5-7a and Fig. 5-7b, respectively.

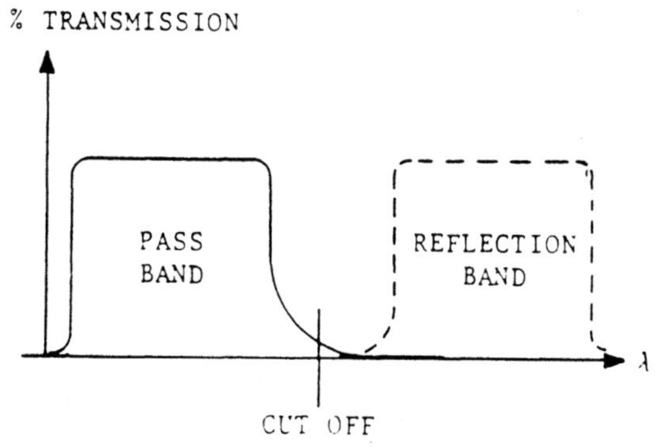

Figure 5-7a. Short-wavelength pass band characteristics of edge interference filter.

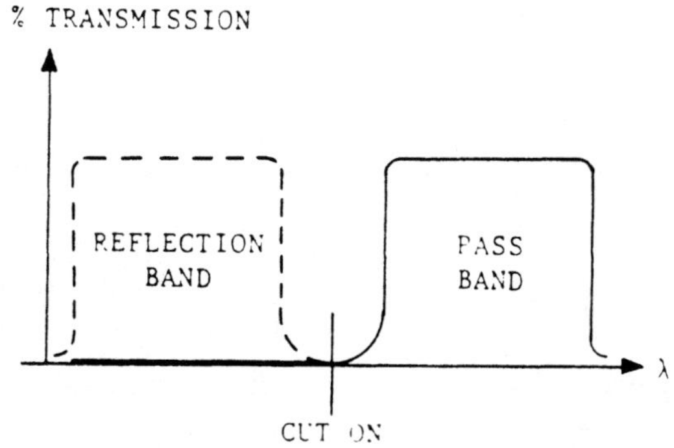

Figure 5-7b. Long-wavelength pass band characteristics of edge interference filter.

The wavelength at which the transition between transmission and reflection occurs is called the "cut off" or "cut on" wavelength. Fig. 5-8 shows how a short-wavelength pass band filter can be used to demultiplex two signals.

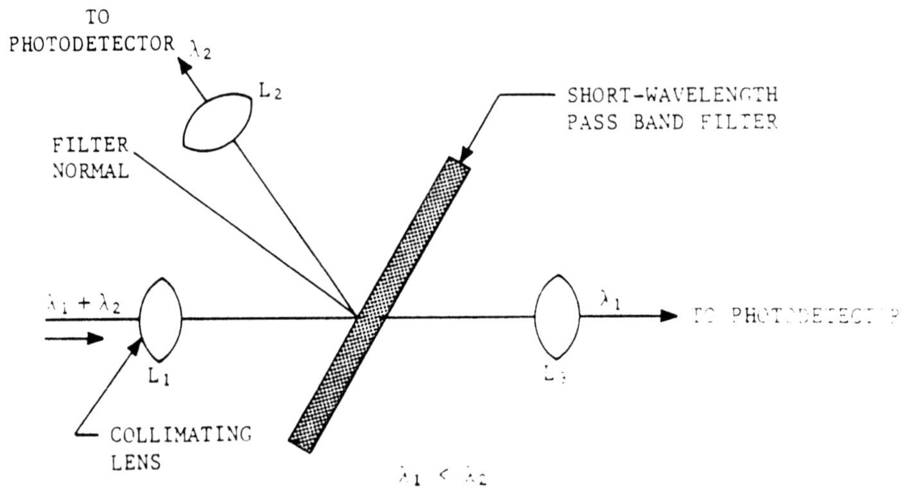

Figure 5-8. <u>Demultiplexing two signals using a short-wavelength edge filter</u>.

Fig. 5-8 shows the edge filter slightly tilted in order to direct the reflected wavelength, λ_2, to the collimating lens, L_2, for detection. The shorter wavelength, λ_1, passes through the edge filter unattenuated to the collimating lens L_3 for detection.

(2) <u>Bandpass interference filter devices</u> -- Bandpass interference filters are constructed similarly to edge filters with respect to the layering of high and low refractive index dielectric films.

However, due to the differences in film thickness and reflective coatings, the bandpass filter allows the transmission of a much narrower spectral band of wavelengths and blocks all others. The spectral transmission window, or bandwidth, of these devices can be less than one nanometer to tens of nanometers. The typical transmission characteristics of a bandpass filter are shown in Fig. 5-9. Since this filter can have a very narrow bandpass, up to six channels can be multiplexed or demultiplexed as shown in Fig. 5-10.

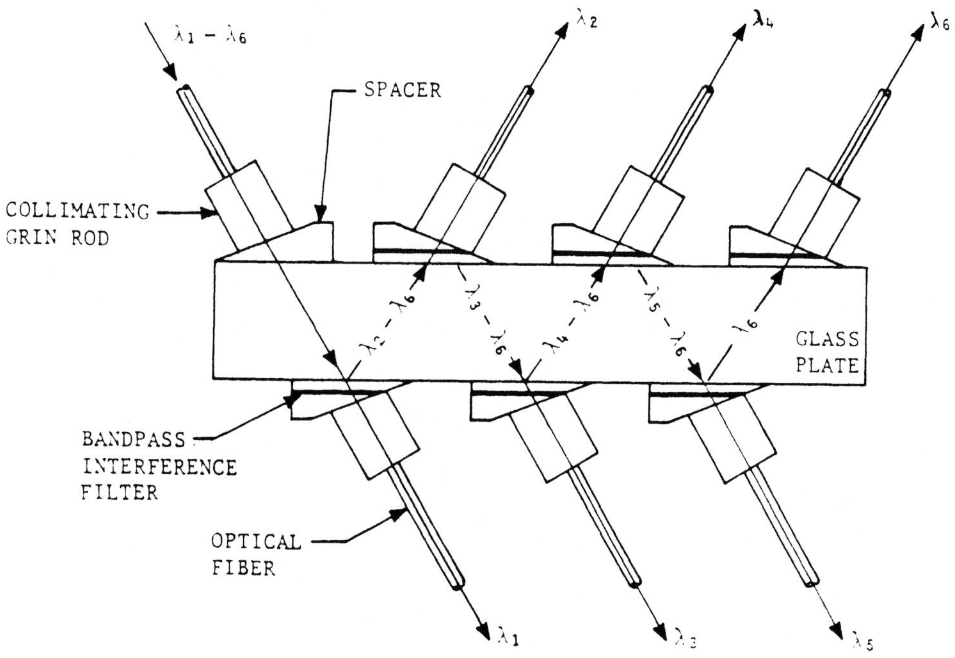

Figure 5-9. Bandpass interference filter characteristics.

The interference filters shown in Fig. 5-10 have narrow enough pass bands which are sufficiently separated such that only one of the six wavelengths will pass unimpeded through each filter. The impeded wavelengths will be reflected to the next filter whose characteristics will allow transmission or reflection. The spectral response of this design will typically resemble Fig. 5-11.

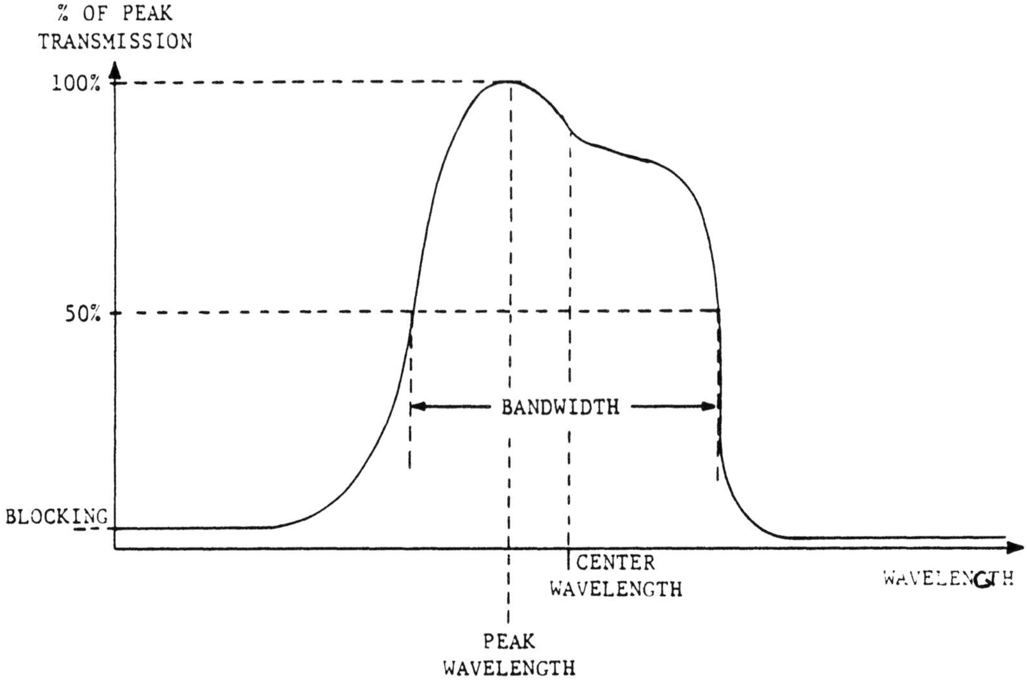

Figure 5-10. Six channel bandpass interference filter demultiplexer.

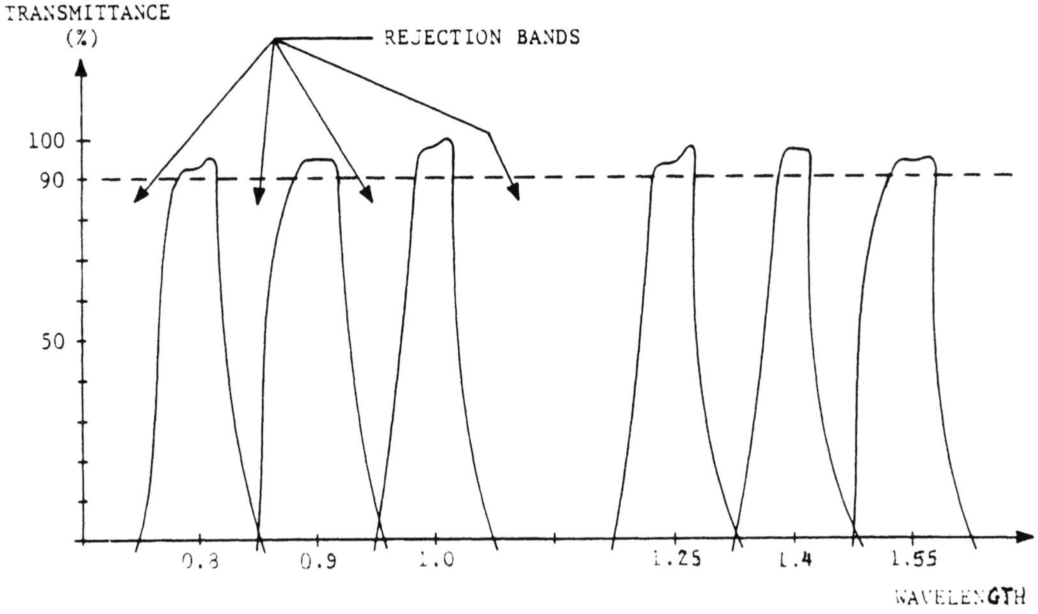

Figure 5-11. Spectral response of the 6-channel demultiplexer shown in Fig. 5-10.

The advantage of this spectral response compared to that of the edge filter shown in Fig. 5-7 are the well-defined rejection bands on each side of the narrower pass bands which allow more channels to be multiplexed over a single fiber for a given range of wavelengths. As can be seen, this is a very complex design which requires extremely critical alignment.

3. EVANESCENT WAVE DEVICES

Evanescent wave multiplexers are very different in their operation than the previously mentioned devices in that these devices do not use optical elements such as gratings, prisms, filters and lenses. Instead, single mode fiber and a unique construction process are used to multiplex or demultiplex optical signals. Evanescent waves are shown pictorially in Fig. 5-12. These waves are that portion of the electromagnetic energy in the mode bound by the fiber core but propagating in the fiber cladding. The evanescent wave energy is typically 20% to 25%[4] of that propagating in the core and is a function of the fiber radius as shown.

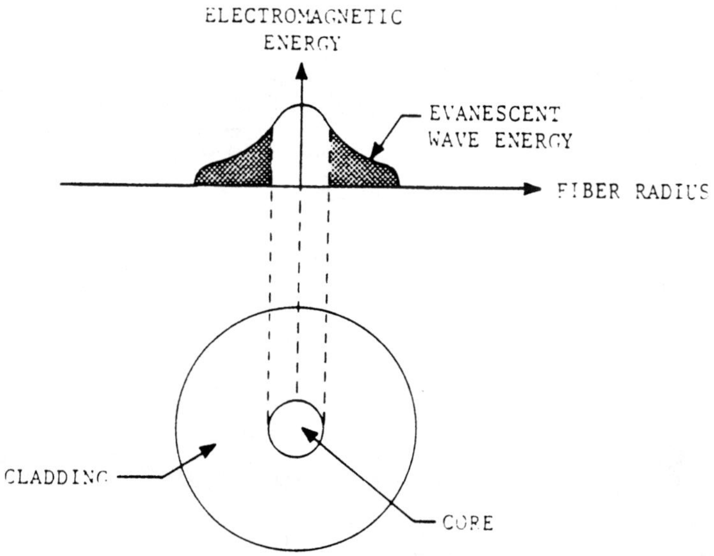

Figure 5-12. Evanescent wave energy.

a. Single mode evanescent wave devices -- Multiplexers and demultiplexers operating on the principle of evanescent mode coupling are fabricated using the fused coupler technique. This is accomplished by fusing two parallel fibers together while they are slowly being pulled under computer control such that a slight taper is formed. Optical energy will be coupled from one fiber to the other at the taper. The amount of power coupled is a function of the relative closeness of the cores, the wavelengths, the core diameters, and the length over which the two cores are allowed to interact.[5]

Fig. 5-13 shows a two channel single mode evanescent wave multiplexer. Proper design of this device will result in a coupling efficiency of almost 0% for λ_1 and 100% for λ_2 at the taper. Therefore, wavelengths λ_1 and λ_2 will be multiplexed onto the primary fiber. The degree of coupling will vary spatially in a sinusoidal fashion as a function of wavelength.

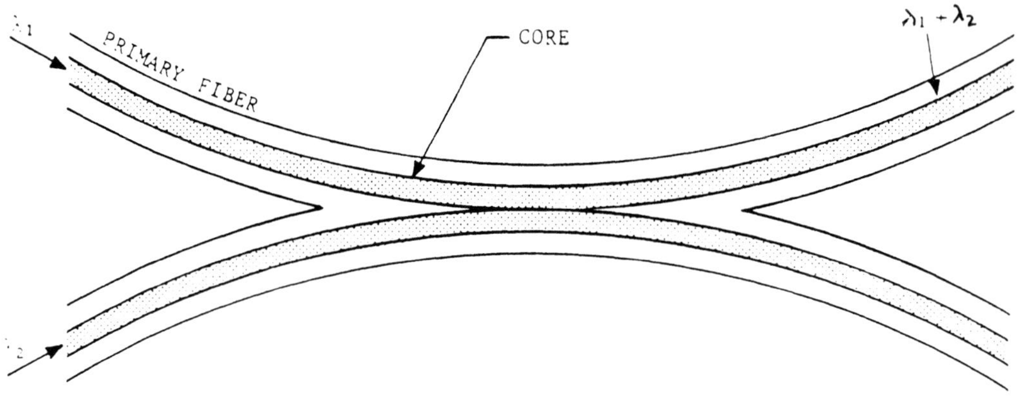

Figure 5-13. Evanescent wave multiplexer construction.

This is depicted in Fig. 5-14. Using this as the response characteristic for the multiplexer in Fig. 5-13, λ_1 and λ_2 would correspond to 1.5 μm and 1.3 μm, respectively.

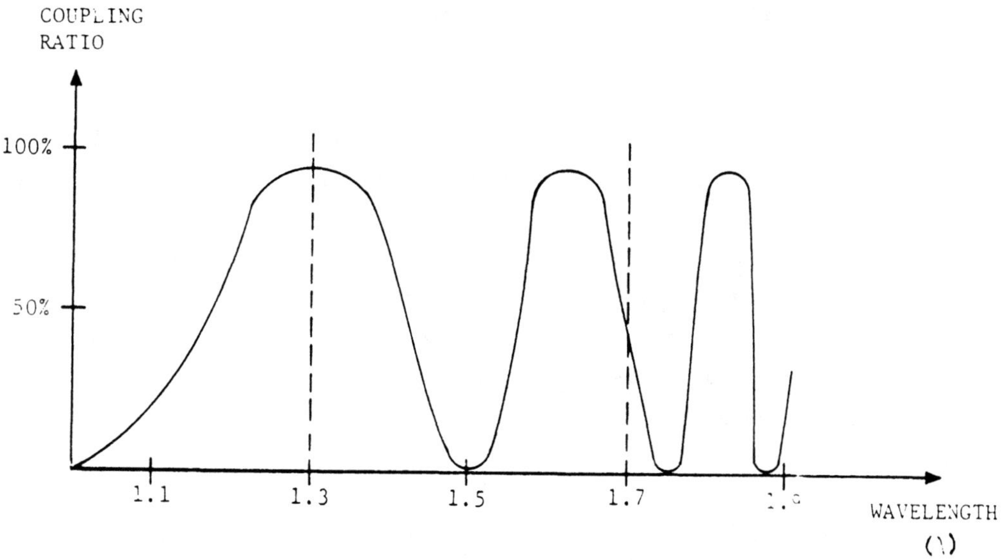

Figure 5-14. Wavelength dependence of the coupling ratio.

II. COUPLERS

Fiber optic couplers are fabricated by one of two techniques: producing an assemblage using bulk optics or altering the physical characteristics of optical fibers to couple electromagnetic fields. The development of couplers by the bulk optics technique occurred first and has since been widely replaced with the latter technique.

Although many different and novel designs have been developed using bulk optics, they suffer from one significant drawback in that the alignment of the components is extremely critical and difficult to realize. Also, maintaining precise alignment during environmental changes, especially temperature and vibration, is very demanding. The optical devices used in these couplers can include prisms, gratings, partially reflective mirrors, curved mirrors and lenses.

Electromagnetic field coupling or mode coupling can be accomplished in various ways but only one method is widely used and that is the Fused Biconical Taper (FBT) technique. This method can be used with both single mode and multimode fiber, however, the distribution of modes is different for the different fiber types. FBTs are fabricated by removing the coating from two or more fibers and twisting them around one another one to two turns. Then the fibers are fused together at the twist at approximately 1500°C. Simultaneously, they are stretched while under computer controlled monitoring. This decreases the diameter of the fused region causing the individual fiber cores to become closer together. The close proximity of the cores enhances the ability of the fibers to couple electromagnetic energy. Fig. 5-15 illustrates the results of this process.

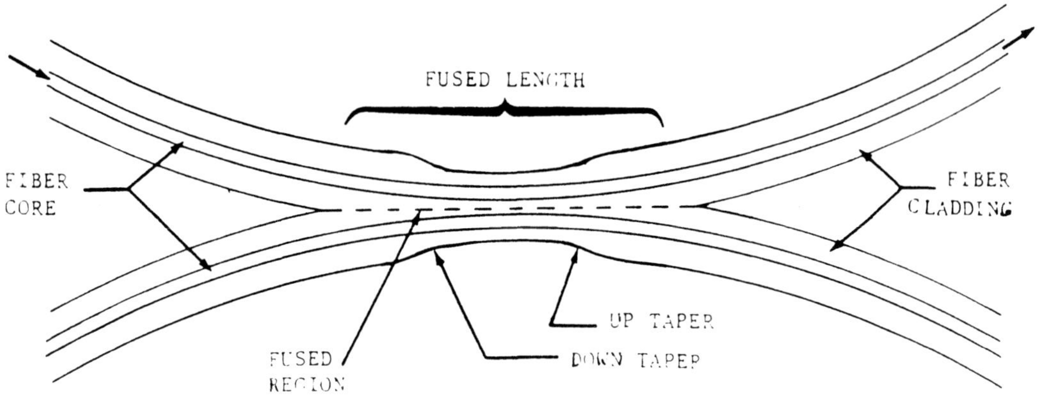

Figure 5-15. Geometry of a fused biconical taper coupler.

1. DIRECTIONAL COUPLERS

Generally, directional couplers connect only three ports; one input to two output ports. This is the definition that is used in this discussion, however, there is not general agreement on the meaning of this term.

a. <u>Splitters and combiners</u> -- Splitters are used to direct the optical signal equally from one input fiber to two output fibers. This can be accomplished using bulk optics or mode coupling. Fig. 5-16 shows a technique using bulk optics and Fig. 5-17 shows a mode coupling technique.

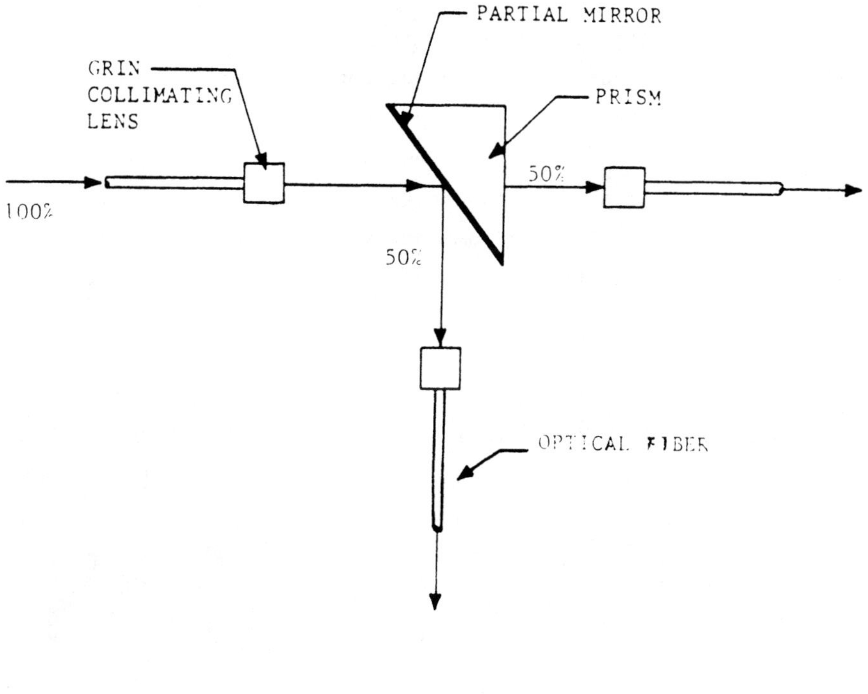

Figure 5-16. <u>Splitter using bulk optics</u>.

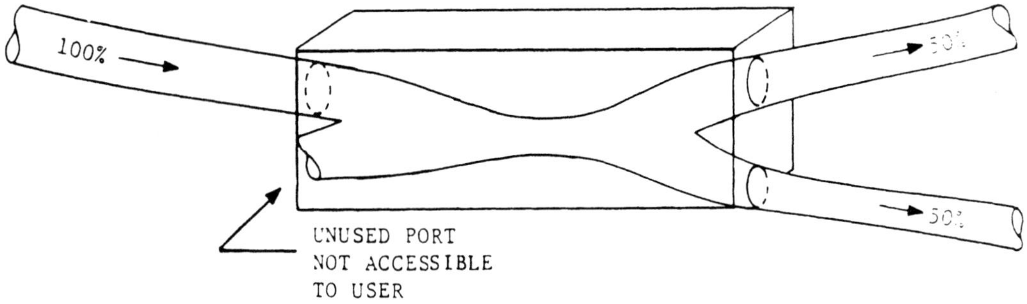

Figure 5-17. Splitter using mode coupling.

A combiner is used to combine two signals, not necessarily of equal strength, onto one fiber. Basically, it is a splitter with the input and output ports reversed.

b. Taps -- The tap is similar to the splitter in that it is a three-port device; one input and two output ports. There is a primary and secondary output port where the secondary port signal is a small fraction, 5% to 10%, of the primary signal. This tapped signal is usually used to monitor the transmitted signal or as an input to feedback circuitry to control a sources' optical output. Like the combiner and splitter, the tap coupler is also fabricated using bulk optics or by using mode coupling techniques.

2. STAR COUPLERS

This class of passive coupler distributes the available optical power from one or more input ports to many output ports. By definition these devices have M input ports and N output ports where N is greater than two and M can range from one to N. They are designated as M x N star couplers.

Star couplers are available in two general configurations; transmissive and reflective. Each type is used extensively in data networks as a central or local distribution point and each type can be manufactured using bulk optics or the fused biconical taper technique.

a. Transmissive couplers -- The configuration of a transmissive coupler fabricated using bulk optics is shown in Fig. 5-18. The purpose of the dielectric waveguide other than to confine the signal is to receive optical power from any or all input ports and evenly distribute this power over the entire surface of the waveguide output face. This allows each output port to receive any and all input signals.

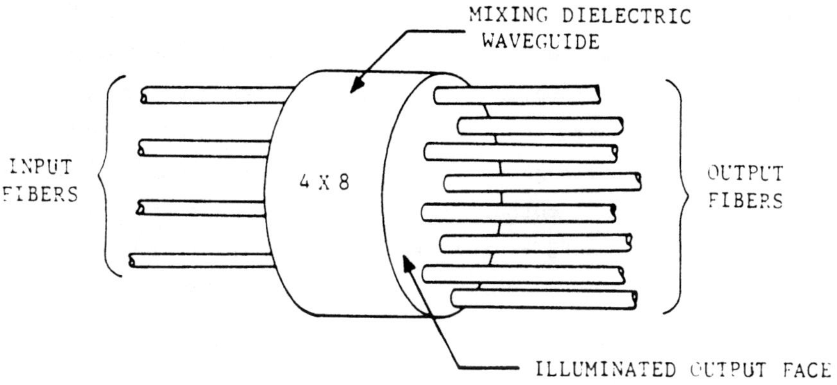

Figure 5-18. 4 x 8 transmissive star coupler using bulk optics.

As Fig. 5-18 shows, all of the light striking the output face does not enter an output port. This results in what is termed a packing fraction loss. The output fibers are bundled or packaged together then butted against the output face. A portion of the illuminated face is not covered with these fibers which allows a fraction of the optical input signal to escape.

Transmissive star couplers can also be manufactured using the fused biconical taper technique which results in an equal number of input and output ports. However, if the device is to be used as a single input port device, the other input ports are not made available to the user. Instead, they will be confined within the device's housing. Because the primary element of the FBT is fiber, these devices are much less sensitive to temperature variations relative to the bulk optic couplers. Also, the lack of discrete components eliminates alignment problems. In addition, there is no packing fraction loss associated with this design.

Unlike the transmissive star coupler manufactured using bulk optics, all of the output ports of a FBT transmissive star coupler do not carry equal amounts of optical power. As already mentioned, FBT devices operate on the principle of coupling electromagnetic energy between fibers due to the close proximity of the excited fibers' core and the claddings to adjacent fibers. As Fig. 5-19 shows, there is a down taper and up taper portion of the coupler. As light enters the down taper region, all of the modes can not be bound by the decreasing volume. This causes the higher order modes to couple to the adjacent fibers where they propagate in bound cladding modes. As the light reaches the up taper region, these modes couple equally from the claddings to the individual output fiber cores. These higher order modes are shared equally among all output port fibers, including that fiber associated with the excited input fiber. However, note that the <u>lower order modes</u> cannot couple into the adjacent fibers. This causes an unequal mode distribution, and therefore, an optical output power distribution non-uniformity. The result is that the output port associated with the excited input port has approximately 10% more power than the other output ports.[7] This must be taken into consideration when performing a system budget analysis.

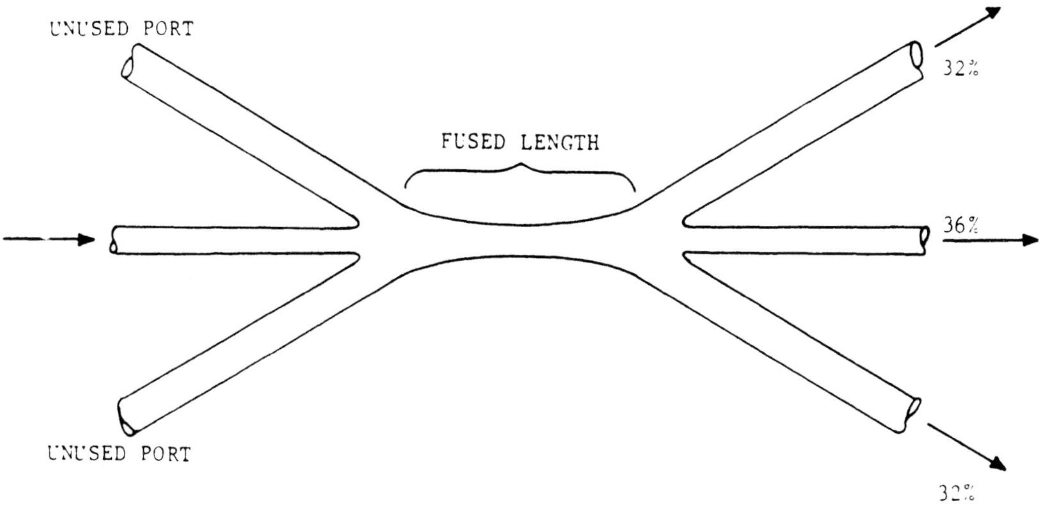

Figure 5-19. Distribution of power in a FBT transmissive star coupler.

b. Reflective couplers -- The characteristics of the reflective star coupler are similar to the transmissive in that the input signal on one port will be directed to all output ports. One significant difference is that the input signal will be directed to all input ports as well. This can result in a significant increase in signal noise due to the reflections into the active optical source, especially when lasers are used.

Fig. 5-20 shows a reflective star coupler which uses bulk optics. In operation, the input signal illuminates the entire reflective face of the coupler. This illumination is then reflected back to the input face where all of the ports receive an equal portion of the reflected signal. This allows every port to function as both an input and an output.

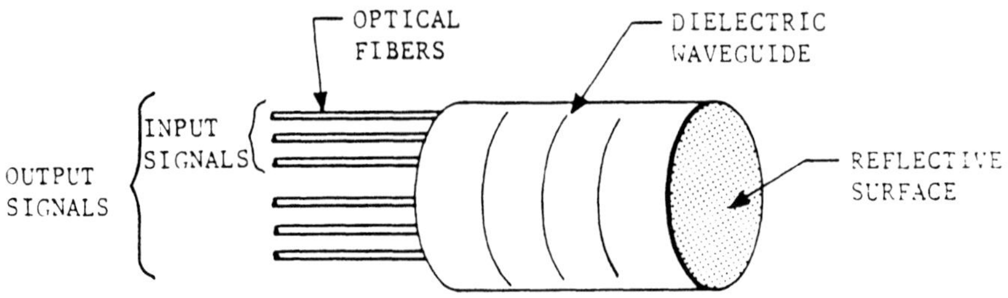

Figure 5-20. Reflective star coupler using bulk optics.

This makes the device very flexible, however, a large portion of the reflected signal is lost to the input fiber ports in a unidirectional link. In a bidirectional link, this loss is not as significant. In both applications there is packing fraction loss.

Like the transmissive star coupler, the bulk optics element is a dielectric waveguide which acts to confine and distribute the input signals over the entire reflective surface.

Fig. 5-21 shows a reflective star coupler manufactured using FBT technology. This device also directs the input signal to all ports just like the bulk optics construction. Although the FBT reflective star coupler does not suffer from packing fraction loss; there is the same non-uniform modal distribution between the ports as discussed for the FBT transmissive star coupler.

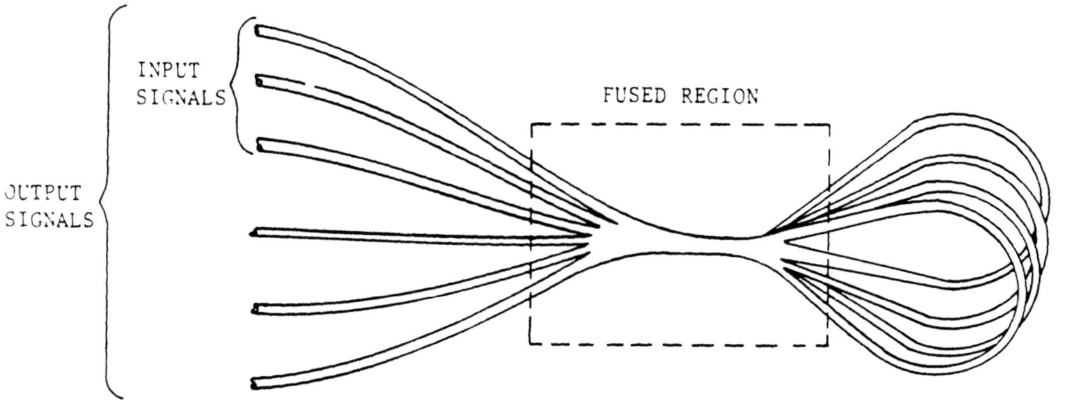

Figure 5-21. FBT reflective star coupler.

III. MEASUREMENT PARAMETERS

There are several measurements which are made on multiplexers, demultiplexers and couplers which characterize their performance. These parameters are typically specified by the manufacturer after monitoring during quality assurance tests. The definitions for these parameters as used in this document are given in the text that follows. This is to avoid any confusion with differing definitions which may exist in the community.

1. WAVELENGTH ISOLATION

Multiplexers and demultiplexers require a certain amount of isolation or separation between each channel to reduce crosstalk. Fig. 5-22 identifies the important spectral response characteristics of a three wavelength bandpass interference filter multiplexer or demultiplexer. The wavelength isolation is most important in demultiplexers for only one wavelength must exit each output port. The degree of isolation between wavelengths is an indicator of how much

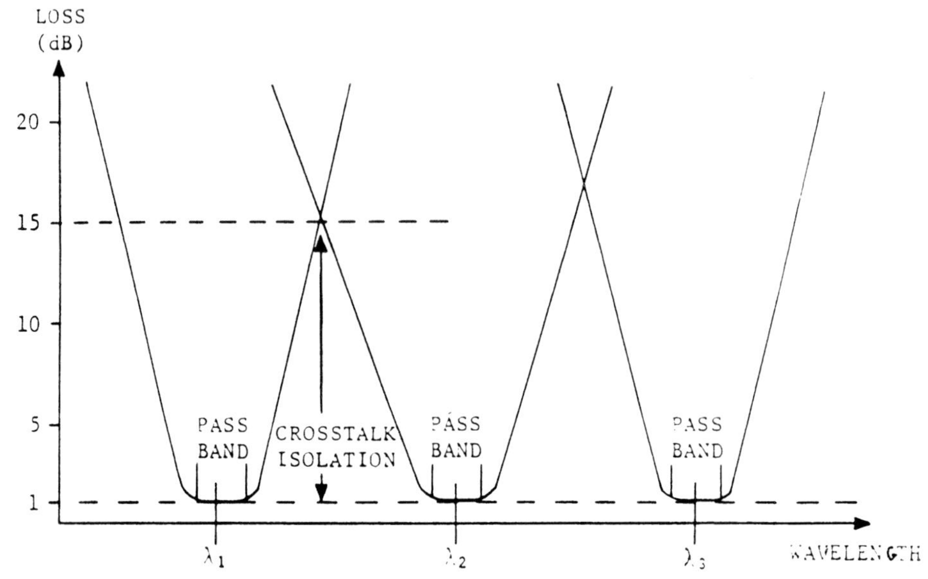

Figure 5-22. Multiplexer or demultiplexer spectral response.

power at one wavelength is measured at an output port that should not be carrying that wavelength. Referring to Fig. 5-23, wavelength isolation can be expressed by equation 5-3.

$$W_I = 10 \log \left[\frac{P_O(\lambda_n)}{P_I(\lambda_n)} \right] \qquad \text{Equation 5-3}$$

where W_I = demultiplexer wavelength isolation

$P_O(\lambda_n)$ = output power at an undesired channel

$P_I(\lambda_n)$ = input power at the selected wavelength

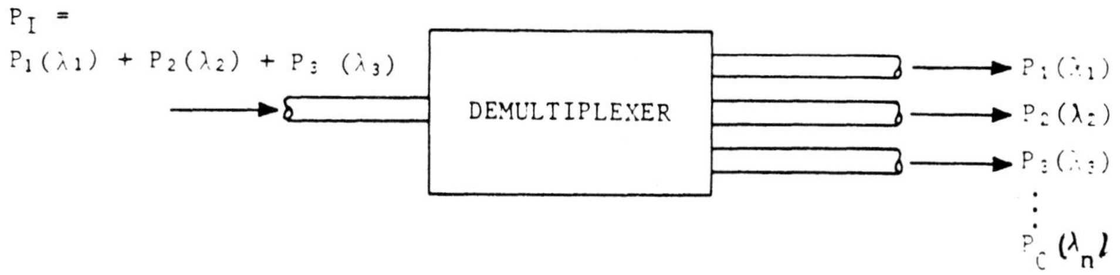

Figure 5-23. Demultiplexer port designations.

Due to the instability of most laser sources with variations in temperature, the source peak wavelength will vary. These variations must remain within the pass band. If not, the wavelength isolation will change with the peak wavelength which can cause crosstalk and will increase the device loss.

LEDs have a relatively wide emission spectrum. Therefore, to prevent a high percentage of this spectrum, and therefore power, from falling outside of the pass band, a wide band is required. Otherwise, a larger portion of the total power will be attenuated and the separation between central wavelengths will have to be increased to achieve an adequate crosstalk isolation. This wider separation decreases the information carrying capacity of the system.

2. INSERTION LOSS

The insertion loss, or excess loss, is a measure of the power lost within the device. For a multiplexer or demultiplexer, this is determined by measuring the amount of input power which is available at the desired output channel. Referring to Fig. 5-24, this is expressed by equation 5-4.

$$I_L = 10 \log \left[\frac{P_O(\lambda_n)}{P_I(\lambda_n)} \right] \quad \text{Equation 5-4}$$

where I_L = multiplexer insertion loss

$P_O(\lambda_n)$ = output power at the selected wavelength

$P_I(\lambda_n)$ = input power at the selected channel

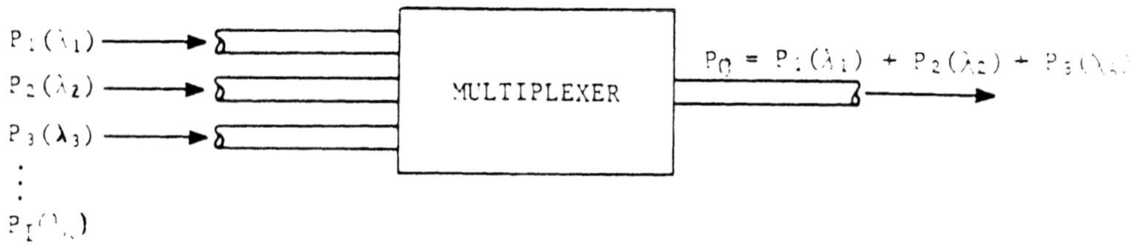

Figure 5-24. Multiplexer port designations.

For a coupler, the insertion loss is a measure of the total power available at the output ports relative to the total input power. Referring to Fig. 5-25 for port designations, the insertion loss is expressed by equation 5-5.

$$I_L = 10 \log \left[\frac{P_3 + P_4}{P_1} \right]$$ Equation 5-5

Figure 5-25. Coupler port designations.

3. COUPLING RATIO

The coupling ratio is a measure of the division of the output power between a coupler's output ports. Common coupling ratios for two output port devices are 50/50% and 10/90%. This parameter can be expressed either as a percentage or as an expression of powers measured in decibels. Equations 5-6 and 5-7 give these expressions, respectively.

$$CR\% = \frac{P_4}{P_3 + P_4} \times 100\%$$ Equation 5-6

$$CR(dB) = \pm 10 \log \left(\frac{P_4}{P_3 + P_4} \right) \qquad \text{Equation 5-7}$$

4. UNIFORMITY

Uniformity of a coupler is a measure of how much difference there is in output port powers as different ports are individually excited. When expressed as a percentage, the uniformity is given by equation 5-8.

$$U\% = \frac{P_{n(Max)} - P_{n(Min)}}{P_{n(Max)}} \times 100\% \qquad \text{Equation 5-8}$$

where U = coupler uniformity

$P_n(max)$ = maximum output power with port P_n excited

$P_n(min)$ = minimum output power with port P_n excited

When expressed as a percentage, there is a uniformity value for each excited port. This leads to a maximum value for the entire coupler.

The uniformity can also be expressed in decibels as the difference between the maximum and minimum values for insertion loss. This results in a power range.

In addition to these methods of defining uniformity, there is the expression given in equation 5-9 which uses the minimum and maximum values for coupling ratio, (CR).

$$U = 10 \log \left[\frac{CR\ (Min)}{CR\ (Max)} \right] \qquad \text{Equation 5-9}$$

5. DIRECTIVITY

Directivity is a measure of a coupler's ability to direct the optical input signal to the desired output ports and not to undesired ports. For example, if P2 in Fig. 5-25 is unused when Pl is excited, this coupler would have good directivity if the power measured at P2 was -40dB to -60dB. That is, very little power would be lost to the unused port. Referring to Fig. 5-25, equation 5-10 gives the expression for directivity. This is a very important parameter in single mode applications since the stability of laser sources is adversely effected by reflected light from external elements such as couplers, multiplexers and connectors. Considering the manner in which reflective star couplers operate, directivity has no meaning for these devices.

$$D = \left[10 \log \frac{P_2}{P_1} \right] \qquad \text{Equation 5-10}$$

IV. MAINTENANCE REQUIREMENTS

The multiplexers, demultiplexers and couplers that have been discussed do not contain active components. The only discrete elements used in the manufacture of these devices are bulk optics and these are mounted in a housing which is not meant to be user serviced. As such, any maintenance to be performed will be that associated with the device pigtails, or optical fiber leads. This amounts to no more than preparing them for termination or fusion splicing. Once this is properly done, there should be no need for further maintenance.

V. CONDITIONS EFFECTING DEVICE PERFORMANCE

The tests performed on these devices are typically used to determine the devices' ability to tolerate external forces. These forces, which can influence the devices' stability, are discussed separately below.

1. TEMPERATURE DEPENDENCE

The coupling ratio of fused biconical taper devices tends to be relatively independent of temperature when temperature cycle tested. Multiplexers, demultiplexers and couplers can be fabricated by the FBT technique with the only difference in manufacture for the different devices being the wavelength sensitivity. During the manufacturing process, if the sensitivity is increased, multiplexers and/or demultiplexers are developed. If the sensitivity is decreased, couplers will result. The implication is that both device type coupling characteristics will be relatively independent of temperature when manufactured using FBT technology.

Tests indicate that FBT devices made from multimode fiber (100/140 μm) is slightly less stable then both single mode fiber FBTs twisted one full turn during manufacture and single mode FBTs having no twist.[8] The variation in coupling ratio was <1% for the single mode FBTs and < 1.5% for the multimode FBTs. The temperature was cycled from 10°C to 90°C to 10°C.

The temperature dependence of devices which use bulk optics is more pronounced than FBT-fabricated devices due to the materials used. This is most significant when they are in an environment which deviates from the nominal operating temperature. Typically, when interference filters are used in temperatures deviating from the specified nominal (usually -60°C to +60°C) the peak wavelength will shift approximately linearly with temperature. Outside of this temperature range, there may be permanent changes in the operating characteristics of the device.[3]

Also, it is extremely important that the individual bulk elements be matched in their coefficients of thermal expansion to prevent separation of the elements or unequal movements resulting in optical misalignment.

2. POLARIZATION

FBTs which are used in applications requiring distinct interference fringes, e.g., in interferometric work, must have good fringe visibility. This is a measure of the quality of the fringes caused by interference. Varying the temperature from 10°C to 90°C while monitoring the outputs of multimode (100/140 μm), single mode twisted and single mode untwisted FBT devices resulted in only the twisted device demonstrating a significant polarization variation with temperature. This is considered to be due to the birefringence induced by the internal strain at the twisted area.[8]

The effects of the input signal polarization on the performance of a device which uses a diffraction grating can be rather significant. Many optical input signals are randomly polarized. This will effect the grating efficiency since light incident on the grating has a different efficiency if it is polarized parallel to the grating rulings than it does if it is polarized perpendicular to the rulings. The efficiency is also a function of the wavelength of the incident light. Therefore, if the efficiency curves are as shown in Fig. 5-26, it is apparent that a significant portion of a randomly polarized signal will be lost to the grating.

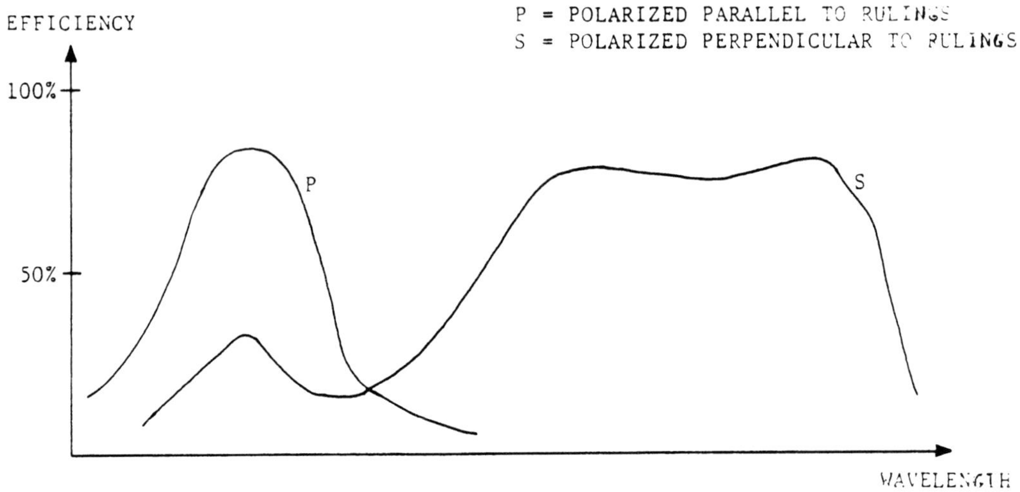

Figure 5-26. Efficiency curves for a hypothetical grating.

3. HUMIDITY

Multiplexers, demultiplexers and couplers are tested for their vulnerability to humid conditions. This entails immersion in water which is slowly raised in temperature. After the maximum temperature is reached, measurements are performed to detect any variations in the coupling ratio. Short duration (13 hour) tests of this nature have been performed on ruggedized FBT couplers with 0.01%/°C deviations in the coupling ratio.[8]

Devices using interference filters make use of an index matching medium or cement on each side of the filter which acts to eliminate the approximate 4% Fresnel loss which is inherent at optical interfaces with air. In addition, this material is necessary to help protect the filter from moisture since these filters are typically fabricated using hygroscopic materials. If moisture comes into contact with the filter, its optical performance will degrade with time. The drawback associated with using cement is the limitation it places on the angle of incidence. The angle of incidence is generally limited to 30 degrees in this case. Devices fabricated early in the development of dielectric filters used chiolite and zinc sulfide (ZnS) which turned out to be very unstable to moisture. Modern devices are fabricated using much more stable materials such as silicon dioxide (SiO_2) and titanium dioxide (TiO_2).

4. VIBRATION AND SHOCK

Devices fabricated using the FBT technique are very stable with respect to vibration and shock. This is apparent from their construction since they do not employ discrete components as do devices constructed using bulk optics. However, it is essential that the packaging of the FBT devices be rugged since the fibers are extremely fragile at the tapered region. This is typically achieved by epoxying a sleeve around the tapered region then mounting this in a protective, environmentally stable housing. This housing is filled with a thermally and vibrationally compatible material or an environmentally safe material is extruded over the sleeve. The fiber jackets are then provided with adequate strain relief and protected against environmental contaminants. This results in a very solid device which withstands high pressures, isolates against shock and vibration, and provides thermal protection.

Multiplexers, demultiplexers and couplers which are designed with bulk optics have the same environmental considerations to contend with as do the FBT type devices, but, they also have the very difficult problem of maintaining proper alignment of the individual elements. This is achieved by the packaging techniques which relies heavily on the proper housing and optical filling compound or epoxy. This is a difficult task and has been met by some manufactures, however, the FBT technique is still more popular.

5. ENVIRONMENTAL TESTS

In an effort to determine how well multiplexers, demultiplexers and couplers will withstand the environmental forces to which they will be exposed, manufacturers of these devices subject them to a battery of tests. However, given the broad variety of possible conditions, ranging from office benign to tactical, the level of stress that these tests impart to the devices will vary in severity. The developer of the FBT technology, Canstar, performs the environmental tests listed in Table 5-1 on their militarized, star coupler. This device is being designed for shipboard applications.[9]

Table 5-1. Environmental Tests for Militarized FBT Star Coupler.

Test Condition	Specification
Vibration	MIL-STD-167-1 Type 1
Mechanical Shock	MIL-S-901 Grade A, Type A Class 1
Thermal Shock	MIL-STD-810 Method 503.1
Humidity	MIL-E-16400 Paragraph 4.8.3.4
Low Temperature	MIL-E-16400 (Operating -28°C, Paragraph 4.8.3.2 Storage -62°C)
High Temperature	MIL-E-16400 (Operating 65°C, Paragraph 4.8.3.3 Storage 71°C)
Temperature, Life	MIL-STD-202 Test Condition H

VI. MAINTENANCE CONSIDERATIONS

No preventive maintenance is necessary for these passive devices nor are there any special logistical requirements other than component availability. The only technical skills required in their installation is that associated with fusion splicing or terminating the fiber pigtails.

VII. FIELD PERFORMANCE

There is very little field data available on the passive devices discussed here. However, one plausible assessment that can be made about these components relates their reliability to the individual elements and compounds which comprise them. The devices are as reliable as the least reliable element contained in them and the reliability of this element must be evaluated in terms of its effect on the other elements. This is not to say that the device lifetime will be equal to the shortest lived element by itself for this depends on the definition of failure. For example, the characteristics of the potting agent or adhesive may change over time but retain the ability to perform its function as a securing agent. If a material's coefficient of thermal expansion changes, then there will most likely be an increase in insertion loss resulting from microbends in FBT devices and misalignment in devices using bulk optics. However, the material still performs its intended function. Also, FBT devices may experience strain induced birefringence which will create a polarization dependence. This dependence will cause variations in the coupling ratio, however, the device will still function. If any one of the above conditions occurs, the device will still function. It will only be considered a failure when it no longer meets the specified system requirements. This means that individual elements of a device may degrade and cause system failure. Therefore, attention should be paid to the discrete elements which make up these passive devices.

Over the last three years, one manufacturer has estimated that their facilities have manufactured approximately 12,000 multiplexer, demultiplexer and coupler devices which have been installed in that three year period. Out of these, less than 1% (~ 120) have failed. Out of these 120 failures, greater than 80% (≅96) were due to improper handling and are termed non-relevant failures. The most frequent cause of failure being the breaking of the fiber pigtail.[10] Therefore, less than 20% (≅24) of the 120 failures were due to other factors which were unknown to the manufacturer. This points out the care that must be taken in handling these devices. Strain relief boots are not provided to prevent the fiber minimum bend radius from being exceeded. Also, the fused section of FBT devices is extremely fragile which makes the methods used to counter tensile loads very important.

Since these devices were installed over a three year period, the length of time the devices have been fielded varies. The installation times are not known in more detail, therefore, it has been assumed that these devices were installed at a uniform rate. This is reflected in Fig. 5-27 which shows the estimated cumulative time that each device has been installed. From this figure, the approximate installed total device-hours can be calculated.

Calculating the area of triangle ABC will give the approximate total device-hours. This is calculated as:

Approximate Total Device-Hours = (1/2)(12,000)(26280)
= 157,680,000 device-hours.

A more accurate figure can be determined by adding the area of the small triangles outside of the large triangle as shown in the inset. This totals:

Area of small Triangles = (1/2)(1)(2.19)(12,000)
= 13,140 device-hours.

Therefore, the sum of these two values gives an estimate of the total installed device-hours which is:

Total Device-Hours = 157,680,000 + 13,140
= 157,693,140 device-hours.

This value was used in calculating the Mean Time Between Failure (MTBF) for the 12,000 installed devices.

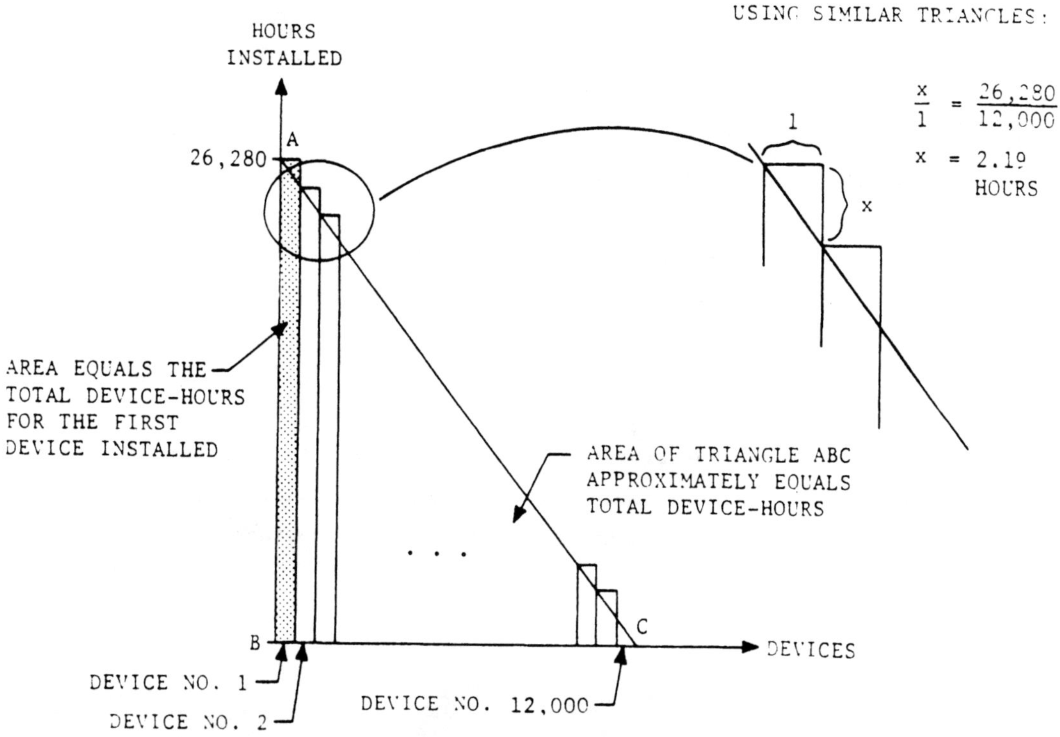

Figure 5-27. Graphic representation of accumulated device-hours.

Understanding that this data was taken from a mixture of component types used under a variety of conditions, the usefulness of the MTBF will be limited. The MTBF was calculated using a 90% confidence interval under the assumption that the components do not experience a time dependent failure mode. The failures were assumed to occur in a continuous and random fashion which allows the Chi Square distribution to be used in determining the MTBF confidence limits. The truncated time form of the Chi Square was used which is given in equation 5-11. The source data and results of the calculations, in which a standard Chi Square table was used, are given in Table 5-2.

$$\text{MTBF Confidence Limits} = \left(\frac{2T}{\chi^2\left(\frac{\alpha}{2}, 2r+2\right)}, \frac{2T}{\chi^2\left(1 - \frac{\alpha}{2}, 2r\right)} \right) \qquad \text{Equation 5-11}$$

where T = total device-hours

α = acceptable risk of error = 0.1

r = number of failures = 120, 96, and 24

2r+2 and 2r = degrees of freedom for lower and upper limits, respectively.

In the case where the number of failures resulted in a degree of freedom which exceeded the Chi Square table values, the approximation given in equation 5-12 was used to determine the value of the denominator to be used in equation 5-11.

$$\chi^2 = \tfrac{1}{2} \left\{ t_\alpha + \sqrt{2r-1} \right\}^2 \qquad \text{Equation 5-12.}$$

where t_α = degree of freedom used for large number of failures

r = number of failures = 120 and 96.

Table 5-2. Reliability Data For Multiplexers, Demultiplexers and Couplers.

Total Devices	12,000
Total Device-Hours (estimate)	157,693,140
Total Failures	120
Non-relevant Failures	96
Relevant Failures	24
MTBF (10^6 Hrs)	
<u>Type of Failure</u>	
Combined Failures	2.16 to 3.30
Non-relevant	2.64 to 4.26
Relevant	4.67 to 9.52
Failure Rate (10^{-6}Hrs)	
<u>Type of Failure</u>	
Combined Failures	0.303 to 0.464
Non-relevant	0.235 to 0.379
Relevant	0.105 to .214

These figures were based on numbers reported as being very rough estimates of actual field performance and should be used accordingly.

VIII. SUMMARY

The passive optical multiplexers, demultiplexers and couplers addressed serve a very important function in photonic systems, namely to control the direction of optical power and to allow the simultaneous transfer of multiple signals on a single optical fiber. This allows a dramatic increase in the information carrying capacity per fiber and gives added flexibility to system configurations.

The two major methods of manufacturing these devices, namely, bulk optics and fused taper method have been available long enough for a predominant trend to be established. This trend is found in the fused coupler devices being much more widely used than the bulk optics devices. This is a result of the complicated manufacturing techniques used in bulk optics. The individual elements used in the bulk optic technique have extremely critical alignment tolerances and the piece part assembly used to construct the final product is very time intensive. In contrast, the fused coupler components are typically manufactured on computer controlled "assembly lines" which allows a high annual volume of products having very consistent optical characteristics.

The absence of laboratory data and scarcity of field data does not allow any more than general statements to be made with regard to the reliability of these devices. Due to the higher number of elements in the bulk optics components, these devices would be expected to have an inherently higher failure rate when compared to the fused coupler components. Also, their alignment requirements are extremely demanding.

The information from field data does indicate that FBT components have demonstrated reliable performance over the short three year time period the data covers. However, this same information demonstrated that proper handling is very important. For it is the fiber that has been most prone to damage and not the packaged coupling mechanism.

Chapter 6. Switches

Solid state switching has been serving the electronics community quite satisfactorily for many years. However, the increasing demands being placed on electronic components in terms of bandwidth has led to the rapid and widespread development of photonic systems where bandwidth is not a problem. Where photonics is being used in large systems which require the use of switching networks, the full potential of the system cannot be met since only the much slower electronic switching is available. To put this in perspective, electronic technology provides switching speeds in the range of 10^{-11} to 10^{-5} seconds while photonic technology is capable of switching speeds of 10^{-11} seconds and faster.[1] Since the laser diode can readily achieve a switching speed in excess of 10^{-12} seconds, it becomes apparent that the development and deployment of photonic switches will be instrumental in allowing the demands of future systems to be met.

The photonic switch serves the same function as the electronic switch and that is to alter the route information is taking. This can be accomplished using a variety of methods, each having their own advantages and disadvantages. Some of the more attractive switch designs are just recently emerging from the developmental stage and being introduced into the marketplace. Therefore, little is known about their performance in the field. However, both these "new" photonic switches and those photonic switches that have been used in the field will be discussed here.

I. SWITCH CONSTRUCTION AND OPERATION

Photonic switches can be broadly classified into three categories; mechanical, integrated electro-optic and optically activated. Mechanical switches are the most widely used as a result of their early availability and relatively simple design. Integrated electro-optic switches have just become available commercially (1987) after many years of intense research whereas optically activated switches are still in the developmental stage. Each type will be discussed in turn.

1. MECHANICAL SWITCHES

There are several types of mechanical switches available for commercial and military applications. The majority of these switch types operate on the principle of fiber alignment, prism/mirror movement or shutter movement.

a. Fiber alignment -- These switches operate by physically moving one fiber with respect to another fiber which changes their alignment, thereby achieving switching. Fig. 6-1 shows two typical configurations for this type of switch. The switch shown in Fig. 6-1a is a single pole-double throw (SPDT) switch which can be activated manually at the switch or remotely. Remote activation is accomplished by applying the proper voltage to the solenoid coil. As shown, no voltage is applied. Upon application of voltage, the moveable plate is attracted by the induced magnetic field which causes the plate to move until it comes to rest against the lower stop. At this point the input fiber is aligned with the second output fiber. This design can be modified to switch numerous fibers simultaneously by securing a silicon chip array having two rows of N fibers to the moveable plate and attach another array having only one row of N fibers to the stationary plate. In one position, the top row of the two row array is aligned with the single row array. When the switch is activated, the moveable plate shifts, causing the bottom row of the two row array to align with the single row array. This is very useful in systems where spare sources are required to prevent appreciable downtime, such as undersea applications. When a failure occurs, the switch can be remotely activated.

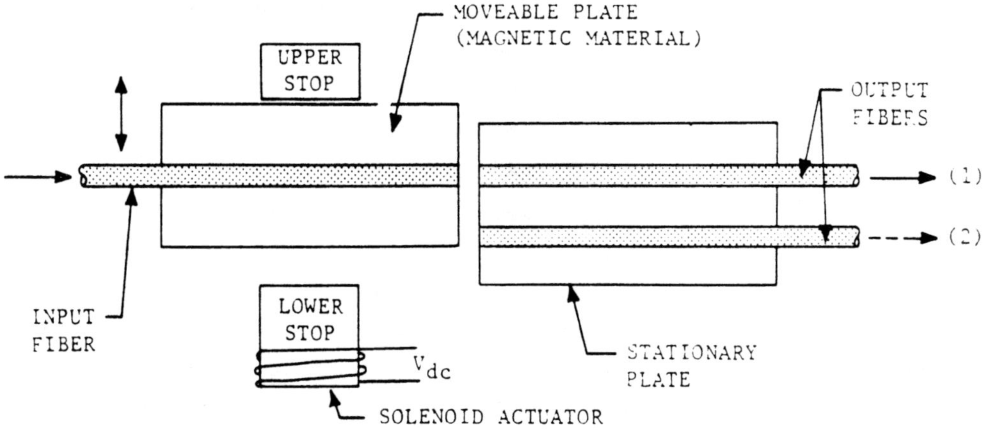

a. SPDT Switch

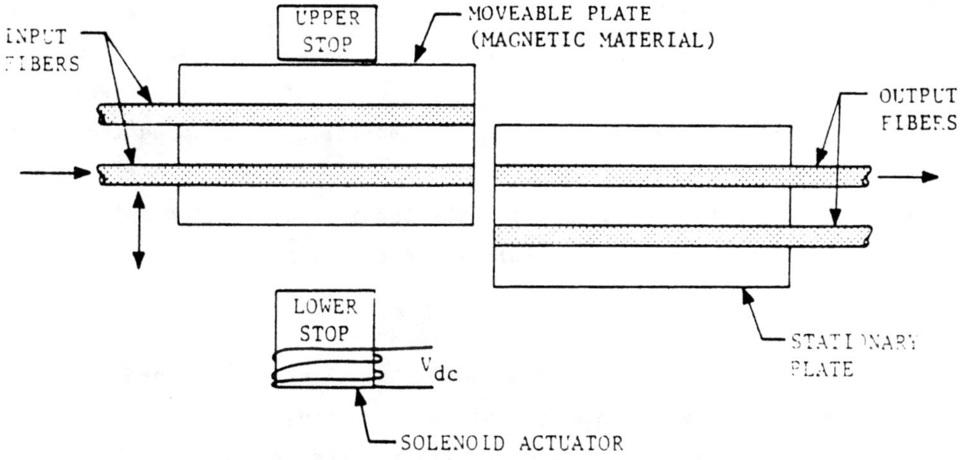

b. Bypass Switch

Figure 6-1. Typical fiber alignment switch designs.

The switch configuration shown in Fig. 6-1b is used extensively in networks having a ring topology. It operates on the same principle as the SPDT switch in Fig. 6-1a, that is, manually or by the application of a voltage to the solenoid which causes the moveable plate to change position. Its use in networks is depicted in Fig. 6-2. As shown, by the solid lines, it is in the normal operating mode. This configuration allows the node to access the ring network. When the switch is actuated, it assumes the position shown by the dashed lines which is referred to as the bypass mode. This configuration bypasses or effectively removes the node from the network while maintaining continuity of the ring. This switch is often designed to bypass the node upon loss of power to that node. Once removed, maintenance procedures can be performed on the node if necessary.

b. Prism/mirror movement -- Mechanical switches which use optical elements such as mirrors, prisms and/or lenses have been available for commercial applications for at least four years. They are actuated by an applied voltage to an electromagnet which causes either the prism or mirror to change position. Fig. 6-3 shows the configuration of a parallelogram prism movement switch which acts as a SPDT switch. As shown, the input beam is intercepted by the moveable prism and reflected to output port 2. When the switch is actuated, the moving stage changes position which takes the moveable prism out of the optical path. This allows the input beam to be reflected by the stationary prism to output port 1. The input port and each output port is fitted with a collimating lens to improve the coupling efficiency.

c. Shutter -- This type of switch is designed such that the endfaces of two optical fibers are separated by a very small air gap. When the actuator is depressed, an opaque shutter passes through the gap which blocks the optical signal as shown in Fig. 6-4. It is necessary to manually operate this switch which does not have a latching mechanism.

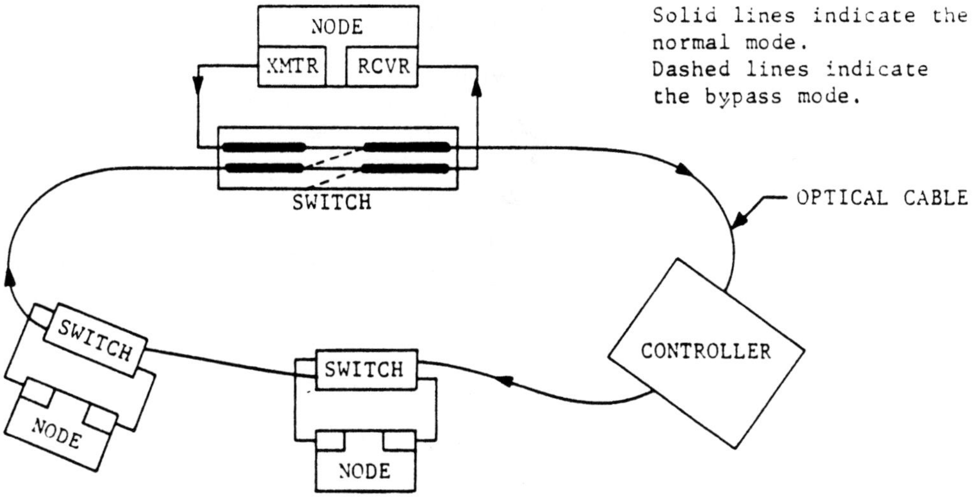

Figure 6-2. Application of bypass switch in ring network.

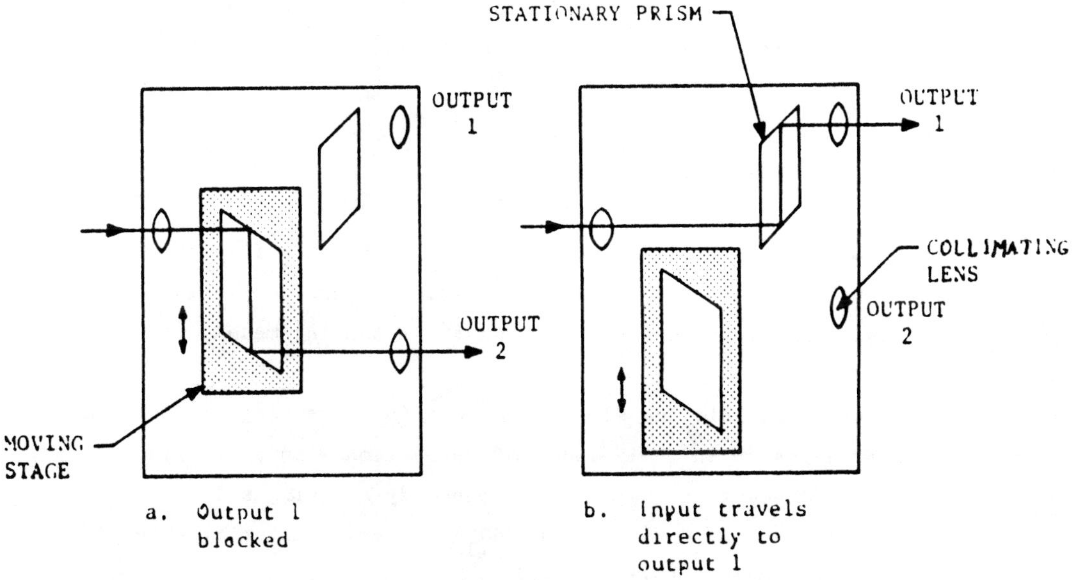

Figure 6-3. Parallelogram prism movement switch.

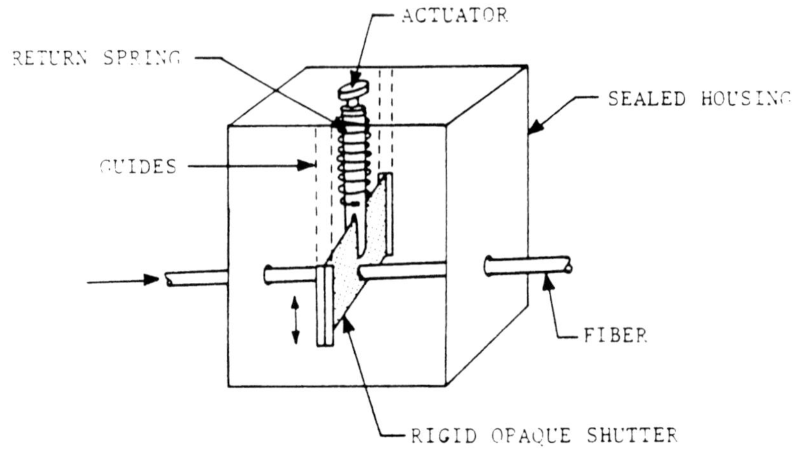

Figure 6-4. Normally open shutter switch design.

This type of switch is suited for use in industrial applications where explosion proof electrical switches are required in hazardous environments.

2. INTEGRATED ELECTRO-OPTIC SWITCHES

Only one switch design in this category has received sufficient attention and was far enough along in the development cycle to consider in this study. This switch uses the electro-optic effect as the actuating mechanism.

Materials which exhibit the electro-optic effect are birefringent and are characterized by a change in their index of refraction when placed in an electric field. The electric field is developed by applying a dc voltage across the material. Lithium niobate ($LiNbO_3$) is such a material which is commonly used in switches of this type.

Integrated electro-optic switches are constructed of a substrate material such as $LiNbO_3$ with two channels formed in the substrate by the diffusion of titanium (Ti). Electrodes are placed on top of the substrate directly over the channels and separated from the substrate by a thin buffer which is usually silicon dioxide (SiO_2). Fig. 6-5 shows the construction of a typical electro-optic switch. This switch uses $LiNbO_3$ as the substrate and Ti as the channel material; as such it is designated Ti:$LiNbO_3$.

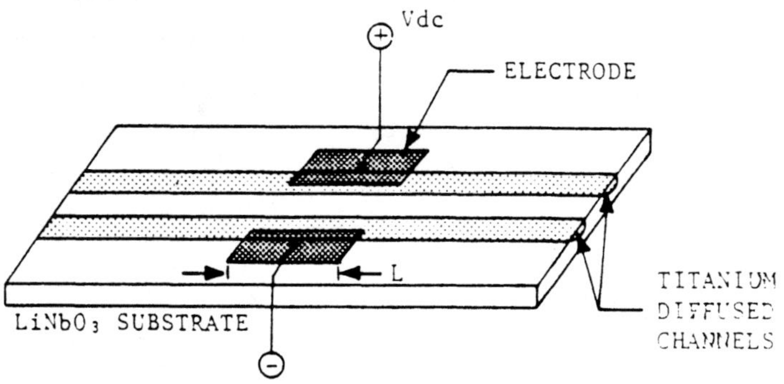

Figure 6-5. TI:LiNbO3 integrated electro-optic switch construction.

With no voltage applied to the electrodes, the index of refraction of the substrate is slightly less than that of the channel material. This results in each channel acting as an optical waveguide by the laws of total internal reflection. In this state (called the bar state), any light which is launched into the input channel will primarily remain in that channel. However, upon the application of a dc voltage to the electrodes, the index of refraction of the channel material in the vicinity of the electrodes will decrease below that of the substrate. Therefore, these areas of the channels no longer have waveguide properties. This allows the coupling of light from the input channel to the other channel, thus achieving switching action. This state is referred to as the cross state.

With no voltage applied, the coupling of optical power occurs when the evanescent field of the launched optical signal overlaps the adjacent channel which is only several micrometers away. The amount of coupling is dependent upon the length o^ the channel section over which there is a transfer of optical power, the propagation constants of the channels and the polarization of the launched light. For there to be complete (100%) coupling, it is necessary that the propagation constants of the channels be equal and that the coupling length be properly dimensioned. How the power is spatially coupled between two channels with no voltage applied to the electrodes is shown in Fig. 6-6. Here it can be seen that the proper coupling length dimension for 100% coupling is $\lambda/2$. Note that the ratio of $2L/\lambda$ determines whether the input signal will remain in the input channel or couple to the adjacent channel with no voltage applied. If the ratio of $2L/\lambda$ is even, as in Fig. 6-6, then the output channel will be the same as the input channel. If this ratio is odd, then the output channel will not be the same as the input channel.

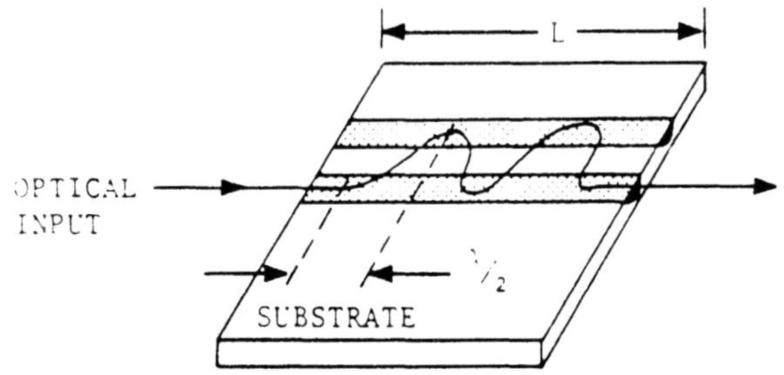

Figure 6-6. Spatial coupling of evanescent field between channels of an integrated electro-optic switch with no voltage applied.

When a voltage is applied across the electrodes, the propagation constant of each channel is altered such that one increases and the other decreases. This results in two effects: 1) the amount of power coupled between channels decreases which precludes 100% coupling and, 2) a phase change occurs in the light which retards its propagation sufficiently to allow switching action as shown in Fig. 6-7.

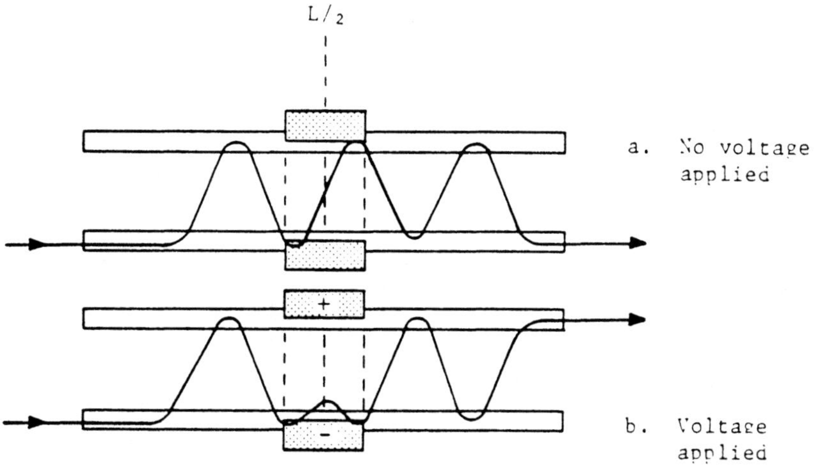

Figure 6-7. Spatial change in light beam upon application of voltage to electrodes with even $2L/\lambda$.

When a voltage is applied across the electrodes, the transfer efficiency, which is the ratio of the output power of one waveguide to the input power of the other waveguide, changes. This relationship is shown for two different channel configurations in Fig. 6-8 for a switch designed such that $2L/\lambda$ is odd. As such, with no voltage applied, the switch will be in the cross state. Therefore, increasing the voltage decreases the transfer efficiency in a rippling fashion. As the voltage increases, the decreasing transfer efficiency corresponds to the crosstalk which will occur as the switch approaches the bar state. Therefore, the voltage V_a in Fig. 6-8a will give a crosstalk of -10dB. As long as the voltage is maintained at this level or higher, the crosstalk will be this level or less.

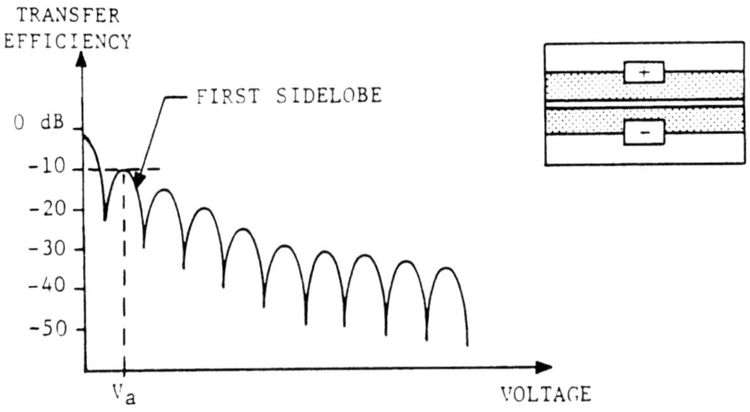

a)

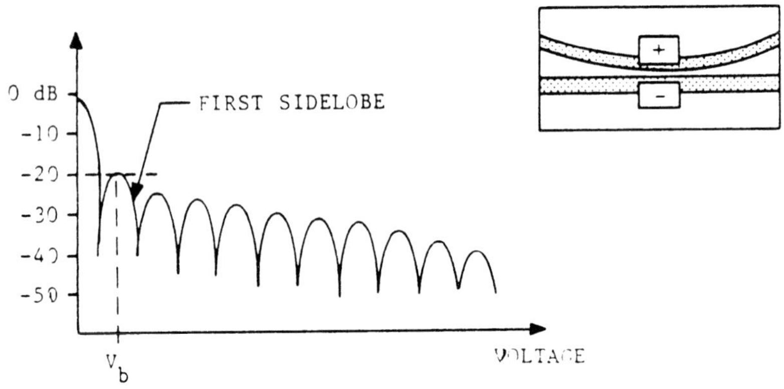

b)

Figure 6-8. Transfer efficiency versus applied voltage for an integrated electro-optic switch with odd $2L/\lambda$.

It is desirable to achieve the least amount of crosstalk with the lowest applied voltage. Fig. 6-8b shows an improved crosstalk level with the same applied voltage. This was accomplished by tapering the interaction region of the channels as shown in the inset of Fig. 6-8b. The result is a significantly lower first sidelobe which corresponds to a much lower crosstalk.[2]

A result of using a birefringent material such as $LiNbO_3$ as the substrate material is the dependency it has on the polarization of the input light. The index of refraction the TM and TE orthogonal linear polarizations experience are different. The TM polarization is perpendicular to the top plane of the crystal while the TE polarization is parallel to the plane of the crystal. This has a significant impact on the applied voltage and, therefore, the coupling efficiency. This is shown in Fig. 6-9. This figure is not to

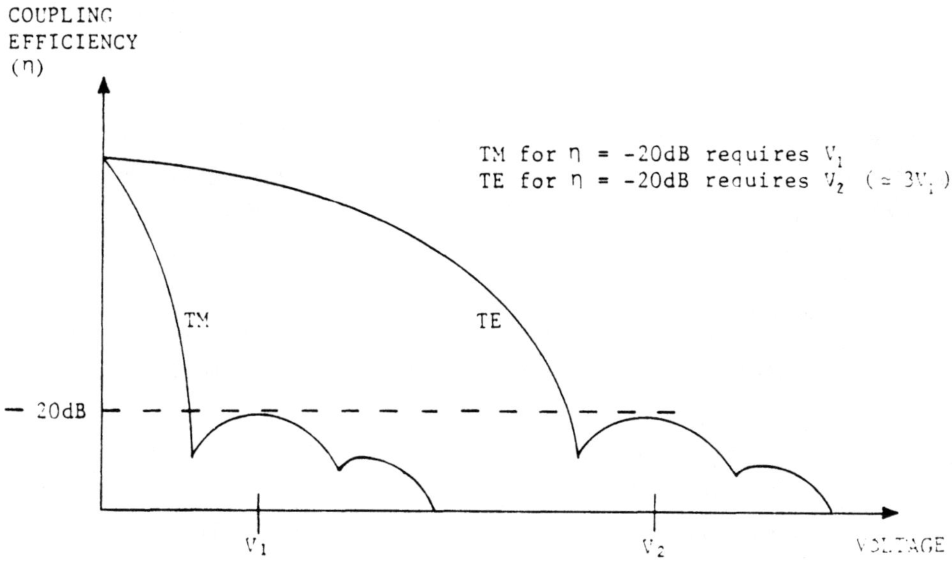

Figure 6-9. Transfer efficiency versus voltage for different polarizations of input light.

scale, however, the voltage normally required by TE polarized light is approximately 3 times that required for TM polarized light to achieve the same transfer efficiency.[2] As indicated in Fig. 6-8, the zero voltage point represents the cross state for this switch. Therefore, as long as the first sidelobe level is sufficiently low, this level will represent the crosstalk in the bar state. As long as voltage V_2 is maintained, the switch will be polarization independent. However, if only TM linearly polarized light is launched into the switch, then the lower voltage, V_1, will provide approximately the same crosstalk level and transfer efficiency.

There is a trade-off that must be made when deciding to use polarization independent switches. That is, the desirable characteristic of a low switching voltage versus restrictions on the type of fiber that can be used.

Only polarization maintaining fibers can be used from the transmitter supplying the signal to the switch. This type of fiber is more expensive (approximately 200 times as of late 1986) than non-polarization maintaining fiber and not as readily available due to the relatively low demand. Therefore, expansion of a system using polarization maintaining fiber will require a significant lead time for purchasing cable. Also, like any other fiber types, polarization maintaining fiber exhibits attenuation due to microbends. However, in addition, microbends degrade the polarization characteristics of this fiber. This means that another form of potential degradation is present when polarization maintaining fiber is used. However, if a polarization independent switch is selected, the fiber restrictions are relaxed in that polarization maintaining fiber is not required. But, the compromise is a switching voltage which is approximately three times higher is required. This places a higher demand on the power supply electronics with respect to thermal dissipation and slew rate.

3. OPTICALLY ACTIVATED SWITCHES

This type of switch is characterized by an optical signal being directly controlled by another optical signal. This switch is still very much in the developmental stages and will only be briefly described. The construction of this style of switch is shown in Fig. 6-10. The principle of operation relies on altering the apparent path between the two mirrors.

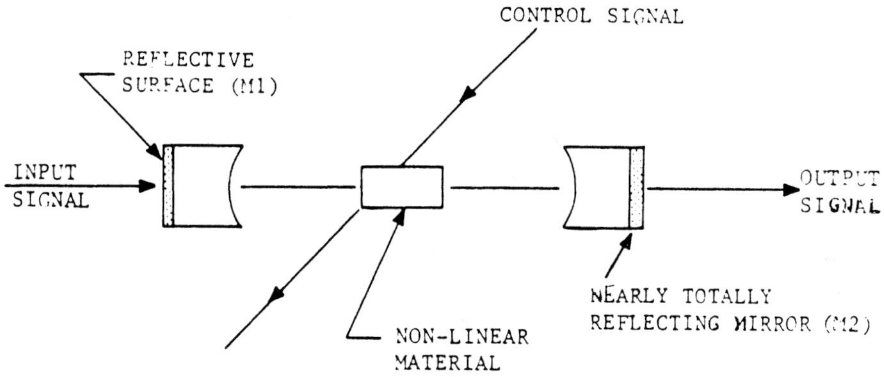

Figure 6-10. Optically activated switch construction.

Two nearly totally reflecting mirrors are situated back-to-back (the reflective surfaces facing outward) and separated by an even number of quarter wavelengths. This produces a Fabry-Perot resonator which allows incident light to be transmitted through mirror M2 as an output signal. This occurs due to the large field which builds up between the two mirrors. Light from this field leaks through mirror M1 which cancels input signal reflections through the mechanisms of interference. In this state, the switch would be analogous to an electrical switch in the closed position. To change the state of this switch, it is necessary to alter the optical distance between the mirrors. This is accomplished by placing a block of non-linear optical material between the mirrors as shown. This material has the property that its index of refraction changes when impinged upon by an optical control signal. The effect of this is to change the apparent length of the block which changes the apparent distance between the two mirrors. As a result, the optical elements no longer act as a Fabry-Perot resonator and the input signal is reflected off of mirror M1 instead of being transmitted. This constitutes switching action.

II. TYPICAL OPTICAL AND MECHANICAL CHARACTERISTICS

The three types of switches discussed thus far are quite different in terms of their optical and mechanical characteristics. These characteristics will be given in this section for the mechanical and integrated electro-optic switch types. Since the optically activated switch is still in the developmental stage, its characteristics are in a state of flux and will not be addressed.

Tables 6-1 through 6-4 give the optical and mechanical characteristics for the individual switch types. These values reflect the range of information obtained from various manufacturers data sheets for mechanical switches currently available. Information presented on integrated electro-optic switches was taken from data sheets and from results published in numerous technical articles.

TABLE 6-1. Fiber Alignment Switch Characteristics.

Parameter	Value
Insertion Loss	0.5dB to 1.5dB
Crosstalk	<< -50dB
Switching Speed	12 msec to 25 msec*
Repeatability**	< 0.1dB
Operating Temperature	-25°C to 65°C
Storage Temperature	-30°C to 70°C

* This switching speed includes bounce time which is the time required for the moved fiber(s) to come to a complete stop upon switch activation.

** Repeatability is the maximum variation in insertion loss over the life of the switch.

TABLE 6-2. Prism/Mirror Switch Characteristics.

Parameter	Value
Insertion Loss	< 1dB to < 3dB
Crosstalk	< -40dB to < -60dB
Switching Speed	5 msec to < 20 msec
Repeatability	±0.02dB to 0.1dB
Operating Temperature	-20°C to 55°C
Storage Temperature	-55°C to 85°C
Relative Humidity	0 to 99% (non-condensing)

TABLE 6-3. Shutter Switch Characteristics.

Parameter	Value
Insertion Loss	3db to 6dB
Crosstalk	N/A
Switching Speed	3 msec to < 5 msec
Operating Temperature	-40°C to 70°C
Storage Temperature	-50°C to 80°C

TABLE 6-4. Integrated Electro-optic Switch Characteristics***.

Parameter	Value
Insertion Loss	3dB to < 7dB
Crosstalk	-20dB to -25dB
Switching Time	several nsec to tens of nsec
Electrode Voltage	
Single Polarization	4 Vdc to 30 Vdc
Polarization Independent	30 Vdc to 100 Vdc

*** Values are for single polarization devices unless otherwise specified.

III. DEGRADATION MECHANISMS

The forms of degradation that the individual switch types experience depends upon the switch design and application. The performance of mirror and prism movement switches is dependent upon their orientation with respect to gravity while this is not a concern for integrated electro-optic switches. However, electro-optic switches can suffer from the effects of photorefraction at high power levels. These and other areas which affect switch performance will be explained.

1. FIBER ALIGNMENT SWITCHES

This switch primarily suffers from the same problems that a connector does since it functions in much the same way. That is, it aligns the fiber endfaces in very close proximity to each other to achieve optical coupling. However, instead of securing the fibers axially as in a connector, they are deflected. Therefore, the main difference is that this switch is an active device which deflects the fiber(s) when activated, whereas, a connector keeps the fibers in continual alignment. This gives the switch a greater sensitivity to vibration and shock. Also, once the fibers are deflected from a colinear position, the effects of angular misalignment become more pronounced. This loss factor can be reduced by increasing the length of the deflected fiber, but this will result in even greater sensitivity to vibration. To reduce the attenuation, the fibers may be immersed in a fluid to control reflection. The fluid must be compatible with the temperature ranges in which the switch will be used.

These devices are commonly operated by solenoids which are highly reliable electrical components. The solenoids perform well in switches, however, tests[3] have shown that repeated false switching occurs when subjected to shock tests. These occurrences may be reduced if the switch is provided with a latching mechanism while in the non-activated mode, that is, while in the non-deflected state. False switching due to shock can also be reduced by mounting the switch such that anticipated shock does not occur in the same plane that deflection occurs.

2. PRISM AND MIRROR MOVEMENT SWITCHES

The materials from which the optical elements are fabricated in these switches determines the wavelength range over which they can operate. Outside of this range, degradation of the optical signal occurs due to aberration. This makes them wavelength dependent devices. The anti-reflection coatings used are also a contributor to this wavelength dependence. As a result, if a switch which is designed to operate at short wavelengths is used at long wavelengths, the switch insertion loss will increase by approximately 2dB[4] (66% to 100%). This dependence reduces the wavelength band over which the switch can satisfactorily operate.

There is a temperature consideration when using these devices, since the optical elements can frost if changes in temperature are too rapid. This will, of course, degrade the performance by increasing the attenuation.

Thermal problems can also arise from continued application of power to the electromagnetic drive mechanism. This problem can be prevented by designing the switch with a latching mechanism which will eliminate the need to maintain the application of voltage.

Due to the mass of the optical elements, these switches must be positioned with the direction of gravity in mind. For a constant voltage device, switching time will be shortest when the element moves in the direction of gravity. Switching time will be increased when changing state in the opposite direction of gravity. The larger the mass of the moving optical elements, the greater this problem will be unless the activating forces are increased also. When the moving element moves in the horizontal plane, the impact of gravity is minimal in either direction of motion.

3. SHUTTER SWITCHES

These switches are simple in design and construction and typically use large core fibers. This gives them good inherent reliability with the primary causes of failure resulting from mechanical problems. Information contained in survey questionnaires mailed to manufacturers[5] indicate the most prevalent problem to be the failure of the actuating spring. Fatigue or wear out of the spring can cause incomplete and sluggish operation of the switch.

4. INTEGRATED ELECTRO-OPTIC SWITCHES

A considerable amount of research has been performed on these switches because they show such high promise for closing the gap between hybrid and all-optical systems. However, they have only recently (1987) been available for applications other than laboratory work. The material presented here reflects laboratory findings and switches used in test bed environments.

The photorefractive effect occurs in birefringent materials as a result of photoconductivity and an applied electric field. When an optical signal excites the carriers, they can migrate into the material and become trapped.

This produces a space-charge which has its own associated electric field. This electric field acts upon the birefringent material by the electro-optic effect which causes a local change in the index of refraction. Therefore, this disturbs the index change intentionally imposed on the material by the voltage applied to the electrodes.

Experiments have determined that the photorefr-ctive effect does not have a significant impact on switch operation when typical power levels are being used, that is <5mW. When the power level is raised to 50 mW, the result is incomplete switching and increased crosstalk.[6]

As shown in Fig. 6-6, the spatial coupling between channels is sinusoidal in nature. The photorefractive effect causes this sinusoid period to increase and to flatten out. Since the interaction length is fixed, there can no longer be 100% coupling. This effect increases with increasing power levels.

Reports addressing this phenomenon indicate that the impact on coupling is a function of time. The relationship between coupling and time most closely fits an exponential. The coupling increases steadily at first then tapers off to a steady state value with a time constant of approximately 100 hours.

It is important that 100% of the power is coupled from one channel to the other while in the cross state or there will be crosstalk. Since there is a change in the refractive index of the channel due to the photorefractive effect, the propagation constant will change and there will be a resultant phase change as discussed in Section I, paragraph 2. This results in an increase in crosstalk which is aggravated by longer interaction lengths. Therefore, the shorter the interaction length, the less susceptible the switch will be to photorefractive induced crosstalk. It has been suggested that doping the material with magnesium may reduce this problem at high power levels.[6]

Optical power has been reported to drift from one output channel to the other when a constant voltage is applied. This is highly undesirable because it constitutes crosstalk. The cause of this drift has been related to an insufficient buffer layer separating the channels from the electrodes. The use of a 200 nm thick layer of SiO_2 has eliminated this problem.[2]

Random variations in the indices of refraction of the channels have been attributed to causing poor crosstalk loss. These variations are considered to be a result of two possible sources: 1) fluctuations in the density or dimensions of the channels and 2) inhomogeneities in the diffusivity of the substrate. The suggestion has been made that annealing the Ti before diffusing it into the substrate to form the channels may provide better crosstalk loss.[7]

IV. LIFETIME DATA

The life of mechanical switches is given in switching operations or on/off cycles. No figures are available from field installations, but, information was obtained from manufacturers which is presented in Table 6-5. Again, considering the very short time that integrated electro-optic switches have been available, there is no lifetime data available. However, since these switches are very much like evanescent wave couplers, except they make use of an applied voltage, it is reasonable to assume that they would have similar lifetimes. But, the fact that there is an applied voltage should alert the reader to the potential for deterioration of the electrodes over time. This may cause a non-uniform electric field resulting in unacceptable crosstalk levels due to output power drift.

TABLE 6-5. Manufacturers Switch Lifetime Data.

Switch Type	Lifetime (10^6 On/Off Cycles)
Fiber Alignment	> 20 - >100
Prism/Mirror Movement	1
Shutter	> 10

V. SUMMARY

This discussion has presented information on the operation and characteristics of the major photonic switch types in use today. The lack of field data makes it difficult at best to provide a suggestion as to the impact they have on fiber optic systems. It is apparent that the switches which are most frequently used, that is the mechanical switches, are only used in applications such as isolating a node from a network, bringing into a system spare sources, switching in a redundant system, laboratory testing when multiple wavelengths are being used individually and providing fault tolerance. In other words, these switches are typically being used as single pole-N-throw switches. They are not typically being used in continuous switching applications since they are not well suited for that type of operation.

As use of the integrated electro-optic switch becomes more widespread, the closer we will be to utilizing the full potential of fiber optic systems. At present, electronic switching is being used in its stead. Until this trend is reversed, it will remain difficult to assess the impact that photonic switching is having on fiber optic communications systems.

Chapter 7. Closures and Organizers

Junction boxes, closures, and splice racks are all enclosure devices which have essentially the same function. They are used to store and protect fiber optic cable, organizers and splice trays. They also provide an access point for system reconfiguration, testing and maintenance. The enclosures are used for aerial, direct buried, duct, and office environments. The type of enclosure used is largely dependent upon the environmental conditions to which the system will be exposed.

An organizer is housed inside an enclosure and secures a series of splice trays in which are held completed optical splices. This provides a means of storing the splices in an orderly fashion for easy access. To accommodate system reconfiguration, spare fiber is coiled inside the organizer.

I. DESCRIPTION OF ENCLOSURE TYPES

1. JUNCTION BOX

The junction box is a light to heavy gauge metal cabinet as shown in Fig. 7-1 which is used to store and protect excess fiber optic cable or to provide a cable access point. It may also be used as a splice box, patch panel, or central point for system testing and trouble shooting. Junction boxes are also called jointing boxes, termination boxes, and cable enclosures. Typically they are intended to be mounted on walls, poles, or inside manholes.

2. CLOSURE

The closure is used to provide protection to splices from the environment. A typical closure is shown in Fig. 7-2. Closures are also referred to as gas blocks, jointing closures, and splice closures. Closure

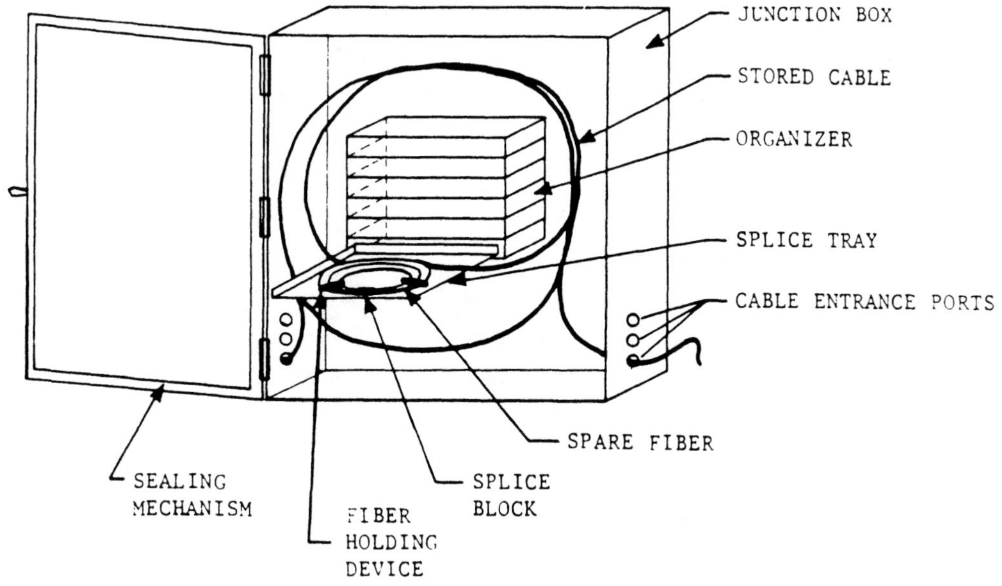

Figure 7-1. Junction box with organizer and splice tray.

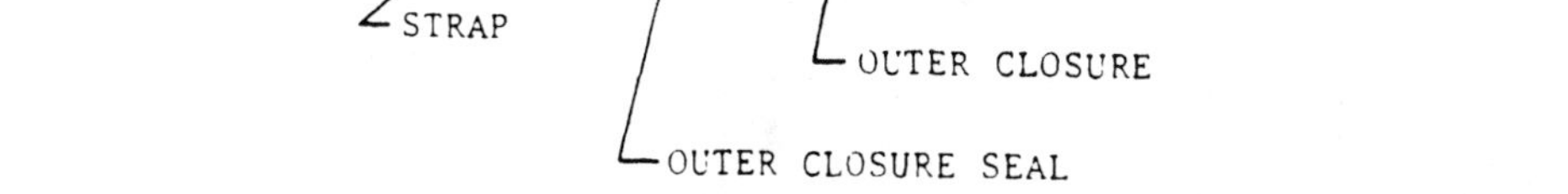

Figure 7-2. Splice closure with inner closure.

designs are available which are re-enterable, pressurized and capable of being filled with a waterproofing compound. For fiber optic applications, two closures are generally used; one inside the other. The spliced fibers are secured in the inner closure using splice trays mounted in splice organizers. The outer closure is then installed around the inner closure with the void between the two closures filled with a waterproofing compound. This approach of using two closures prevents the compound from directly touching the fibers which can cause microbends. It also allows for easy re-entry to the splice point in the event of repair or reconfiguration. Three types of closure cable entry are available; straight, branch, and butt as shown in Fig. 7-3.

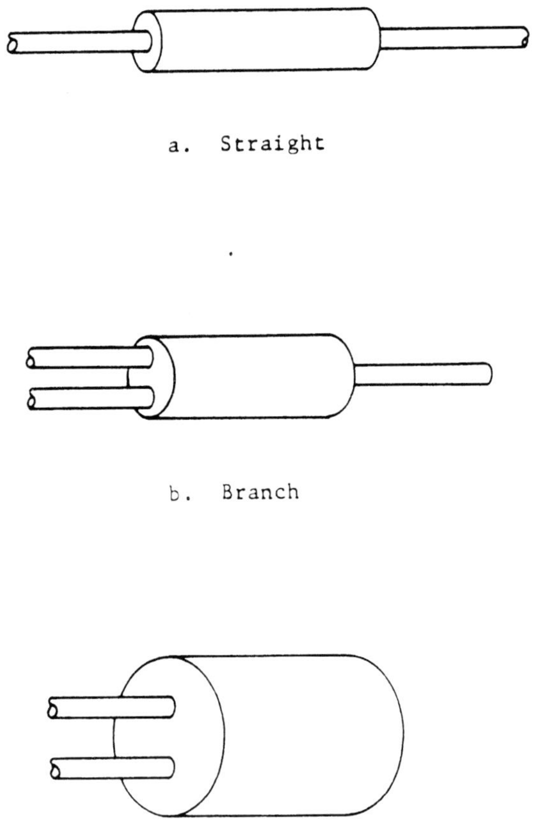

a. Straight

b. Branch

c. Butt

Figure 7-3. Closure cable entries.

3. SPLICE RACK

The splice rack is used to organize and provide mechanical support to fiber optic splices or connections, cables and fibers, see Fig. 7-4. The mechanical support adds strain relief and prevents the application of potentially damaging stresses. If the splice rack is used with connectorized fibers which can be rearranged, it is generally referred to as a connector panel. Splice racks are available in a wide range of sizes and configurations and for use inside junction boxes and 19" and 23" electronic racks.

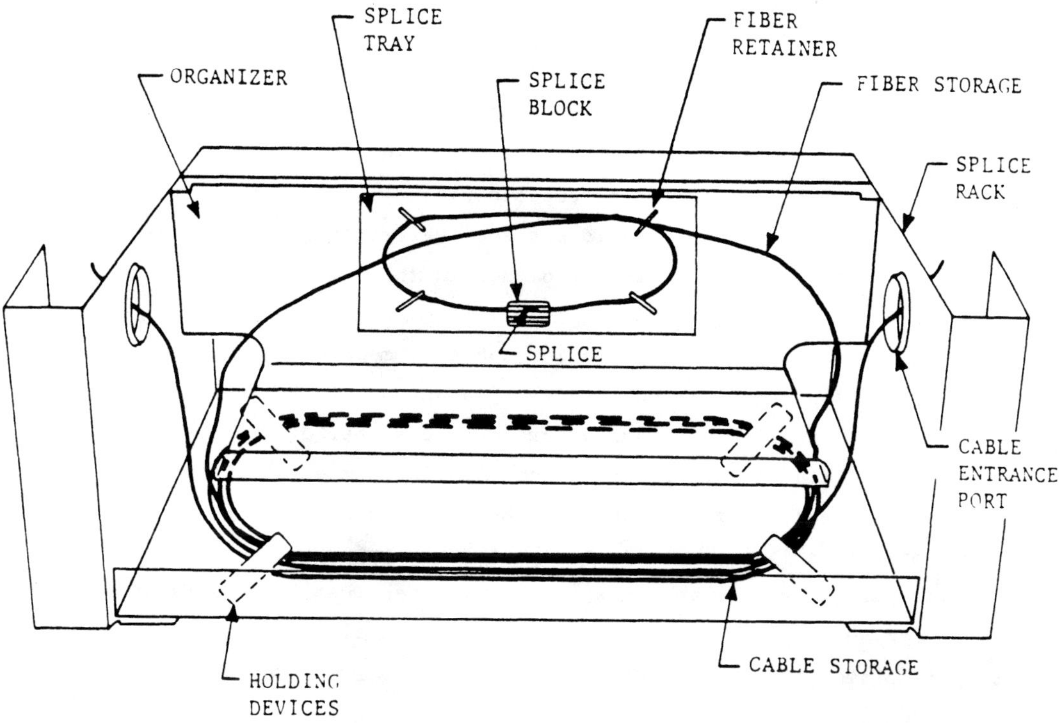

Figure 7-4. Splice rack with organizer and splice tray.

II. DESCRIPTION OF SPLICE ORGANIZERS AND SPLICE TRAYS

Splice organizers are generally made of a lightweight metal or plastic and are used as racks to secure splice trays as Fig. 7-5 shows. They are capable of securing from 1-12 splice trays and are designed for use with a specific splice tray usually made by the same manufacturer. The splice tray is the unit that actually holds the optical splice. These trays are also generally made of a lightweight metal or plastic.

Splice trays are designed to house the three most common splice types; fusion, mechanical, and silicon array. Generally, trays that can accommodate fusion splices can also accommodate mechanical splices. This is not true for all manufacturers though, and the type of splice to be used must be checked to ensure compatibility. Silicon array splices are unique and require a different style splice tray.

Splice trays are designed to house either single or multimode fibers or both and either loose or tight tube cable constructions. Some splice trays can accommodate either type of cable but they generally require minor modification in the field to switch from one to the other.

The number of splices an organizer can accommodate is dependent upon the number of trays the organizer can hold. In a survey of 12 different splice organizers from six different manufacturers, the number of trays that each organizer could hold ranged between 1 and 12. The number of splices that each tray can hold ranged between 1 and 24. The completed organizer with the maximum number of splice trays can hold between 1 and 144 splices. There has been preliminary work done on a new fiber organizer that is capable of accommodating 800 splices. This work is developmental and is expected to be capable of holding 160 five-fiber ribbons for a total of 800 splices[1] However, the more splices that are held within an organizer, the more difficult it is to remove a splice for repair without tangling the stored fiber. Proper storage of the splices is important as it will impact maintenance. If too many splices are stored on one splice

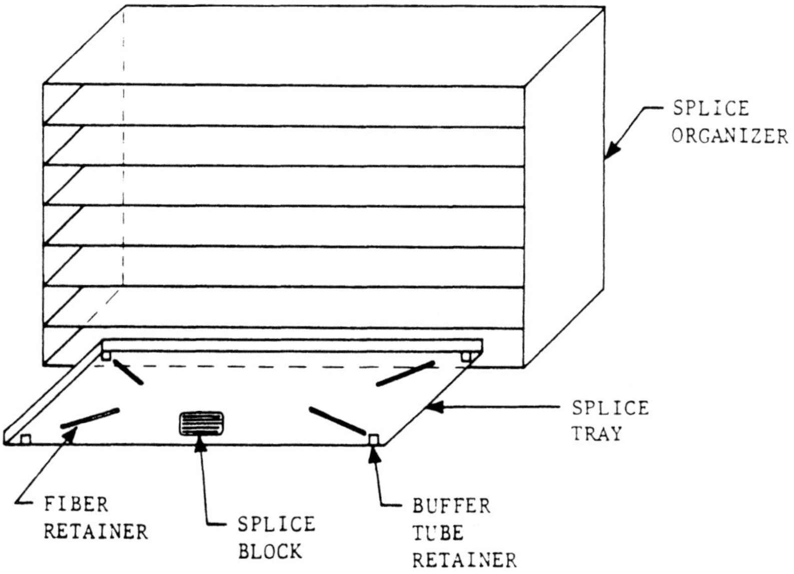

a. Splice Organizer & Tray

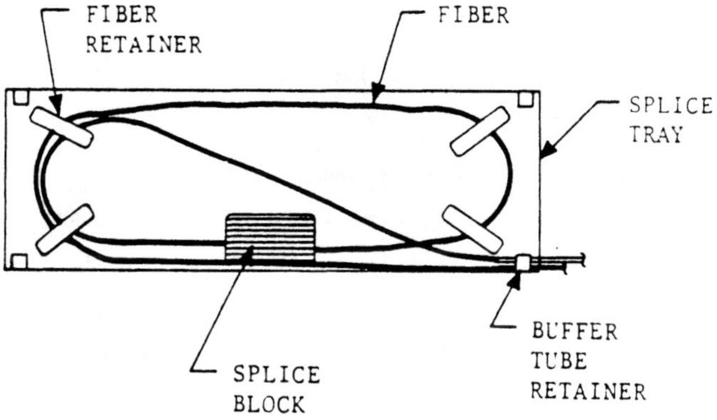

b. Splice Tray

Figure 7-5. Splice organizer and tray.

tray, it is more difficult to remove one splice for repair without damaging the other splices or excess fiber. When a splice is removed from the tray, the excess fiber, which is very prone to tangling with other stored fiber coils, must also be removed.

Storage of extra fiber, and extra fiber encased in buffer tubes, is a feature that is provided by most organizers in the event a splice needs to be reworked. Fiber storage lengths range from 10 in. to 36 in. per fiber with 30 in. being a typical length. Buffer tube storage is useful for future splicing applications where the system design may be modified, however, the spare fiber generally provides enough excess length so that the spare buffer tube is rarely used.

Splice trays must also limit the bend radius of the fiber while it is stored to prevent excess loss due to microbending. According to fiber manufacturers, a minimum radius of approximately 3 inches is desirable for storing fiber without inducing stress on the fiber, however, splice trays are designed which successfully allow a 1.5 inch minimum bend radius. These trays have been used for many years without any report of excess loss due to the minimum bend radius. The 1.5 inch minimum bend radius places minor stress on the fiber but not enough to induce significant loss.

Providing strain relief is very important to insure that the fibers are not under strain and to prevent lateral movement during temperature cycling. This is generally provided in the form of strain relief devices applied to the buffer tubes. The splice is protected with a glass, plastic or metal sleeve which is snapped or held in place on the tray by friction fit in combination with a strapping mechanism such as Velcro[R].

III. COMMON FAILURE MECHANISMS AND THEIR CAUSES

1. CLOSURES/SPLICE RACKS/JUNCTION BOXES

These devices have an estimated life of 30-40 years[2,3]. The feature that is most susceptible to wear is the sealing mechanism, though it is not

[R] Velcro is a trademark of Du Pont Company.

prone to failure until after the expected lifetime. Premature failure of the seal may occur if it is exposed to chemicals or weather extremes that it is not rated to withstand.

Junction boxes, closures and splice racks can often be used interchangeably. They can serve as splice points and patch panels. The selection will depend on the specific application, number of terminations or connections, cable sizes, fiber count, required accessibility, and the environment to which it will be subjected. The selection will also depend on whether the device will lay on the ground, be buried, be pulled through duct work, or be suspended.

2. SPLICE TRAYS/ORGANIZER

Physically, the splice organizer and splice tray are not prone to failures that would impact the overall reliability of a fiber optic system. They could experience problems such as cracking or corroding due to environmental conditions over time, however, there will be a very small chance of this occurring when the organizer is mounted within a properly selected outer enclosure. Failures which have occurred within the organizer or splice tray were found to be related to the stored spare fiber and to the fiber splice itself. These failures can be mechanically or environmentally induced, installation induced or design related. Typical mechanical failures caused by improper installation include storing the fiber in a state of residual tension or exceeding the minimum bend radius which can cause microbends, propagation of microcracks and breaks in the fiber[4]. All of these conditions increase the overall system loss and can cause degradation of the system over time. Environmental factors that can cause failure if not protected against include temperature extremes, excessive temperature cycling, moisture, humidity and dust. Excessive temperature extremes or cycling should not adversely affect the fiber if properly coiled and mounted but may affect the adhesion properties of the epoxy used in the splice or splice protection sleeve. Excessive moisture may enhance the propagation of any microcracks existing in the fiber. These are actua failures of the fiber or splice, however, the chance of failure depends upon proper storage of the optical fiber and the splice within the organizer. One cannot quantitatively state that the failures are a result of improper use of the trays and organizers but the fiber/splice/tray/organizer are so integrated when used with an optical system that they must be perceived as units and if possible, the exact cause of failure should be determined. To prevent these problems, the outer device in which the organizer and tray is mounted must be chosen to provide adequate environmental protection.

The splice trays and organizers must be designed to store the fiber in a relaxed state with no residual tension or induced stress. This can be done by insuring that the fiber is not placed in a bend radius of less than 1.5 inches, preferably 3 inches. The strain relief devices for the buffer tube and the fiber must adequately relieve any stress and keep the buffer tube and fiber from having lateral motion during temperature cycling. It is very important that these same considerations be recognized and their adherence insured during the installation process.

IV. MAINTENANCE

Junction boxes, closures and splice racks require no routine maintenance when properly selected for the environmental conditions to which they will be exposed. Splice trays and organizers also require no routine maintenance. The only time that an organizer and splice tray would need to be accessed is when a problem occurs with a splice or when a fiber is being rerouted. Trays and organizers should be selected which allow easy access to the splices and which include adequate room to manipulate the fibers. This makes repairs less prone to create additional problems and less time consuming.

V. SUMMARY

Junction boxes, closures and splice racks provide a means of storing or terminating fiber optic cable and fibers. Junction boxes and closures can be used for multiple purposes, such as splice points, patch panels, and distribution centers. They can be used in aerial, buried and duct installations.

Fiber optic splice trays and organizers protect and organize fiber splices within an outer protection device such as a junction box or enclosure. Reliability of the splice trays and organizers have not been found to be a problem. The most common problems that occur within splice trays/organizers are related to the minimum bend radius of the fiber being exceeded and inherent failure of the optical splice. The devices must be designed and installed to alleviate any residual tension or induced stress on the fiber and the splice.

Preventive maintenance of the trays and organizers is not needed and corrective maintenance is seldom required. In order to reduce the probability of damaging the fibers and splices, the splice trays and organizers should be moved as little as possible. However, if a splice needs to be accessed, the organizer and trays should be selected such that their design allows easy access.

Section IV

Systems

Chapter 8. AN/FAC-2A, 2B & 3 Fiber Optic Communications Sets

I. SYSTEM DESCRIPTION

The AN/FAC-2A, -2B and -3 are fiber optic communications systems that provide multichannel, point-to-point, digital data transmission. The information is transmitted in baseband Non-Return-to-Zero (NRZ) format. Data channel inputs and outputs are compatible with existing MIL-STD-188-114, 128 ohm balanced option. These systems consist of one to six Transmitter Channel Units (TCU), one to six Receiver Channel Units (RCU), two frame assemblies, and a fiber optic cable. The direction of data flow in any channel may be reversed simply by interchanging the TCU and the RCU in that channel. A typical system configuration is shown in Fig. 8-1 where three channels carry data in one direction and three channels carry data in the opposite direction.

The primary difference between the AN/FAC-2 and the AN/FAC-3 is that a clock circuit has been incorporated into the AN/FAC-3. Both sets can transmit two independent digital signals over one fiber by optically multiplexing the 830nm and 1060nm wavelength signals at the TCU and optically demultiplexing them at the RCU. A maximum of six TCUs may be transmitting at one time over six independent fibers. This allows for 12 signals to be simultaneously transmitted; six signals at 830nm and six signals at 1060nm. This is considered the DATA/DATA mode in which there are two independent data streams transmitted over a given fiber. The AN/FAC-3 has added a CLOCK/DATA mode in which two digital signals and an associated clock signal are transmitted in a single optical wavelength multiplexed data stream. This mode provides total data rate transparency because a bit synchronizer is not required at the RCU which is needed when operating in the DATA/DATA mode. Twelve signals can be simultaneously transmitted in this mode. There is an adjustable electronic delay in the data signal path of the RCU which compensates for the difference in the channel propagation times of the two wavelengths.

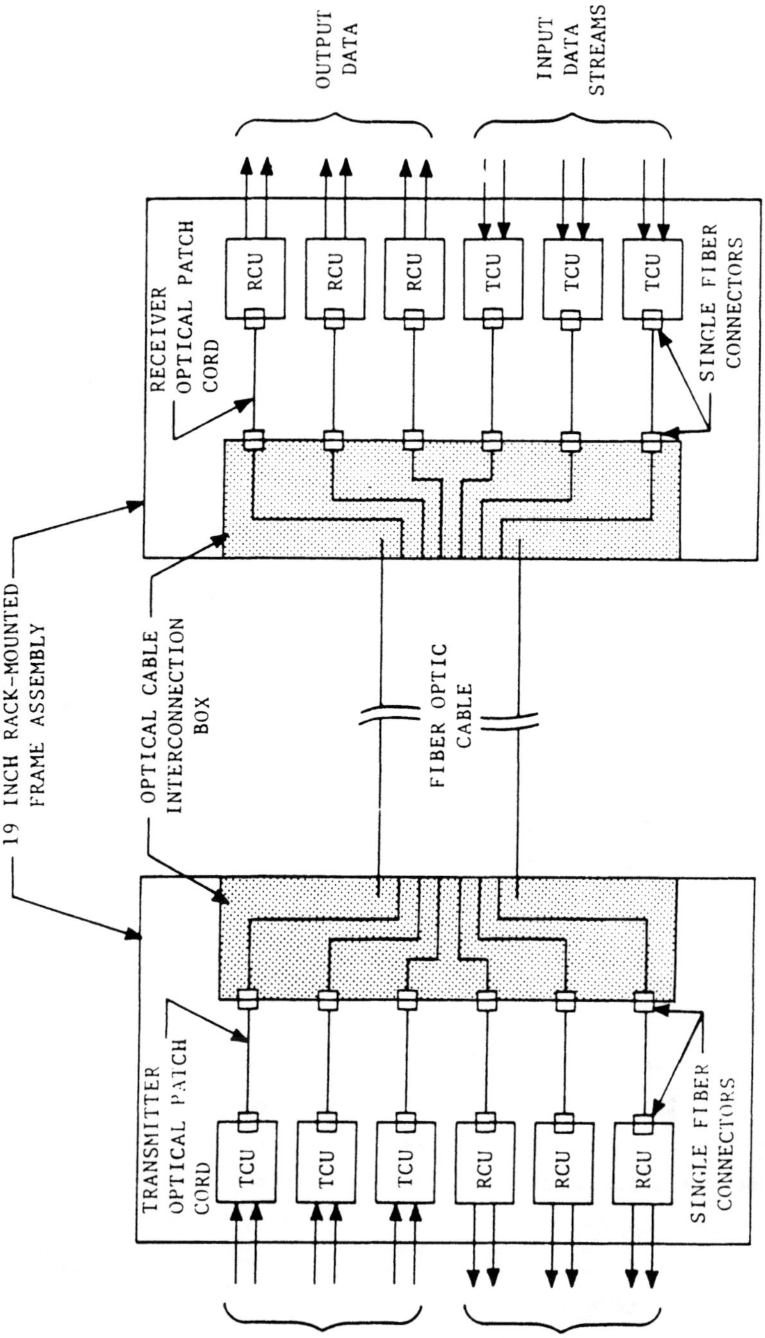

Figure 8-1. System configuration-typical AN/FAC communication set.[1]

Differences also exist between the AN/FAC-2A and the AN/FAC-2B. The -2A was basically a limited edition development model which was used primarily by the U.S. Army. The -2B is a TRI-SERVICE production model which is currently in use. Minor electrical changes were made to the -2B TCU and RCU which had no impact on the fiber optic portion of the system. Also, the fiber optic cable was color coded. The -2B TCU and RCU can be backfit to the -2A.

These systems were originally designed as fiber optic systems and thus, there is no counterpart electrical system. The AN/FAC-2A has been in existence since 1978, the -2B since 1981, and the -3 since 1981. We are not aware of the -3 being used by military agencies. The unit was designed and specified by the National Security Agency (NSA).

II. ASSEMBLY OF THE SYSTEM

The AN/FAC sets are easy to install and no special tools or equipment are needed for installation into standard equipment cabinets at the selected sites. The first steps are to unpack the sets from the shipping containers and visually inspect them. The sets are then installed in the 19-inch rack mounted assemblies. Once the sets are installed, an installation checkout procedure is performed. This is a performance test in which optical power output is measured, analog and digital signal receipt is verified, and any adjustments to the propagation delay are made. The optical patch cords are then connected to the TCU, RCU and the Optical Cable Interconnection Box (OCIB). The connector plug and jack must be connected by first mounting the connector jack on the OCIB and then rotating the plug onto the jack. The installation instructions indicate that great care must be taken to prevent rotation of the fiber. This is accomplished by holding the fiber against the palm of the hand with the little finger while screwing the plug into the connector jack with the thumb and forefinger of the same hand. Once the plug is screwed tight, the technician is to gently pull the fiber out of the plug about 1mm and then release the fiber. This causes a spring action which permits the ferrule to seat properly in the connector. There have been no reports of failure due to over-extending the fiber when trying to seat the ferrule in the connector.

Once the cable has been installed in underground ducts or aerial poles between the two equipment locations, it must be terminated. Termination of the fiber optic cable is the most difficult part of the installation. To install the ITT single fiber jeweled connectors on the cable, an optical connector installation kit is required. The piece parts of this connector are shown in Fig. 8-2. The cable must be stripped and the individual fibers terminated in a jeweled ferrule.

The fiber is then polished after which it must pass a visual examination with a 150X microscope. Single fiber jeweled connectors are installed at the other end of the OCIB. An attenuation test is then performed in which the loss should be less than 3dB. If the loss is greater than 3dB, the polishing procedure and visual exam must be repeated.

The fiber termination is time-consuming and labor-intensive but once complete and done properly, there should be no need to reterminate the connector.

The operations and maintenance manual indicates that uncertainty exists on loss measurements made in the field due to the inability to strip the cladding modes while making the measurements. The manual indicates that laboratory losses below 3dB measured without stripping cladding modes correspond to losses below 2dB with the cladding modes stripped.[1]

III. OPERATIONAL FEATURES

The AN/FAC sets provide up to six identical and independent transmission channels for data (and associated clock for the -3) over distances of 1,000 ft to 15,000 ft for the -2A and -2B respectively, and 6600 ft for the -3. The two digital signals are simultaneously transmitted over an optical fiber by multiplexing the signals at 830nm and 1060nm. A data channel consists of a fiber optic cable, two optical patch cords, four ITT single fiber jeweled connectors, one fiber optic TCU, and one fiber optic RCU. A functional block diagram of the entire system is detailed in Fig. 8-1. Each of the items in the diagram will be described in the following paragraphs in regards to its function, reliability and maintenance features.

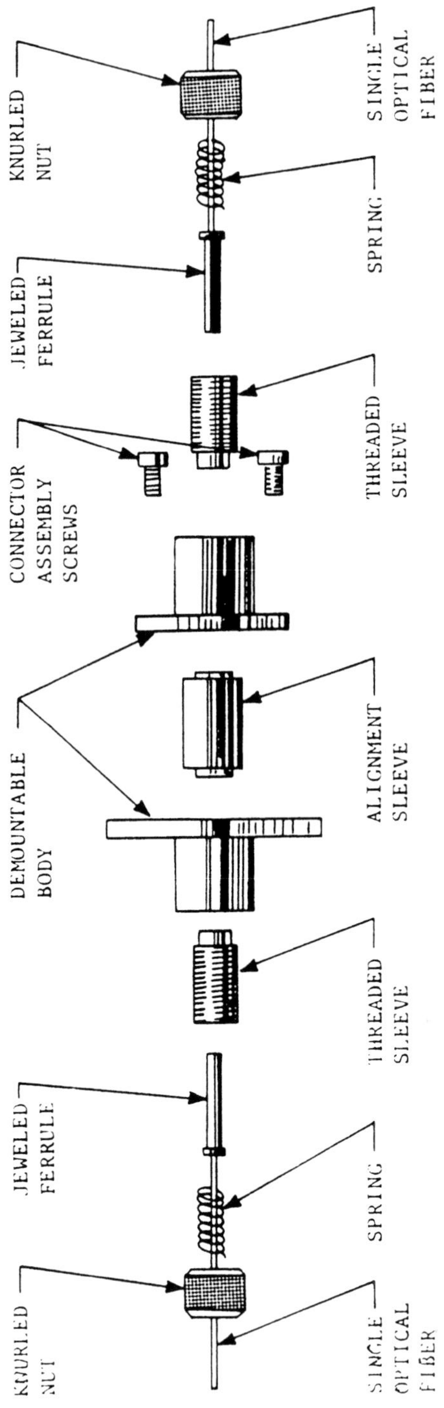

Figure 8-2. Components of ITT single fiber jeweled connector.[1]

The TCU detailed in Fig. 8-3 accepts signals, modulates the two light sources at 830nm and 1060nm wavelength and then optically multiplexes the two signals onto a single fiber. This unit contains three sub-units: the power supply, the transmitter interface card and the optical transmitter module. The power supply and transmitter interface cards are purely electronic.

The optical transmitter module consists of an optical coupler assembly and two LED drive circuits. There are two pigtailed LEDs in this module with one operating at 830nm and one operating at 1060nm. The coupler assembly contains a Lensed Dichroic Coupler (LDC) and a box in which the coupler is mounted. The optical outputs of the LEDs are fed into the LDC where the signals are multiplexed onto one transmission fiber. The LDC also filters the LED out-of-band radiation which improves the performance.

All sets are capable of operating in the DATA/DATA mode in which two independent data channels are transmitted over the same fiber. The operating data rates are 10Kb/s to 20Mb/s for the -3, and 20Kb/s to 20Mb/s for the -2A and -2B. Only the -3 is capable of operating in the CLOCK/DATA mode which is activated by a switch on the front of the unit. In this mode, one of the incoming signals is clocked in a retiming circuit which transmits the two signals in the correct phase relationship. Use of this mode eliminates the need for bit synchronizer electronics at the RCU which is a reliability problem with the -2A and -2B electronics. The optical transmitter module and the power supply are considered non-repairable in the field. The transmitter interface card can be repaired in the field.

The RCU, which is detailed in Fig. 8-4, demultiplexes the intensity modulated optical signals, reconverts them to electrical form and amplifies the signals. The RCU contains three sub-units: the power supply, receiver interface card and optical receiver module. The power supply and receiver interface card are purely electrical. The optical receiver module consists of an optical coupler assembly and two Avalanche Photodiode (APD) receiver circuits. There are two pigtailed APDs in this module with one APD sensitive to 830nm signals and one sensitive to 1060nm signals. The coupler assembly contains a Lensed Dichroic Coupler (LDC) and a box in which the coupler is

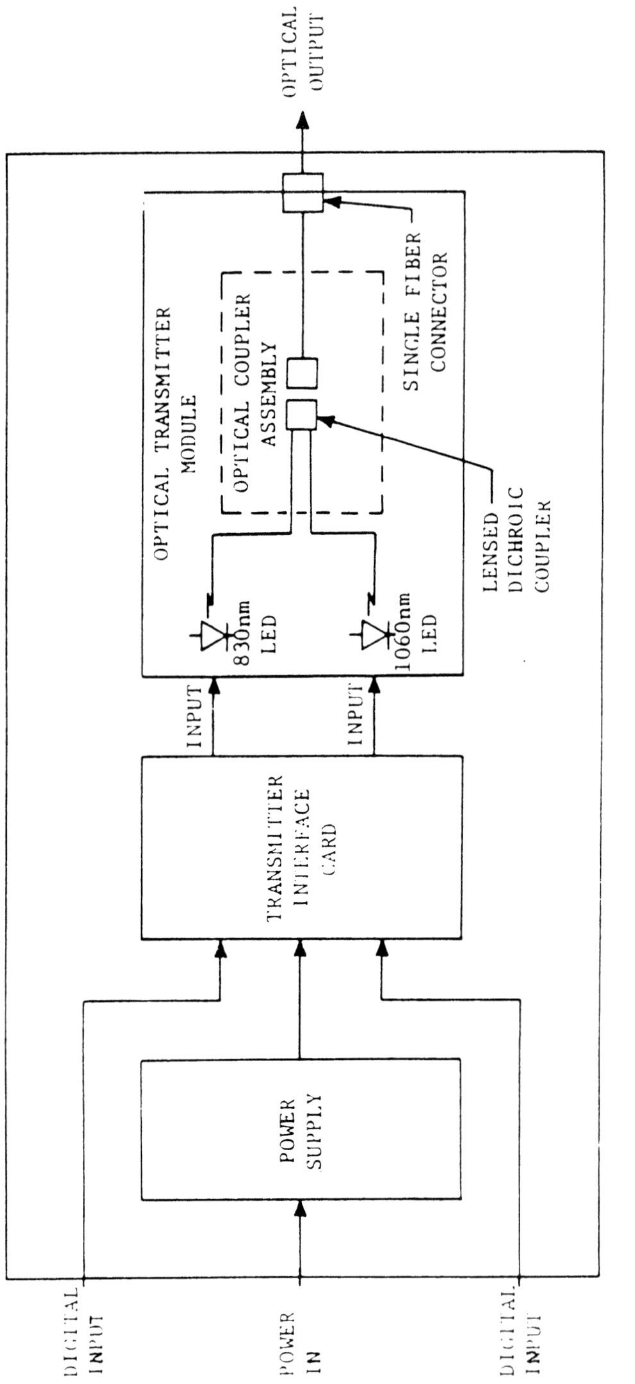

Figure 8-3. Transmitter channel unit (TCU).[1]

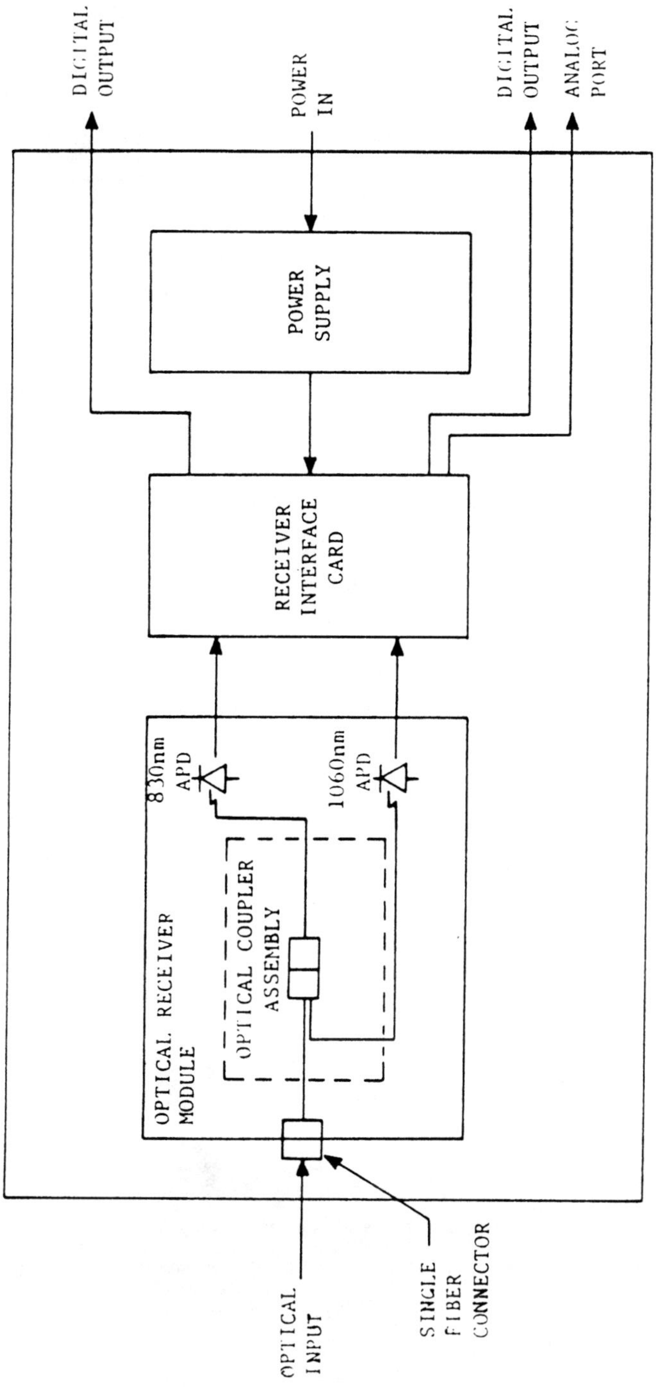

Figure 8-4. Receiver channel unit (RCU).[1]

mounted. The received optical signal is coupled to the LDC through a fiber pigtail which is terminated in a single fiber connector on the receiver module. The LDC splits the received light into 830nm and 1060nm signals and routes each signal to the appropriate APD which converts the optical signals into electrical signals. The electrical signals are then amplified, fed through a low-pass filter to an automatic gain control amplifier, an output buffer amplifier and a voltage comparator which provides a digital output. The digital output uses Transistor Transistor Logic (TTL) and is compatible with MIL-STD-188-114.

The intensity modulated signals are reconstructed into a digital format by level detection circuitry. Analog ports are also available to drive external equipment. Each digital signal is routed through a variable delay element which compensates for the inherent difference in the propagation velocities of the two signals due to the different wavelengths. The delay is adjusted by a knob at the rear panel of the receiver. The optical receiver module and power supply unit are considered non-repairable in the field. The receiver interface card can be repaired in the field.

The optical connectors used in the system are single fiber jeweled connectors which have a precision machined, stainless steel ferrule. A jewel bearing is inserted into the ferrule and the fiber is inserted into the precision hole in the center of the jewel. Assembly of the connector may be performed in the field by trained technicians using a fiber optic connector installation kit with an instruction manual.

The optical fiber cable used in the various AN/FAC sets varies with each model. The characteristics of the fiber and cable are listed in Table 8-1. The optical fiber cable is constructed with external strength members between an internal and external jacket as shown in Fig. 8-5. The cable is designed to be installed in conduit and ducts and is rated to withstand tensile forces up to 300 pounds. The outer and inner jackets are made of polyurethane which provides abrasion resistance and protection against water.

TABLE 8-1. Fiber Optic Cable Specification.

Parameter	AN/FAC-2A	AN/FAC-2B	AN/FAC-3
Fiber Core Size (mm)	0.0558 ±0.0050	0.050 ±0.005	0.050 ±0.005
Fiber Cladding Size (mm)	0.127 ±0.0076	0.0125 ±0.005	0.125 ±0.005
Number of Fibers	8	7	7
Cable Diameter (mm)	6.985 ±0.64	6.985 ±0.64	7.00 ±0.64
Tensile Strength (kg)	136	136	140
Isolation Between Fibers (dB)	≥ 50	≥ 50	≥ 50
Minimum Bend Radius (cm)	*	*	5
Multimode Dispersion (ns/km)	5 (between 3dB points)	4 @ 900nm	≥ 2
Attenuation (dB/km)	Range 1: 6 to 8 @ 820nm Range 2: 4 to 6 @ 820nm Range 3: 2 to 4 @ 820nm	3 to 6	≥ 6

* Not Known

Ref: 1,2

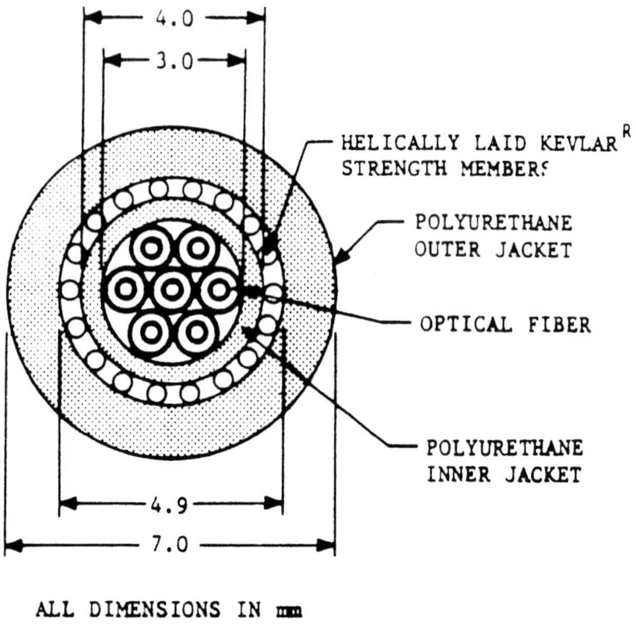

Figure 8-5. Optical cable cross section.[2]

RKevlar is a registered trademark of Du Pont Company.

IV. ENVIRONMENTAL DESIGN CONSIDERATIONS

AN/FAC-2A, -2B, and -3 communication sets are not designed for portable field conditions but for a rather benign, indoor, office-type environment. The operating temperature of the sets is from 0 to +50°C. There are no environmental tests run by the manufacturer and there is no indication of any special optical or electrical subassembly or component designs.

V. PERFORMANCE OF FIBER OPTIC SYSTEM

The AN/FAC sets have been in various stages of operation for six years. The installation at Ramstein Air Base in Germany was operating for a couple of years when new buildings were constructed in which the existing AN/FAC-2A sets were installed.[3] A new fiber optic cable was installed and the existing TCU and RCU 19 inch rack mounted frame assemblies were moved into the new buildings. One of the optical receiver modules was updated by replacing the existing Positive Intrinsic Negative (PIN) photodiodes with Avalanche Photodiodes (APD). There was an ample power margin in those channels using the PIN photodiode and those using the APD photodiodes, thus the remaining receivers were not updated with APDs. The AN/FAC-3 was originally designed using APDs.

Reliability predictions were calculated in accordance with Section 2 of MIL-HDBK-217C. The Mean Time Between Failure (MTBF) and the failure rate were calculated for all components. The reliability information contained in this report is for the AN/FAC-3.[4] Primarily, only the fiber optic components will be addressed on an individual basis. The failure rate for the fiber optic connector is an estimate based on the manufacturer's operating experience in many different systems. There have been very few fiber or cable failures reported in six years of operating experience for all of the AN/FAC models combined. The cable failure rate assigned is an estimate based on this background. The predicted and actual failure rate data is presented in Table 8-2.

TABLE 8-2. AN/FAC-3 Reliability Parameters.

Individual Predicted Parameters for Optical Assemblies Only

Cable Assembly	Failure Rate (10^{-6} Hrs)	MTBF* (Hrs)
Connector	1.52	657,900
Single Fiber Cable	2.08	480,800
Multifiber Cable	1.04	961,600
	4.64	215,500
Receiver Channel Unit		
Optical Receiver Module	11.88	84,200
	11.88	84,200
Transmitter Channel Unit		
Optical Transmitter Module	8.78	113,900
	8.78	113,900
Optical Coupler Assembly (Receiver)		
Optical Coupler	2.00	500,000
Optical Connector	0.76	1,315,789
APD, 830nm	0.26	3,846,154
APD, 1060nm	0.26	3,846,154
	3.28	304,878
Optical Coupler Assembly (Transmitter)		
Optical Coupler	2.00	500,000
Optical Connector	0.76	1,315,789
LED, 830nm	5.00	200,000
LED, 1060nm	2.00	500,000
	9.76	102,459
Combined Predicted Parameters For Optical Assemblies Only		
	38.34	26,082
Actual MTBF Reported For The Entire System		
		24,000 - 30,000

Ref: 5

* MTBF = Mean Time Between Failure

The actual numerical failure data for all of the AN/FAC sets combined is unknown. From phone conversations with various organizations, it has been reported that the fiber optic portion of the sets are very reliable. After six years of operatior of 300 AN/FAC-3 units, there have been no optical failures in the cable, fiber, connectors or couplers. The AN/FAC-3 design MTBF for the entire system (electrical and optical) is 8,000 hours and the actual MTBF is 24,000 to 30,000 hours.[4] Reliability indicators have been calculated for the 300 AN/FAC-3 units which have been operational for six years. The total operating hours were calculated assuming all units have been continuously operating which gives 15,768,000 total system-hours. The data was considered to be continuous and random which allows the Chi Square distribution to be used. The MTBF at a 90% confidence interval was calculated using equation 8-1. The source data and calculated results, in which a standard Chi Square table was used, are given in Table 8-3.

$$\text{MTBF Lower Limit} = \frac{2T}{\chi^2(\alpha, 2r + 2)} \qquad \text{Equation 8-1}$$

where T = Total System-Hours
α = Risk of error = 0.1
r = Number of failures
χ = Chi Square distribution

TABLE 8-3. Calculated AN/FAC-3 Data Summary.

Total Number of Systems	300
Total System-Hours	15,768,000
Total Number of Reported Failures	0
Calculated MTBF Lower Limit (10^6 Hours)	6.85
Calculated Failure Rate (10^{-6} Hours)	0.146

Any problems that have occurred have been electrical problems associated with the transmitter and receiver. A major problem that ultimately was the basis for the new design of the AN/FAC-3 was the periodic failure of the AN/FAC-2B bit synchronizer electronics, which reestablishes the clock for the timing. A cable failure was experienced with the AN/FAC-2B at a location in Norfolk, VA. In this incident, a rat chewed the cable in two and proceeded to bite and damage the cable for approximately 250 ft. The cable was installed in underground ducts to which the rat gained access via a manhole cover which was not secured. The facility has since installed locking mechanisms on the manhole covers and the manholes are regularly inspected.

There have been some failures due to shipping of the AN/FAC sets. On long air shipments to Australia, power supply problems have occurred; solder joints have broken loose and one epoxy bond which secured an optical fiber to a coupler broke.[5] However, no special requirements due to the fact that the sets contain fiber optics, are currently required for packaging or handling. This simplifies logistics considerations.

VI. MAINTENANCE REQUIREMENTS

Baseline temperature testing is done on ten percent of all manufactured sets. Manufacturer operational tests check every function, indicator and switch, plus signal symmetry, delay, rise and fall times, distortion and bit error rate at 5Mbits/sec and 20Mbits/sec.

Installation testing includes checking the optical power levels at the transmitter optical patch cords to insure that they are within the specified values. The optical power output from the receiver patch cords is also measured to insure that the power level is within the specified values. If the power level is above 2mW, the transmitter power can be reduced. Reducing the power of the transmitter greatly extends the lifetime of the transmitter LED but still allows adequate optical signal power at the receiver.[1] The propagation delay must be checked and set.

Preventive or scheduled maintenance is not required for the fiber optic portions of the AN/FAC sets. In fact, scheduled maintenance is strictly prohibited on most of the sets. The users are instructed not to touch the system unless it fails. The maintenance manual for the AN/FAC-2A & 2B has a note which says not to remove the fiber optic cable unless it is absolutely necessary as damage to the fibers may occur. No preventive maintenance is suggested because experience has shown that the more the connectors are handled, the more likely it is that a failure will occur.[2]

It was anticipated that corrective maintenance procedures may occasionally be required on some components of each transmitter and each receiver, therefore, detailed corrective maintenance procedures were given to check the following:[1]

1. Signal to Noise Ratio (SNR)
2. Crosstalk
3. Analog signal amplitude, pulse spread, rise/fall times
4. Signal monitors/signal loss alarms and,
5. Bit Error Rate

Repair of the models is basically performed by swapping out TCUs or RCUs. The units are then sent to the return repair bin. The depots repair or checkout the units using the 50% return repair rule which follows the premise that repairs are made until the cost of making the repairs reaches/exceeds 50% of the initial cost at which point the unit is scavenged. Once the TCU and RCU have been replaced and the system is still not operating properly, the connectors are cleaned and the system is retested. Then the optical loss of the malfunctioning channel is measured. The operations manual indicates that power readings made directly on the optical output of the transmitter may be inaccurate by as much as $\pm$3dB. The readings taken this way should only be used as a guide to determine 'proper' operation of the transmitter. The readings should be taken through an optical patch cord to insure a higher degree of accuracy. The single fiber jeweled connectors can be repaired in the field. A list of tools required for connector installation/repair is detailed in Table 8-4.

TABLE 8-4. Optical Connector Installation Repair Kit.

Required Parts	
Aluminum Adpato-Case	Flat wooden toothpicks
Power cord, 115 volts ac	Clear rubber tubing
Cable stripper	3 cc disposable syringe
Knife and blades	10 cc disposable syringe
Scissors	30 cc disposable syringe
Screwdriver	Paper discs, 600 grit
Tap hammer	Polishing pads
Tweezers	Polishing compound (aluminum)
Jewel punch	Plastic bottles, 2 ounce
Micrometer	Epoxy, resin and hardener
Ferrule polishing fixture	Solvent
Polishing fixture adapter	Masking tape and dispenser
Aluminum plates, 4 x 4 inches	Deionized water
Aluminum weighing pans	Instruction manual
Optional Parts	
Microscope (150 X)	Optical Power Meter
Microscope light and batteries	

Following the hands-off maintenance philosophy, fiber optic system failures have only occurred due to uncontrolled events such as rodent damage rather than operator induced failures. If a routine maintenance philosophy were followed, a slow degradation of the system may be noticed when periodically measured bit error rate results are compared. Generally, a degradation would be the result of degraded input data rather than poor fiber optics performance. In any event, there have been no reports of system failure due to degradation of the system.

Personnel that perform the maintenance on these sets are not required to have any special training or instruction. The personnel must have basic electrical test background to perform the corrective maintenance tests. Maintenance requires the use of an oscilloscope, pattern generator, error detector, pulse generator, voltmeter, and other common electronics test equipment. The operation and installation manuals contain very detailed instructions on how to install the fiber optic connectors. Stripping the cable and buffer material from the fiber, cleaving the fiber, installing the connectors, polishing and inspecting the fiber endface are all described. However, thorough instructions cannot compensate for hands-on experience. Because the engineers had obtained more experience with the components in the laboratory, when the relocation of the AN/FAC-2A was performed in Germany, a team of engineers rather than a team of technicians was on-site to install the connectors and the APD.[3]

VII. LOGISTIC INFLUENCES

Information on the number of spare parts required is unavailable. The major spare parts that would be required are the TCUs and the RCUs. This is due to the fact that maintenance follows a swap unit philosophy. The only portion of both of these units that are field repairable are the interface cards and the single fiber connectors.

No special packaging, handling or storage requirements are currently required for the AN/FAC sets. The sets are shipped as standard communications gear. Failures have occurred in both the fiber optic and electrical portions during long air shipments which indicates that perhaps special handling or packaging requirements should be implemented.

Repair equipment and spare parts should be located at each building where an equipment rack containing an AN/FAC set is installed. This will allow immediate repair of any failure that may occur and will allow testing of the completed repair from both ends. This location of equipment was suggested in documentation concerning the relocation of AN/FAC-2A units in Germany.[3]

VIII. SUMMARY

The AN/FAC-2A, -2B and -3 have proven to be successful fiber optic communications sets based on operating performance. The advantages inherent to fiber optics allow the sets to be separated by longer distances and the all dielectric cable is immune to EMI, has small size and is light weight. Therefore, the use of fiber optics in this system can only be considered beneficial and reliable since there have been no operational failures resulting from its use. Special tools are required for connector installation, but this is not unique to fiber optics. These special tools simply replace the analogous tools that would be required for a specific electrical connector. Logistics is not a problem for this system.

Overall, the entire system has proven to be very reliable with a new model currently being designed. The new model may utilize 62.5 micron core fiber which will add to the existing inventory of cable types.

Chapter 9. AV-8B Harrier Fighter Aircraft

I. SYSTEM DESCRIPTION

The fiber optic system on board the AV-8B Harrier is a point-to-point, unidirectional, single channel data link used to transmit serial Communication, Navigation and Identification (CNI) information at a low data rate of 125 kbits/sec over a 0.5 MHz bandwidth. Typically, altitude warning, display data, and control communication systems parameters data is transmitted.[1] As Fig. 9-1 shows, there are two twenty foot sections and one three foot section of optical cable which run concatenated from the cockpit to the aft of the airplane.[2] The front half of the aircraft is assembled by British Aerospace in England and the rear half is assembled by McDonnell Douglas in St. Louis, MO. The two halves join at the bulkhead which must be penetrated by the fiber optic link. An SMA style connector with a bulkhead feed-thru is used at the interface of these two halves. Also, an SMA style

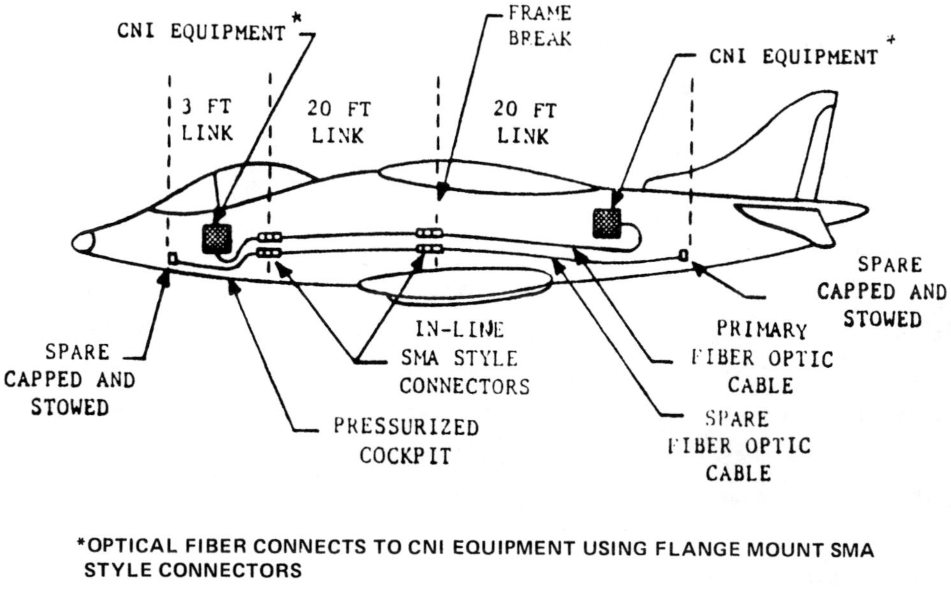

Figure 9-1. AV-8B fiber optic cable configuration.

connector with a bulkhead feed-thru is used to penetrate the bulkhead at the interface of the pressurized cockpit and the unpressurized fuselage. The selection of the SMA style connector was driven primarily by what was available on the market.[3] When the CNI optical system was developed in the early 1980's, very few optical connectors were being manufactured. Out of those which were available, the SMA style connector was the most popular and its reputation was good. A spare fiber optic cable is installed alongside the primary cable. This cable is identical to the primary cable including the two bulkhead SMA connectors. The spare cable remains concatenated with the extreme ends capped and stored near the CNI equipment.

Ninety-eight AV-8B Harriers have been produced since November 1981. The CNI system aboard these aircraft has been in operation for more than five years which represents over 44,394 flight hours[4] as of 30 June 1987. The installation of fiber optics on the AV-8B was done for experimental purposes; there were no particular advantages associated with its use or requirements that could not be met by electrical components. The intent of the experiment was to determine whether fiber optics could operate reliably aboard an aircraft.[3] The results of this effort have been overwhelmingly in favor of the use of fiber optic technology, for this system has not experienced a single optical failure.

II. OPERATIONAL FEATURES

The CNI system is comprised of two main units: the Communication, Navigation and Identification Data Converter (CNIDC) and the Auxiliary Communication, Navigation and Identification Panel (ACNIP). As the functional block diagram of Fig. 9-2 shows, the electrical link provides two-way data transmission and can fulfill all of the functional requirements of the CNI system without the fiber optic link. Therefore, with both links, this system has inherent redundancy.

The ACNIP simultaneously transmits the same information over both the optical and electrical links to the CNIDC. However, during normal operation, only the fiber optic data received by the CNIDC is used. All information transmitted from the CNIDC to the ACNIP is only transmitted over the electrical link.[1]

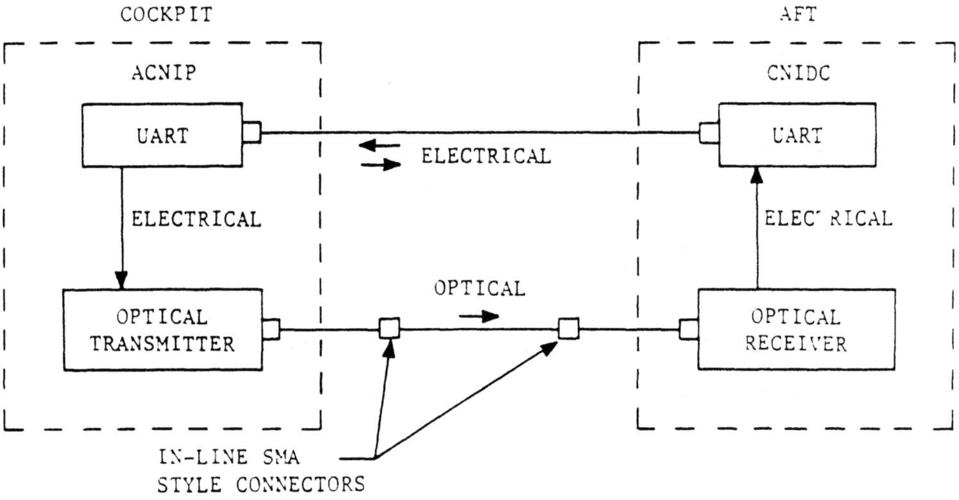

Figure 9-2. Functional block diagram of AV-8B single channel fiber optic system.

The ACNIP unit is located in the controlled environment of the cockpit. It has an electronic Universal Asynchronous Receiver/Transmitter (UART) and a fiber optic transmitter which uses a Light Emitting Diode (LED) that operates in the pulsed mode. Data is transmitted over 100/140 micrometer multimode fiber[3] in an 80 microsecond long word followed by a 500 microsecond gap, another eighty microsecond long word, another 500 microsecond gap and then another 80 microsecond long word. This sequence is repeated every 50 milliseconds.[5]

The LED operating wavelength is 820nm and its typical power output is specified at 1100 microwatts at a forward current of 100 milliamps. This GaAlAs LED is a planar double heterostructure device fabricated using liquid-phase epitaxy techniques. It is contained in a hermetic, metal TO-46 style package mounted in a SMA style flange mount receptacle.[6]

The optical data is received at the CNIDC which is located in the aft of the airplane in an uncontrolled environment. A Positive Intrinsic Negative (PIN) photodiode fabricated using planar, oxide-nitride passivated, bipolar silicon processing techniques is used in the optical receiver. It is specified to have a responsivity of 0.30 ampere/watt and a dark current of 0.05 nanoampere at a reverse bias voltage of 30 volts. It is housed in a hermetic, metal TO-46 package mounted in a SMA style flange mounted receptacle.[6] The electrical output signals from the PIN photodiode are sent to the UART located in the ACNIP.

Considering the typical power output of the LED and the responsivity of the PIN photodiode, the maximum allowable cable assembly attenuation was specified at 20 dB.[1] The cable assembly includes the three sections of optical cable, two in-line SMA style connectors and two SMA style connectors for mating with the active device flange mounts at the ACNIP and CNIDC.[2]

The actual cable assembly attenuation was measured at 7-9 dB. After 3-4 years of operation, the cable attenuation had not deteriorated below this initial value. This provided a worst case optical power margin of 20 - 9 = 11 dB.[1]

In the event that a failure occurs in the optical system, an indicator lamp will light in the cockpit to alert the pilot. The use of an indicator lamp to identify a cable fault is not normal Air Force practice. This lamp was unused when this system was designed and therefore, available for this purpose.[5] When a failure does occur, the electrical link must be forced to transmit data from the ACNIP to the CNIDC. Two designs have been used to accomplish this while a third is under development.[1,5]

The first design consisted of a manually operated switch. Activation of the switch transferred transmission of the data from the fiber optic link to the electrical link. This design was only used on the flight test planes.[1]

The second design performed the same function as the first but required a hardware modification to implement. This consisted of grounding an electrical pin in the ACNIP.[1,5] This design is currently used on all AV-8B aircraft produced after the flight test planes.

The third design, which is under development, automatically initiates a switch-over to the electrical link when the CNIDC suspects a problem either in the fiber optic link or in the ACNIP. A problem in the ACNIP can be either electrical or optical. The function of the developmental system is outlined below. The CNIDC directs the ACNIP to begin built-in-testing. Then the CNIDC waits for a response which normally takes about 3-5 seconds. If the CNIDC does not receive a response within 30 seconds, it is assumed that there is a failure either in the ACNIP or in the fiber optic cable. At this time all transmission is automatically transferred to the electrical link.[1,5]

The integrity of the transmitting side of the fiber optic system is continuously assessed by electrically monitoring the forward current of the LED in the ACNIP to insure that it stays within the limits specified for proper operation. If the forward current goes outside these limits, a failure is reported in the cockpit.[5] This monitoring method was selected in lieu of monitoring the spectral output of the LED because of its simplicity.

III. FIBER OPTIC COMPONENT CHARACTERISTICS AND SPECIFICATIONS

1. CONNECTOR

Single channel SMA style fiber optic connectors are used in the CNI system. These metal connectors are terminated using the epoxy/polish method and are used with feed-thru adapters. This method of termination is labor and time intensive (30-45 minutes)[7] and undesirable for field repair. Epoxyless terminations are most desirable, however, they are currently unavailable. An epoxyless termination would reduce the probability of contamination and significantly decrease the time required to complete a termination. The requirements and specifications the current connector is required to meet are given in Table 9-1. Table 9-2 gives the requirements and the specifications which the fiber optic feed-thru adapter must meet.

2. FIBER AND CABLE

The single channel, high temperature, aerospace 100/140 μm optical cable selected for the CNI system has the construction shown in Fig. 9-3. This cable was selected because it can be wrapped around a 1/8 inch diameter mandrel without difficulty. This mandrel size was considered to be representative of the flexing which would be encountered on the AV-8B.[3] Initially, designers of the system specified a fiber with a 200 μm cladding but no fibers of that size could meet the flexibility requirements. The 200 μm size fiber was specified due to the fact that it would make the termination process easier.[3] The 100/140 μm fiber is specified to have an attenuation rate 10 dB/km and a bandwidth 100 MHz-km when operated at 850 nanometers. The fiber numerical aperture is 0.29 +/- 0.02 and it is proof tested to 100,000 pounds per square inch.[8] The cable characteristics and the military tests to which it is subjected are given in Table 9-3.

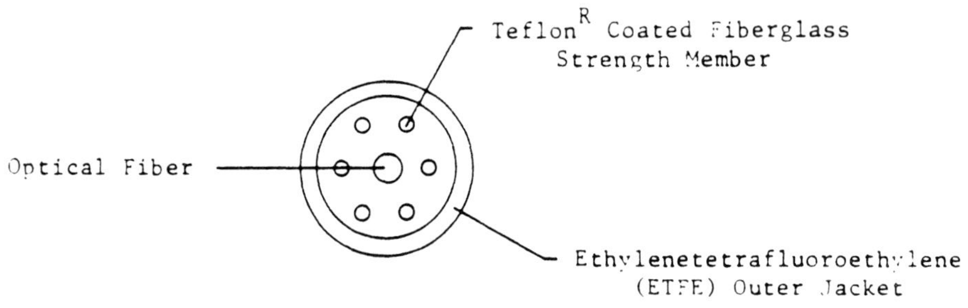

Figure 9-3. Optical cable construction.

[R]Teflon is a registered trademark of Du Pont Company.

Table 9-1. Fiber Optic SMA Style Connector Characteristics.

Condition	Military Specification or Requirement
Temperature Rating	-54°C to +93°C
Altitude	0 - 70,000 ft
Optical Loss When Connected to Feed-thru Adapter	1 dB Maximum
Durability After 500 Matings and Dematings	1 dB Maximum and 0 - 70,000 ft
Finishes	MIL-E-5400 and USAS B46.1
Threaded Parts	MIL-E-5499
Corrosion	MIL-E-5400
Weight	5 grams
Temperature Altitude*	4.1 of MIL-T-5422 Class 2, MIL-E-5400
Vibration*	4.2 of MIL-T-5422 Procedure I
Shock*	4.3 of MIL-T-5422 Part I
Humidity*	4.4 of MIL-T-5422
Salt Fog*	4.5 of MIL-T-5422 Energized Item
Temperature Shock*	4.9 of MIL-T-5422 Energized Item

* During this test, the connector must maintain a maximum 1dB loss with a fiber optic cable terminated with an SMA connector which is attached to the opposite end of the adapter.

Ref: 6

Table 9-2. Fiber Optic Feed-Thru Adapter Characteristics.

Condition	Military Specification or Requirement
Temperature Rating	-54°C to +93°C
Altitude	0 - 70,000 ft
Optical Loss When Connected to Feed-thru Adapter	1 dB Maximum
Durability After 500 Matings and Dematings	1 dB Maximum and 0 - 70,000 ft
Finishes	MIL-E-5400 and USAS B46.1
Threaded Parts	MIL-E-5499
Corrosion	MIL-E-5400
Weight	5 grams
Temperature Altitude*	4.1 of MIL-T-5422 Class 2, MIL-E-5400
Vibration*	4.2 of MIL-T-5422 Procedure I
Shock*	4.3 of MIL-T-5422 Part I
Humidity*	4.4 of MIL-T-5422
Salt Fog*	4.5 of MIL-T-5422 Energized Item
Temperature Shock*	4.9 of MIL-T-5422 Energized Item

* The Feed-thru adapter must maintain a maximum 1dB loss during this test with an SMA connector attached to both ends of the adapter.

Ref: 6

Table 9-3. CNI System Single Channel Optical Cable Characteristics.

Test	Military Specification	Requirement
Core Size	N/S*	100 μm +/- 4 μm
Cladding Size	N/S	140 μm +/- 6 μm
Protective Acrylate Coating	N/S	500 μm +/- 25 μm
Pull Strength	N/S	150 newtons
Minimum Bend Radius	N/S	5.0 centimeters
Weight	N/S	6.72 kg/km
Temperature Rating	N/S	-55°C to +150°C
Flammability	MIL-W-22759/16	N/S
Chemical Resistance	MIL-W-22759/16	N/S
Flex	DOD-STD-1678 METHOD 2010	10,000 cycles
Crush	DOD-STD-1678 METHOD 2040	200 newtons
Impact	DOD-STD-1678 METHOD 2030	0.5 kg at 150 mm
Vibration	MIL-STD-202 METHOD 204D	20 G at 20-2000 Hz
Constant Acceleration	MIL-STD-202 METHOD 212A	100 G
Mechanical Shock	MIL-STD-202 METHOD 213B	100 G

* Not Specified

Ref. 8

3. EPOXY

Two types of epoxy have been used in the termination of the fiber. The first type had a relatively high viscosity which was applied to the end of the fiber which was then slid onto the terminus. The use of this epoxy was abandoned because the viscosity was too high for there to be any wicking action, therefore, no strain relief was provided inside the terminus.[3] The epoxy currently in use is much less viscous and, due to the wicking action, it provides strain relief to the fiber. The characteristics of this epoxy as specified by the manufacturer[9] are summarized in Table 9-4.

Table 9-4. Epoxy Characteristics.

Parameter	Value
Number of Components	2 part mix
Pot Life	24 hours
Curing Schedule	120°C/1 hour
Viscosity	320 centipoise
Glass Transition Temperature (Tg)	106°C
Maximum Service Temperature	160°C
Thermal Coefficient of Expansion (Below Tg)	0.000063 in/in/°C
Thermal Coefficient of Expansion (Above Tg)	0.000125 in/in/°C

Ref. 9

IV. SYSTEM PERFORMANCE

Ninety-eight AV-8B aircraft have been produced since November 1981.[5,10] Four of these were test planes which were produced between November 1981 and June 1982.[5] In the 68 months which have elapsed between November 1981 and 30 June 1987, more than 44,394 CNI system flight-hours have been accumulated.[4] Typically an AV-8B is in flight 1-2 hours per mission.[5]

As of 30 June 1987, there were no known electrical or optical operational failures in the CNI system. To determine the MTBF, the flight-hours must be converted to operational hours. Based on a Rome Air Development Center study which addresses this conversion, one flight-hour equals 1.3 operational hours for fighter aircraft. Therefore, the CNI system has logged,

$$(1.3 \text{ op-hrs/flt-hrs})(44{,}394 \text{ flt-hrs}) = 57{,}712 \text{ op-hrs.}$$

The Chi square distribution was assumed in calculating the MTBF lower limit at a 90% confidence interval. The Chi Square expression used is given in equation 9-1 and the source data and results are given in Table 9-5.

$$\text{MTBF Confidence Limits} = \left(\frac{2T}{\chi^2\,(\alpha, 2r+2)}, \infty \right) \qquad \text{Equation 9-1}$$

where

- T = total operational-hours
- α = acceptable risk of error = 0.1
- r = number of failures = 0
- 2r+2 = degrees of freedom

Table 9-5. Reliability Data For AV-8B CNI System.

Total Systems	98
Total Flight-Hours	44,394
Total Operational Hours	57,712
Total Failures	0
MTBF Lower Limit (Operational Hours)	25,065

The electrical and optical portions of the CNI system combined were predicted to have a mean time between failure of 1,000 operational hours and a lifetime of 10,000 operational hours. Actual performance indicates a tentative MTBF lower limit which is 25 times the predicted lifetime or 25,065 operational hours.

1. EMITTER AND PHOTODETECTOR RELIABILITY TESTS

The individual active fiber optic components have been tested by their manufacturer using military specification methods. These tests are summarized in Tables 9-6 and 9-7. The failure rates were calculating using the device hours and test results provided by the manufacturer. The test results indicating 100% passed has been interpreted to mean no failures.

Table 9-6. LED Reliability Tests.

Stress	Military Specification	Quantity Tested	Device Hours And Test Results	Failure Rate @ 90% Confidence (10^{-6} Hrs)
Operating Life I_f= 100 mA T_c= 25°C	N/S*	40	200,000 2 Failures	between 1.78 and 31.46
Operating Life I_f= 100 mA T_a= 25°C	MIL-STD-750 METHOD 1027	287	48,200 0 Failures	between 47.78 and ∞
Temperature Cycle -55°C to +125°C	MIL-STD-750 METHOD 1056 CONDITION A-1 (+125°C)	157	6,550 Cycles 0 Failures	between 350 and ∞
Thermal Shock 0°C to +100°C	MIL-STD-750 METHOD 1056 CONDITION B	82	4,700 Cycles 0 Failures	between 490 and ∞
High Temperature Storage T_a= +150°C for 24 hours	MIL-STD-750 METHOD 1032	124	1,776 100% Passed	between 1,296 and ∞

* Not specified

Ref. 6

Table 9-7. Photodiode Reliability Tests.

Stress	Military Specification	Quantity Tested	Device Hours And Test Results	Failure Rate @ 90% Confidence (10^{-6} Hrs)
Temperature Cycle -55°C to +125°C	MIL-STD-750 METHOD 1056 CONDITION A-1 (+125°C)	75	1,875 Cycles 100% Passed	between 1,228 and ∞
Thermal Shock 0°C to +100°C	MIL-STD-750 METHOD 1056 CONDITION B	75	1,875 Cycles 100% Passed	between 1,228 and ∞

Ref. 6

The failure rate data for the LED was calculated during the useful operating life of the devices, that is, in the constant failure rate portion of the failure rate versus operating life curve for these devices. This eliminates failures due to infant mortality and wear out as indicated in Fig. 9-4. For the LED tests, the operating life is defined as the time at which the output power drops to one half of its initial value in 50% of the LEDs under test.[6]

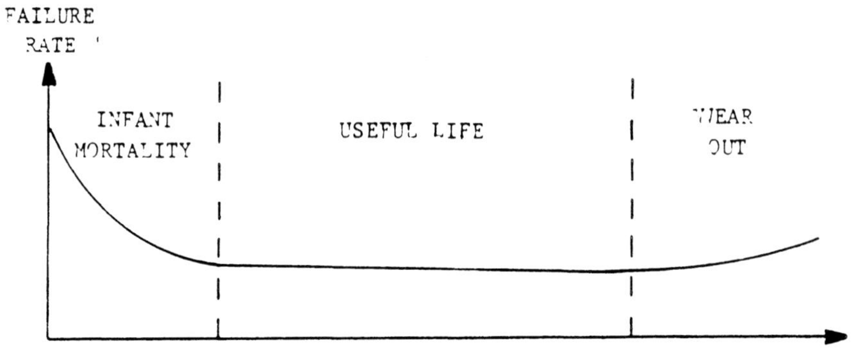

Figure 9-4. <u>Characteristic device failure rate curve</u>.

2. EXHIBITED FAILURES

The tentative MTBF of the CNI system has exceeded the 1,000 hour MTBF predicted for the total CNI system by a factor of 25. There have been no operational failures of the fiber optic or electrical portions of this system and no field failures of any kind.[5,7,10] All cable repairs have been performed by installing new cable from inventory instead of using the spare cable. Therefore, the spare optical cable has been removed since it was never used.[1]

At the time this system was first installed, there was no information on the reliability of optical fiber in an aircraft. Therefore, in anticipation of failures, the optical cable was routed close to the aircraft skin and near the engine compartment in an effort to make maintenance easier to perform.[10,11] Installing the cable this way is not normal procedure and although it is easier to access for repairs, it is also more vulnerable when other unrelated repairs are being made. As a result, induced failures were experienced while maintenance was being performed on other sections of the aircraft. These failures were all mechanical in nature and were the result of fiber breakage due to it being walked on with heavy boots when exposed.[3,5,7,10] The fiber is exposed when the wings are removed to gain access to the engines for major repairs or removal. To remove the engine, it is necessary to detach the wings which come off simultaneously as a single unit. This exposes the engine compartments along with the optical cable. These mechanical failures are not due to a fiber optic technology shortcoming; they are a result of improper cable routing techniques and/or inadequate cable crush resistance requirements. As indicated in Table 9-3, the specified crush resistance is 200 N which is equivalent to 44.96 pounds. This is insufficient considering that the cable is exposed to the weight of repair crew members. However, had the cable been routed alongside the electrical cable, which was routed following normal procedures, it would not have been exposed to this level of crush force and, therefore, would not have failed. The performance of the optical cable when properly routed has been proven in the field, for in the production of the most recent AV-8B aircraft, the same type optical cable was routed under the electrical cable, completely out of sight, and no mechanical or operational failures have occurred.[9]

However, as a result of the failures that have occurred, and since the fiber optic portion of this system was not meant to be a permanent installation, this portion of the CNI system is being removed under a Class 2 Change due to the expense associated with the maintenance of replacing the cable.[4,5,9,10] Each time a cable link was replaced, the new cable had to be terminated with an SMA style connector at each end which takes approximately 30-45 minutes to complete per connector. The time required to perform this repair and the associated costs including parts are the primary factors leading to the removal of the cable.[7] McDonnell Douglas is currently developing a connector which does not require the time and labor intensive epoxy and polishing assembly technique but instead, merely requires a cleave and crimp technique. Removal of the fiber optic portion is the prescribed solution to this maintenance problem resulting from improper routing, not as a result of problems with the technology. This is a logical response since the electrical portion can also perform the function of the optical link and the purpose of the experiment has been accomplished.

V. COMPARATIVE PERFORMANCE OF OPTICAL AND ELECTRICAL LINKS

Neither the optical nor the electrical link in the CNI circuitry has experienced an operational failure. Although the optical link has experienced mechanical failures, this should not be considered a weakness with the technology. It has been shown that had this cable not been specially routed, these failures would not have occurred. With this in mind, the exhibited operational performance of the optical link can be considered to perform as well as its electrical counterpart when its specified mechanical limitations are not exceeded as has been done with the optical cables crush resistance.

VI. MAINTENANCE REQUIREMENTS

Although the fiber optic link on the AV-8B has been operational for more than five years, there have been no preventive maintenance instructions developed for the system. The only tests performed on the optical portion of the CNI system are unscheduled maintenance tests. There are no scheduled maintenance procedures for cleaning the connector termini or measuring the optical power received at the ACNIP.[12] The AV-8B's Organizational Maintenance Wiring Repair Technical Manual (A1-AV8BA-WRM-000)[13] calls out three activities which deal with the fiber optic system. These are the repair of the cable and connector, and a description of how to use the optical attenuation test set to test the CNI system.

1. OPTICAL CABLE REPAIR

The equipment needed to repair the fiber optic cable is listed in Table 9-8. If the cable is found to be defective other than at a connector,

TABLE 9-8. AV-8B Fiber Optic Cable Repair Materials List.

Optical Attenuation Test Set (OATS)
Fiber Optic Repair Set
1,1,1 Trichloroethane Solvent
Optical Lens Tissues

the Wiring Repair Technical Manual (work package number 056 00) instruction is to replace that entire segment of cable using the spare cable which is routed adjacent to the primary cable. Since no failures have occurred in the field, only at repair facilities during other maintenance activities, this has never been done. The cable has always been replaced with new cable from inventory. In the event that the cable breaks at the back of the connector or in the connector backshell, the connector is to be replaced. If a bulkhead adapter or alignment sleeve is broken, they are to be replaced. High attenuation due to contaminated interfaces is to be taken care of by cleaning the connector with 1,1,1 Trichloroethane solvent, a toxic substance. It is not specified whether scratched fiber endfaces should be repolished or reterminated.

These are quick solutions as long as the particular problem can be identified. Considering that an Optical Attenuation Test Set (OATS) is the only test equipment used and this unit does not identify the location of a fault within a cable, it would not be possible to assess which component is at fault. The problem could lie in any one of the three cable segments or at any one of the connectors. The three cables could be tested individually and the fault isolated to a specific cable. But fault location could not go beyond this point with only an OATS unless the damage could be located visually. The only available instrument that is capable of specifying fault locations is the

Optical Time Domain reflectometer (OTDR) which costs many times more than an OATS. Therefore, it is most likely more cost effective to test each cable link separately instead of testing the entire link at once and having to pay for an OTDR. However, this is not explicitly stated in the technical manual.

One benefit that location of a fault within a cable using an OTDR would provide is the ability to repair the fault and not replace the cable. The source data given in Chapter 2 of this study indicated that fusion splices are the most reliable splice type. However, this requires the purchase of a fusion splicing machine which is also very expensive.

Therefore, it appears that the maintenance philosophy of replacing short cable segments is cost effective as long as replacements do not occur frequently.

2. OPTICAL CONNECTOR REPAIR

The procedure outlined in the Wiring Repair Technical Manual (work package number 056 01) when replacing the optical connector is no different than the manufacturer's directions. The materials needed to perform this replacement are listed in Table 9-9. The epoxy cure times are specified to

Table 9-9. AV-8B Fiber Optic Connector Replacement Materials List.

Fiber Optic Repair Set
Crimping Tool (nomenclature M22520/5-01)
Crimping Tool Die Set (nomenclature M22520/5-37)
Wire Stripper
Wire Stripper Blade
Heat Shrink Sleeving
X-Acto Knife Handle with No. 11 Blade
1,1,1 Trichloroethane Solvent
Polyethylene Squeeze Bottle
Optical Lens Tissue

be one half hour or one hour using 149°C or 121°C temperatures, respectively with EPO-TEK 377 epoxy. The fiber is to be polished with five different grit sizes of polishing paper (60, 40, 15, 3, and 1 micrometer) until the smallest grit size gives a mirror finish to the fiber endface.

3. OPTICAL ATTENUATION TEST SET OPERATION

The use of the OATS is given in terms of testing the fiber optic CNI system along with a basic description of its operation. This information is given in Wiring Repair Technical Manual (work package number 056 02). This unit is described as being a ruggedized, compact, lightweight and portable field tester. A digital liquid crystal display reads out the attenuation over a range of 45 dB or the absolute power level from +3 dBm to -65 dBm. The additional materials needed to perform the attenuation tests are given in Table 9-10.

TABLE 9-10. AV-8B Optical Attenuation Test Set Support Materials.

40 ft. Optical Test Cable
6 ft. Optical Test Cable
Bulkhead Adapter
Optical Lens Tissue
1,1,1 Trichloroethane Solvent

The OATS is initially zero calibrated before any measurements are taken. Then, one end of the 6 ft. Optical Test Cable is connected to the receive port of the OATS while the other end is connected to the ACNIP optical port using the bulkhead adapter and a plastic alignment sleeve. One end of the 40 ft. Optical Test Cable is connected to the CNIDC optical port using a plastic alignment sleeve. The OATS is then turned on and the attenuation measurement is taken in decibels from the display.

VII. LOGISTICS INFLUENCES

The number of spare units kept by the Government is minimal for this system. There are 1 or 2 CNIDC spare units and 1 or 2 ACNIP spare units kept in inventory.[5] No individual active components are kept in inventory. Spare cabling, connectors and epoxy is kept in stock by McDonnell Douglas for cable replacement[7] which requires the termination of the cable with SMA connectors.

VIII. SUMMARY

The fiber optic system used in the AV-8B CNI system is very simple in its design. Even though this is the case, its impact can still be assessed and valuable information extracted. The major problem area associated with its use has been mechanical in nature and is not related to the physics of the technology. During maintenance on other areas of the plane, the optical cable has been broken when walked on. The forces exerted on the cable when walked on meet or exceed the cable manufacturer's specified crush resistance. However, when the optical cable was rerouted beneath the electrical cable following normal procedures, there were no failures. This indicates that the fiber optic technology used on the AV-8B was not a source of problems once proper routing techniques were used. In fact, the CNI system has demonstrated a MTBF of 25,065 operational hours. This is approximately 25 times the predicted MTBF and 2.5 times the predicted lifetime. However, one problem area associated with the maintenance of the optical system was once again brought to light. That is the lengthy and costly procedures that must be followed when using the available epoxy and polish connectors. This problem is widely experienced in the fiber optics industry.

No special emitter monitoring techniques were used with this system that are unique to photonics. All monitoring and developmental built-in-testing is performed in the electrical domain and is very conventional in its operation.

This system has clearly demonstrated that the current fiber optic technology has the robustness needed to reliably operate in a fighter aircraft. As an indication of the success of the optical technology displayed by this CNI system, future AV-8B aircraft are scheduled to use fiber optics in the Digital Map Display system.[2,7]

Chapter 10. Fiber Optic Transmission System—Long Haul Army Communications System

I. SYSTEM DESCRIPTION

The Fiber Optic Transmission System (Long Haul) (FOTS (LH)) will be used by the Army in the field for replacement of CX-11230 dual coaxial cable with fiber optic cable. The FOTS (LH) equipment will be used with various existing communications systems such as the AN/TCC-73 Telephone Terminal, AN/TCC-138 Radio Repeater and the AN/TRC-151 Radio Terminal[1]. These systems are intershelter, radio remoting and long haul cable systems and are part of the standard Army Tactical Communications Systems (ATACS). FOTS (LH) is being designed to operate with a MD-1026 Digital Data group modem modified for fiber optic use for future applications in the Army Integrated Tactical Communications Systems (INTACS).

The main reason for choosing fiber optic cable as the transmission medium was that the metallic coaxial cable currently used does not meet operational requirements of high mobility, rapid deployability, dependability, and survivability in a tactical environment. The metallic cable emits RF signatures and can be damaged by an Electromagnetic Pulse (EMP) during a nuclear scenario. The FOTS (LH) fiber optic cable is not affected by Radio Frequency Interference (RFI) and Electromagnetic Interference (EMI), is immune to EMP, is nuclear hardened and is lighter in weight and more rugged than the metallic cable it is replacing. Major savings in weight and deployment time are obtained by using fiber optic cable as was mentioned in the referenced article by Mondrick, et al. The weight reduction is approximately 80%. For a typical 64 km coaxial cable link, 5 trucks are needed to deploy the cable. For the same length using fiber optic cable, only one truck is needed for deployment. Also the number of repeaters required are reduced from 39 to 10 by using fiber optics. Table 10-1 provides for a comparison of the attributes of the different cable types for the specified 6 and 64 km links and the goal of an 8 km link.

TABLE 10-1. CX-11230 Metallic Cable Links vs FOTS(LH) Links.

	CX-11230			FOTS(LH)		
				Spec	Goal	Spec
Distance (km)	6	8	64	6	8	64
Data Rate (MBPS)	20	20	2.3	20	20	20
No. of Repeaters	14	19	39	0	0	10
Cable Cost $(000)	21	28	224	12	16	131
Repeater Cost $(000)	11	15	36	0	0	50
(*) Total Link Cost $(000)	32	43	260	12	16	181
Cable Weight (lbs)	1081	1442	11,536	192	256	2048
Transportation Required (2 1/2 Ton Trucks)		5			1	

*Does not include cost of electro-optic converters.

Ref: 2

II. SYSTEM COMPONENTS

The FOTS (LH) system is basically used in two configurations--a repeatered link of 64 km and a repeaterless link of 6 km. There is a design goal of increasing the repeaterless link length to 8 km. The primary component of the system is the Fiber Optic Cable Assembly (FOCA). The FOCA is a one kilometer two-fiber cable which is terminated at each end by dual fiber hermaphroditic lensed optical connectors. The FOCA is reeled on existing twin coaxial reeling equipment so that it may be deployed in one kilometer sections by two soldiers walking the route. The Remote Optical Assembly (ROA) converts incoming electrical signals to optical signals for transmission and converts incoming optical signals back to their electrical format. The Fiber Optic Repeater (FOR) provides full-duplex regeneration of the optical signals. Repeaters are required every 6 km for this system. An Electronic Interface Assembly (EIA) was designed to allow interfacing with up to four different optical communications links. The primary ATACS or INTACS communications equipment are set up in shelters at various points on the battlefield. The ROA is mounted on the shelter wall and all incoming signals are converted to electrical signals. From that point, all signals within the shelter are electrical and will not be addressed in this report. See Fig. 10-1 for a typical repeaterless link and a typical repeatered link of 64 km.

III. OPERATIONAL FEATURES

The FOTS (LH) system provides two identical and independent transmission channels for data over distances of 1 to 64 km. The electrical data is transmitted at rates from 72 Kb/s to 20 Mb/s (4 to 1144 channels)[1] for the INTACS equipment and from 2.304 to 4.9152 Mb/s for the ATACS equipment. Once converted to optical signals, the standard transmission rates are 18.944 and 20.206933 Mb/s[3,4]. The overall system must meet a Bit Error Rate (BER) of 10^{-9} over the specified temperature range over 6 km repeaterless or 64 km with 10 repeaters prior to a nuclear event and over 4 km repeaterless or 64 km with 15 repeaters after a nuclear event. Each individual link must meet a BER of $6x10^{-11}$[4].

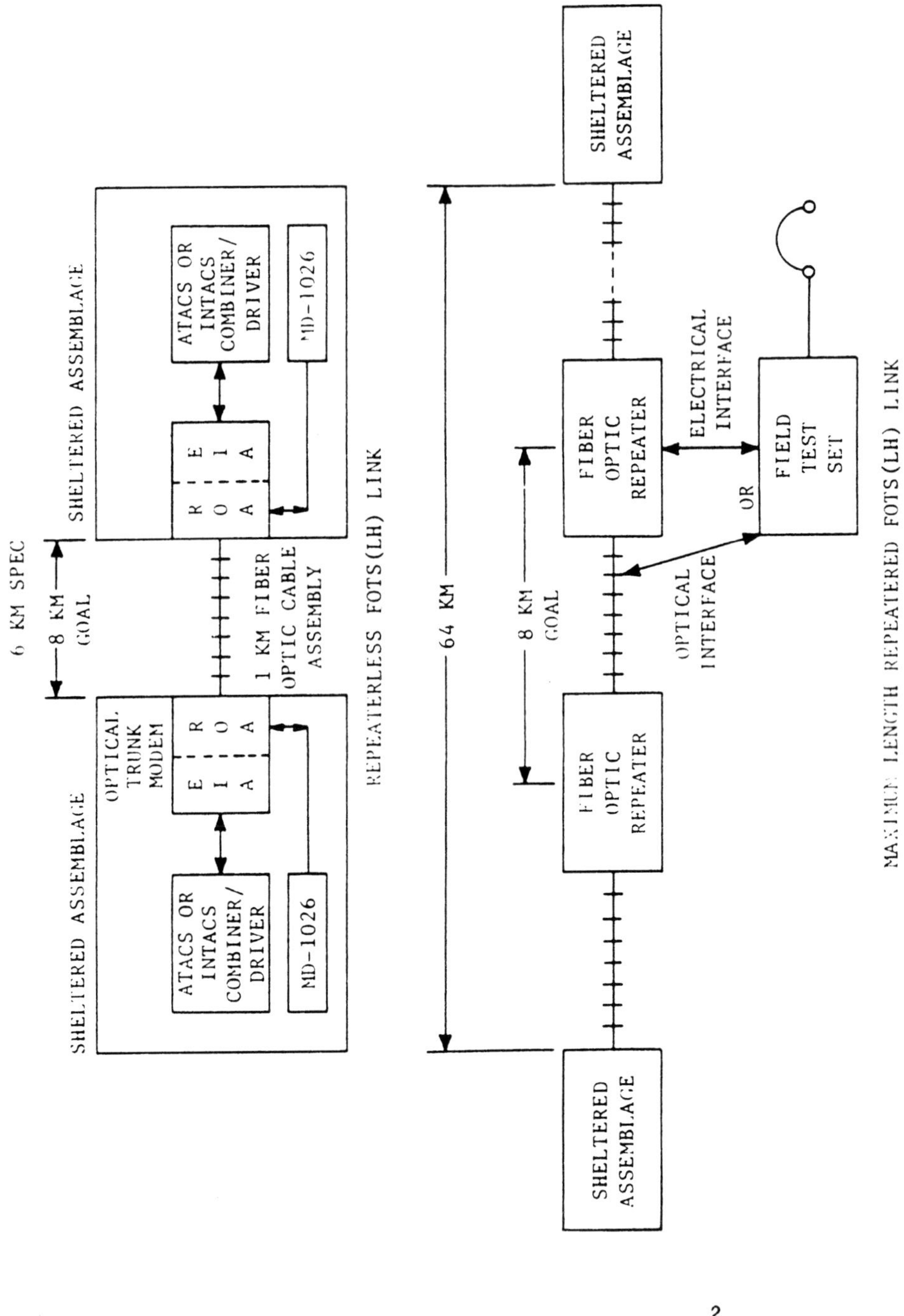

Figure 10-1. Fiber optic links for FOTS(LH)[2].

IV. ENVIRONMENTAL DESIGN CONSIDERATIONS

The environmental conditions that the system must meet are detailed in Table 10-2.

Table 10-2. FOTS(LH)/Environmental Requirements.

CONDITION	REQUIREMENT
TEMPERATURE	MIL-STD-810
o Storage and Transit	-46°C to 71°C
o Starting	-46°C
o Continuous Operating	-46°C to 52°C with solar radiation of 360 BTU/SQ FT/HR
ENVIRONMENTAL	AR-70-38 MIL-STD-454 MIL-STD-810
ELECTROMAGNETIC COMPATIBILITY/ ELECTROMAGNETIC INTERFERENCE	MIL-STD-461A

Ref: 2

The environmental and mechanical conditions that the FOCA must meet are detailed in Table 10-3.

Table 10-3. FOCA Requirements.

CONDITION	REQUIREMENT
Diameter	5.5 ± 0.2 mm
Tensile Strength (operating/ultimate)	200/1780 Newtons Minute
Flex	2000 cycles from -46°C to 71°C
Radial Compression	2000 Newtons
Overhand Knot	100 Newtons
Impact	100 impacts from -46°C to 71°C
Flammability	Self extinguish in 30 seconds with less than 4 in. of flame travel
Weight	72 lbs/km off reel 112 lbs/km on reel
Humidity	Per Quality Control Test Procedures--information not available.
Altitude	"
Vibration	"
Bench Handling	"
Drop	"
Shock	"
Fungus	"
Dust	"
Static Load	"

Ref: 2, 4

V. PERFORMANCE OF FIBER OPTIC SYSTEM

The FOTS (LH) system is still in the developmental stage. A contract was awarded to ITT Defense Communications Division in August 1982 to develop the system. ITT let a major subcontract to Standard Telephone and Cables (STC) in England for development of the electrical portions of the system. A pre-production system has been delivered to Fort Huachua, AZ for field tests.

During the development of the system, there were several design challenges that had to be met, as well as various systems engineering trade-offs, in order to develop a system that would meet the Army's requirements. The primary challenges in the design were nuclear hardening, flammability, low attenuation connector over a -46°C to 71°C temperature range and the development of test methods[4]. One of the key trade-offs was the decision to have the system operating wavelength be 1300nm[3]. The component development for operation at 1300nm was less mature and therefore a higher risk than 850nm components. The higher wavelength increases the dynamic range and lowers the attenuation of the system relative to operation at 850nm. It is also advantageous to use 1300nm because the fiber is more resistant to the effects of nuclear radiation. Another key issue was the fact that by operating at 1300nm, LEDs were used in the transmitters and a PIN/FET was used in the receiver. These components require less power and are more reliable than laser sources or APD photodetectors. The connector design was chosen to be an expanded beam type rather than a precision alignment type connector. The expanded beam connector was considered to be easier to maintain and clean in the field and that it would be more immune to the dirt and dust that would be experienced in a tactical environment. The connector loss was optimized at 1300nm wavelength to obtain a better system operating margin at the higher wavelength.

During the development of this system, there were several design objectives that could not be met[4]. A primary objective of the system was to meet the specified error rate performance with a repeaterless 6 km link with an incentive to go 8 km. An 8 km link was developed although the specified optical power budget was not met[4].

The measured optical power budget was -1.9dB at both low and high temperatures and was 3.4dB at the ambient temperature of 25°C for the pre-production system[4]. The low power budget was attributed to the poor LED output (7.4dB below budget values) at low and high temperature. Poor LED performance necessitated adding input amplitude limiting circuitry and increasing LED operating bias to favor more linear operation. The rise time and the bandwidth requirements on the LED were reduced which resulted in the LED drawing more power, exhibiting slower rise and fall times and having less optical power output than originally designed. A waiver was granted allowing the use of LEDs with outputs up to 2dB below the component specification. During temperature cycling tests, the LED failed as exhibited by a reduction in optical power output. This was due to movement of the fiber supporting structure and to temperature induced stress in the fiber. During the 1,950 hour reliability test, LEDs followed a pattern of gradual degradation resulting in failures. Twelve LEDs failed during the reliability test and four failed the post reliability test[4]. The selected LED did not meet the overall design objectives. See Table 10-4 for reliability requirements, predictions and allocations.

TABLE 10-4. Failure Prediction Results - FOTS(LH).

ITEM	REQUIREMENT	PREDICTION	ALLOCATION
		PROBABILITY OF SUCCESS	
FOTS(LH) System (4 channels - 20 km)	0.953	0.693	0.9685
		MEAN TIME BETWEEN FAILURE (hours)	
FOTS(LH) System (4 channels - 20 km)	500	770	750
Optical Trunk Modem (EIA/ROA pair)	7,000	10,084 (typical) 9,750 (worst case)	7,000
Field Test Set	5,000	14,671	7,500
Remote Optical Assembly	Not Specified	125,173	85,000

Ref: 4

The PIN/FET detector met the design objectives. During the environmental testing, there was one failure due to the fiber attachment process which was remedied[4].

The fiber optic cable was required to have one fiber color coded red. A dye was used which caused an increase in attenuation upon repeated coiling of the cable. For the pre-production model sent to Fort Huachuca, two filler strands were used instead of the dye to identify the fiber. A design or process improvement plan prior to production is being developed to eliminate this problem.

The bandwidth of the cable was too low which prompted the proposal of a process to improve this by automating the control of the fiber preform diameter, using larger preforms and faster pulling speeds.

There was a problem with attenuation at low temperature which was caused by longitudinal compression of the sheath and also by differences between the two fibers. A process control change was implemented to cool the extrudate and to improve the tension control on the fibers.

During the environmental tests as reported in the referenced article by Magnus, et al., one FOCA failed due to incomplete curing of the epoxy; the epoxy had contaminated the sylgard lens material of the connector which caused clouding at cold temperatures which in turn increased attenuation. The epoxy also caused improper seating of the optical fibers secondary coating in the inner ferrule which caused broken fibers after temperature cycling. The large increase in attenuation of the FOCA at cold temperatures has also been attributed to out-of-control steps in the final manufacturing process of the cable. During the reliability test of 1,950 hours, only one FOCA failed. The attenuation of the FOCA when tested at 850nm exceeds the specification requirement of 1.25dB typical and 3.0dB maximum. A waiver was granted allowing the attenuation to exceed the specification by 2.0dB due to connector optimization at 1300nm. The weight of the cable on the reels was typically 110 pounds, with a few weighing more than the requirement of 116 pounds. Four of the completed FOCAs were less than 1 km long (900m) due to retermination after removing cable faults. Overall, the FOCA exceeded the design objectives[4].

The two channel hermaphroditic connector experienced a few problems during the environmental testing. The nuts on the connector mouldings cracked during the testing and during routine handling. The material was changed, the edges were rounded and then heat-treated. No further problems with the nuts were experienced. The connector alignment pin seized occasionally. This was resolved by adding a layer of nickel plating on the pin. The shape of the chamfer also had to be changed to eliminate burrs from occuring during repeated matings.

The EIA, FOR and Field Expedient Splice Kit (FESK) were downgraded to test equipment status in June 1984 by government directive because the repeaterless link satisfies the majority of potential applications[4]. Only a few of these items were built and tested. The program was never funded to obtain ATACS interfaces so these components never advanced beyond the conceptual stage. Since these components were deleted, the ROA now directly interfaces with the external equipment. The ROA Mean Time To Repair (MTTR) depends on the fault recognition and isolation capability of the equipment with which it interfaces. Thus, the reliability and maintainability data contained in this report are only valid for the testing of this particular pre-production system. If the implemented system does not contain these components, the failure and maintainability predictions will have to be recalculated.

VI. MAINTENANCE REQUIREMENTS

The system was designed to be maintained with specially designed kits. The mean time to fault isolate to a single line replaceable unit (LRU) is required to be less than 10 minutes. See Table 10-5 for Maintainability Prediction Results.

TABLE 10-5. Maintainability Prediction Results.

PARAMETER	REQUIREMENTS (Min)	PREDICTION (Min)
Mean Time To Fault Isolate to LRU	10	2.96
MTTR System (Organizational)	15	9.12
MTTR, ROA (Organizational)	Not Specified	5.53

Ref: 4

The Field Test Set (FTS) optically interfaces with the FOCA or electrically interfaces with the FOR. The FTS is used for transmission system set up and maintenance. It detects data traffic faults, identifies and locates faults in optical cables and provides maintenance order wire communication services during field testing. Failure predictions for the equipment are indicated in Table 10-4. The FESK is used by field personnel to make temporary repairs to a damaged fiber optic cable to restore communications. The FESK was developed by TRW and was tested at Fort Gordon, GA in December 1983. The splice used is a no-epoxy-or-heat low loss splice. During the testing, it was determined that temporary repair with FESK is feasible. As was mentioned in the reference by Magnus, et al., the average MTTR for cable, including both day and night repairs, is 22 minutes. 37% of the splices showed losses of 3dB or lower. The average loss of the splices was 0.80dB[4].

A Permanent Cable Repair Kit (PCRK) is used by maintenance personnel to make permanent repairs to the fiber optic cable in a sheltered environment. The kit allows removal of a damaged section of cable and splicing a new section of cable in place or splicing on a replacement connector. Repaired cables are required to be at least as strong as the original cable and the insertion loss can't increase by more than 0.5dB per splice. The kit can be used to make repairs in temperatures from 0°C to 50°C. The kit contains an Optical Time Domain Reflectometer, fusion splicer, attenuation test set and equipment and enough supplies to repair 10 connectors and 12 splices. Total cable repair time is required to be less than 90 minutes and includes fusion splices that have losses less than 0.5dB. The PCRK weighs less than 100 pounds.

A Fiber Optic Cable Simulator (FOCS) is used during training or maintenance in place of the FOCA. It is used to interconnect ROAs, FORs and FTSs and to simulate unrepeatered cable links of up to 8 km in 1 km increments (4.5dB/step). It is a passive optical device which is manually operated and does not require any electrical power. It has 2 hermaphroditic optical connectors.

FOTS also has Built in Test Equipment (BITE). Fault detection is accomplished by continuous monitoring of the transmission activity. The BITE must detect greater than 98% of the faults with less than 1% error.

VII. LOGISTICS INFLUENCES

Information on logistics is unavailable. The major spare parts that would be required are FOCAs, ROAs and EIAs if a swap unit philosophy is used. Spare components for the various assembly and repair kits should also be kept on hand. It is not anticipated that logistics will be a problem.

VIII. SUMMARY

FOTS is not yet operational and it may be a few years until it is deployed by the Army. There have been several problems encountered throughout the design of the pre-production model and the vendors and designers involved have worked together to change process controls or design processes to improve system performance. The system does not yet meet all of the original design requirements. Work is being accomplished to improve the system.

Section V
Testing

Chapter 11. Optical Sources and Power Meters

Optical Sources (OSs) and Optical Power Meters (OPMs) are designed to launch and measure the optical power transmitted over optical cables. These devices are the primary test equipment used to check out a system once it has been installed and to measure connector and link loss.

I. DESCRIPTION

1. PHYSICAL CHARACTERISTICS

OSs and OPMs range from small, lightweight, inexpensive hand-held units to very expensive laboratory units which have the capability to interface with computers. Typical dimensions for hand-held field portable devices are listed in Table 11-1.

TABLE 11-1. Typical Optical Power Meter and Optical Source Physical Characteristics.

Length:	15 cm
Width:	10 cm
Height:	3 cm
Weight:	300 grams

2. OPTICAL SOURCES

OSs utilize either LED or laser sources to launch light into the fiber under test. It is very important to have a stable source because the value of fiber loss measured is a function of the source wavelength. Therefore, if the wavelength deviates much from the actual wavelength at which the system will operate, the loss measurement may be very misleading. This is why the source used for testing should operate at the same wavelength at which the system is designed to operate. If the system operates at more than one wavelength, then it should be tested at each wavelength.

3. OPTICAL POWER METERS

OPMs must have the capability to detect light at various wavelengths to accommodate the wide range of wavelengths different systems employ. Generally, manufacturers utilize solid state photodiodes to detect the optical signal being measured. A listing of photodiode types and their associated spectral ranges is shown in Table 11-2.

TABLE 11-2. Photodiode Types and Spectral Ranges.

Type	Spectral Range (nm)
Silicon (Si)	400-1100
Germanium (Ge)	800-1800
Indium Gallium Arsenide (InGaAs)	900-1800
Mercury Cadmium Telluride	1000-1350

Ref: 2

Most manufacturers offer the different wavelength photodiodes as "plug-in sensors". The sensors are equipped with a spectral calibration trimpot. When the sensor is plugged into the OPM, the trimpot can be adjusted to obtain a calibrated power readout at any desired wavelength within its calibration range.[3]

Silicon photodiodes have a higher quantum efficiency than Germanium photodiodes which is the primary characteristic making silicon sensor heads suited to low light levels, collimated beams and high density sources. Germanium sensor heads are longer wavelength devices which are useful for high power sources such as laser diodes. Germanium sensors are more economical than InGaAs sensors, but InGaAs sensors have better noise characteristics to allow measurements at lower power levels. When using OPMs, the sensor head should be selected such that each wavelength at which the system operates will be tested.

OPMs are compatible with most systems due to the development of connector adapters. When an OPM is purchased, generally these adapters are included for the connector that has been specified.

4. OPTICAL TEST SETS

Power meters and optical sources are often sold as a test set, the typical contents of which are listed in Table 11-3.

TABLE 11-3. Typical Test Set Contents.

- Optical Power Meter
- Sensor Heads
- Optical Source
- Connector Adapters
- Launch Cables & Splice Bushings
- Microscope
- Connector Cleaning Wipes
- Attenuators
- Spares and Supplies
- Case

Ref: 1

Launch cables are used to establish consistent test conditions which allows repeatability and enhances the accuracy of the measurement. Splice bushings or sleeves are used to connect the launch cable to the cable under test. For some tests, a receive cable and splice bushing is used at the far end of the cable under test. Microscopes and cleaning supplies are included to allow the check-out of connector end-faces before testing. Generally, all connectors need to be cleaned prior to testing to insure that excess loss due to dirt and lint are minimized.

II. MEASUREMENT UNITS

OPMs measure power in units of watts (milliwatts or microwatts) or in dBm or dB . If the power is measured in watts, the attenuation in decibels is given by the following equation:

$$dB = 10 \log (P_{out}/P_{in}) \qquad (11\text{-}1)$$

where P_{out} is the power measured by the OPM, and P_{in} is the reference launch power. P_{out} and P_{in} must be in the same units. Decibels are the logarithmic transformation of the basic power unit, watt, and submultiples of watts. These units are useful in compressing power measurement data that has a wide dynamic range. If the reference power launched into the cable is 1mW, equation 11-1 becomes:

$$dBm = 10 \log (P_{out}/1mW) \qquad (11\text{-}2)$$

If the reference power launched into the cable is 1μW, equation 11-1 becomes:

$$dB\mu = 10 \log (P_{out}/1\mu m) \qquad (11\text{-}3)$$

where the subscripts m and μ in these equations denote the reference power level. To obtain the power loss in units of power, P_{in} is subtracted from P_{out} regardless of what units of power were used to measured the power.[2]

III. OPERATION

There are various types of tests that need to be performed in the check-out of an optical system. Table 11-4 lists these common tests and the equipment that can be used.

TABLE 11-4. Equipment Associated with Check-out Tests.

Test Parameter	Required Test Equipment OS	OPM	OTDR	Attenuator	Launch Cable
Emitter Output Power (Launch Power)		X			X
Fiber Loss	X	X	X		X
Connector Loss	X	X	X[1]		X
Receiver Power		X			
Receiver Sensitivity		X		X	
Fault Location			X		

[1] Not Recommended

As can be seen from the table, the OPM when used with an optical source, is a very versatile testing device which has the capability to perform all of the common tests with the exception of fault location where an Optical Time Domain Reflectometer (OTDR) is needed.

To measure emitter output power, an OPM and a launch cable are needed. To measure fiber loss, an OPM and OS or an OTDR is needed. For measuring connector loss, an OPM with the system emitter or an OTDR alone can be used. To measure receiver power, an OPM is needed. For measuring receiver sensitivity, an OPM can be used with an attenuator to simulate cable loss and the bit error rate can be determined as a function of the received optical power.[3] The primary use of an OTDR is to locate faults in an installed cable system. If a cable has been severed or damaged, an OTDR can be used to determine the location of the failure. The procedure for performing each test with an OPM and OS will now be discussed.

1. FIBER ATTENUATION MEASUREMENT BY CUTBACK METHOD

This test method is the standard way of measuring attenuation of unterminated optical fibers and is used by all fiber manufacturers. A source is used to illuminate the fiber under test and an optical power meter is used at the far end as shown in Fig. 11-1. For multimode fibers the launch conditions cannot be consistent unless the fiber is modal conditioned using a mode scrambler. The amount of light at the far end is measured and then the fiber is cut back to a point just past the output end of the mode scrambler. Then, the amount of light at this point is measured. The difference in power measured at one end and the other divided by the length of the fiber before cutback gives the loss per unit length.

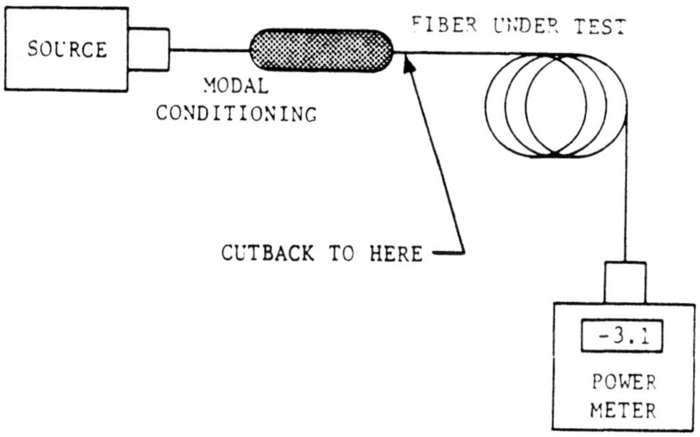

Figure 11-1. Fiber attenuation measurement by cutback method.

2. CONNECTOR INSERTION LOSS TEST

Most connector manufacturers utilize Electronic Industries Association test EIA-455-34 to test the standard insertion loss of connectors. The test set-up is shown in Fig. 11-2. A source is connected to a launch fiber which is modal conditioned, and is then attached to a test fiber which is connected to an OPM. The length of the test fiber is relatively unimportant and can therefore be a conveniently short length. The total amount of power is measured. The fiber is then broken just after the mode scrambler and a connector is installed and mated. The power is measured again. The two power measurements are then used to determine the loss associated with the insertion of the connector in the link.

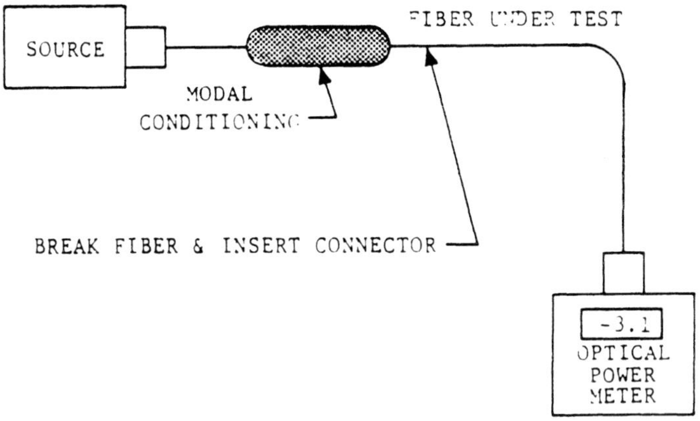

Figure 11-2. Connector insertion loss test.

3. BASIC SINGLE END CONNECTORIZED CABLE LOSS TEST

This test procedure combines the fiber attenuation test by the cut-back method and the basic connector insertion loss test to test single end connectorized cables. Light of a known power level is launched through a launch cable into the cable under test. At the far end of the cable under test, an OPM is connected and the amount of light received at the OPM is measured and recorded, as shown in Fig. 11-3. The launch cable should contain the same type of fiber that is in the cable under test and the connectors must be compatible to eliminate excess loss due to extrinsic factors. The launch cable is used so the launch conditions can be consistently repeated when performing this test. The launch cable also simulates the way that the cable will actually be connected when used. If a launch cable is used, all cables are tested under similar test conditions which allows for comparison of results.

This test procedure is the same procedure that is described in the Electronic Industries Association (EIA) document EIA-455-171. The method simulates a system where the cable goes from the source to the detector which is the primary application of fiber optic systems. The insertion loss of the connector at the launch end and the cable are measured with this test. The connector insertion loss value and the loss of the cable under test is obtained by subtracting the known launch power from the measured power.

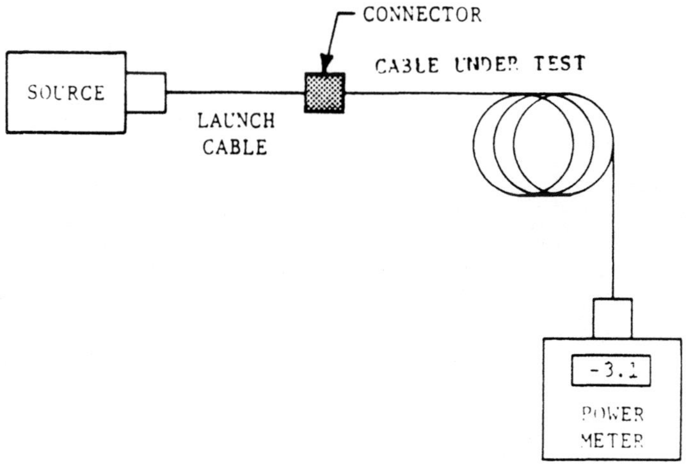

Figure 11-3. Basic fiber optic cable loss test.

4. DOUBLE-ENDED CABLE LOSS TEST

This test is used when the cable under test is connectorized at both ends as may be the case for long-haul applications. The set-up for this test is depicted in Fig. 11-4. This test differs from the basic single end connectorized cable test in that a receive cable is attached to the cable under test before the OPM is attached. The receive cable may either have the same fiber parameters as the cable under test or it may be a larger sized fiber. A larger fiber is often used in testing single mode systems and is commonly called a "bucket" cable. Before the cable is tested, the launch and receive cables are connected and the power meter is set to give zero dB referenced to a microwatt out of the launch cable. The reading taken from the OPM reflects the insertion loss of the cable under test and both connectors. The loss value of the cable under test and both connectors is obtained by subtracting the known launch power (which is generally preset to zero) from the power measured at the receive cable.

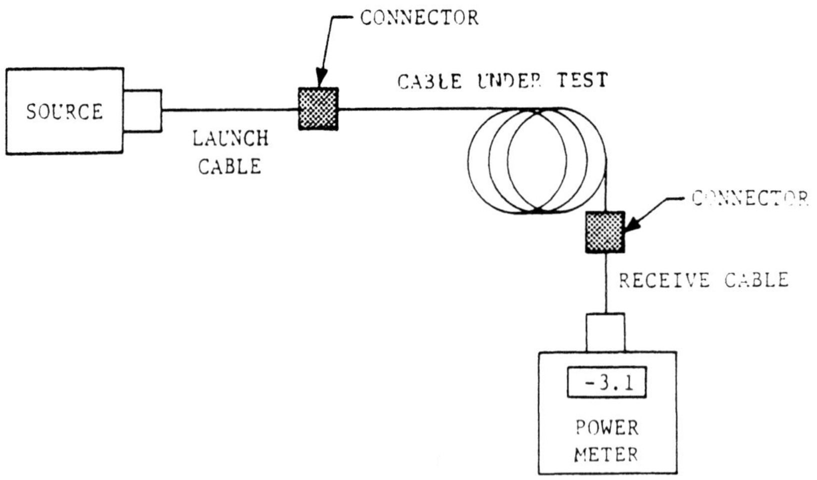

Figure 11-4. Double-ended cable loss test.

5. LOOPBACK CABLE LOSS TEST

The loopback test method is a common method of testing trunk cables from one end. This simplifies testing since the OPM and OS are located at one end of the installation and therefore only one technician is required which eliminates the need for communications between cable ends. However, each end of the cable under test must be visited to complete the set-up. To perform this test, two fibers in the cable under test are used. At one end of the cable, the OS is attached to one fiber through a launch cable and the OPM is attached to the other fiber through a receive cable. At the far end of the cable under test, a jumper cable is used to connect the two fibers together. This effectively makes a continuous loop from the OS through the two fibers and jumper cable to the OPM. Therefore, the loss measured by the OPM will be that of two fibers. Care must be taken when assuming the loss of each fiber is one-half the loss measured for both fibers. It is quite possible that one fiber can have a very high loss while the other has a very low loss. Thus, one-half of the total loss may provide an acceptable value when calculated, but the actual values can be unacceptable. If a cable pair does not meet the system requirements, each individual fiber should be tested in a one way test with a power meter at one end and optical source at the other end. This will isolate the faulty fiber.

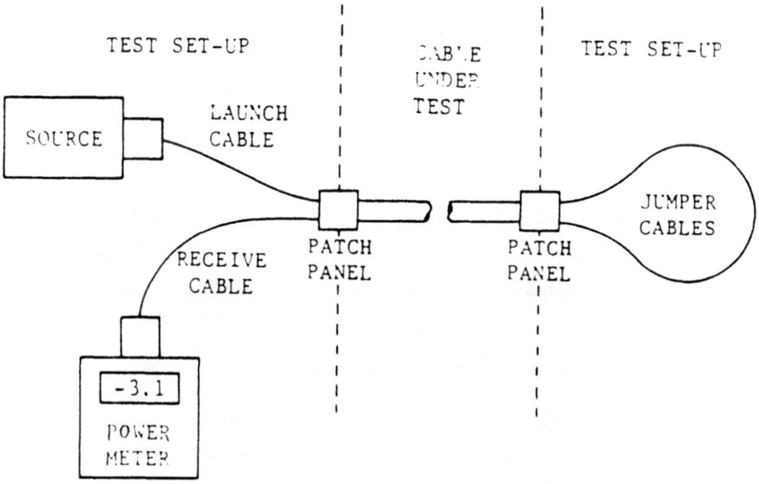

Figure 11-5. Loopback cable loss test.

IV. PREDICTED PRODUCTION LOT FAILURE RATE AND MAINTENANCE REQUIREMENTS

Survey questionnaires were sent out to all known manufacturers of optical power meters. Responses were received on thirteen models of power meters.[5] Although actual failure rate data was not available, the predicted production lot failure rate ranged from 2.5% to 10% with a mean of 6.08%. The number of OPMs used to calculate this data was 2250. The age of the data was 6 months for one model and one year for the remaining twelve models. The elements expected to fail the most were given as follows:

- LCD readout
- Keyboard membrane
- TE cooled Ge detector
- Detectors

The manufacturers recommended that preventive maintenance be performed on a yearly basis. The maintenance actions required include cleaning and recalibration.

V. ADDITIONAL FEATURES

Some OPMs have calibrated lock-in capability which permits autophasing, autoranging, and autozeroing. They can also be operated without an external synchronization signal which allows power measurements over a long cable span without running another cable to supply an external synchronization signal.

Another feature of power meters that should be considered is the inclusion of an IEEE-488 and IEC 625-1 General Purpose Interface Bus (GPIB). These buses are designed to perform automated testing by computer.

VI. SUMMARY

OPMs and OSs are widely used throughout the industry in testing fiber optic systems. Most of the devices are very diverse, hand-held units which are available with many different sensor heads for various applications. When used in conjunction with an optical source, an optical power meter can be used for all standard installation tests.

Chapter 12. Optical Time Domain Reflectometers

The optical time domain reflectometer (OTDR) is a piece of test equipment which is used to perform distance and loss measurements on optical fiber and associated components. Two primary features differentiate the OTDR from the optical power meter/optical source test set. These are the ability to perform all tests from one end of a fiber and the ability to identify the location of a fault. However, these desirable features are obtained through highly sophisticated and complex circuitry and significant bulk.

The operation of the OTDR, interpreting its trace and the accuracy considerations related to its use are discussed in this chapter.

I. DESCRIPTION

1. PHYSICAL CHARACTERISTICS

OTDRs are designed using state-of-the-art optical and electronics technologies. This instrument is constructed for both field and laboratory use with different units able to tolerate various degrees of mechanical and environmental abuse. The dimensions and weight of a typical ruggedized OTDR for field use are listed in Table 12-1.

TABLE 12-1. Typical Optical Time Domain Reflectometer Physical Characteristics.

Length (inches)	16.0
Width (inches)	16.0
Height (inches)	11.0
Weight (pounds)	38.0

2. OPERATING PHYSICS

Briefly, the OTDR operates by sending a pulse of light down a fiber and monitors the light reflected by Rayleigh scattering and Fresnel reflection. The intensity of the reflected light controls the vertical displacement of a

Cathode Ray Tube (CRT) trace while the time associated with the distance the pulse has traveled down the fiber controls the horizontal displacement. The end result is a trace of power versus distance for the length of the fiber measured. A more detailed description follows.

The basic functional configuration of an OTDR is shown in Fig. 12-1. During operation, a pulse of light is launched into the fiber through the beamsplitter. As the pulse propagates down the fiber, light is reflected back towards the OTDR by the mechanisms of Rayleigh scattering and Fresnel reflection.

Rayleigh scattering (backscatter) occurs continuously and is caused by composition and density variations which are "frozen" into the fiber during the manufacturing process. This scattering is dependent on the wavelength and is proportional to λ^{-4}. Therefore, as the wavelength increases, there is less backscattering which results in a reduction in the optical power returning to the OTDR.

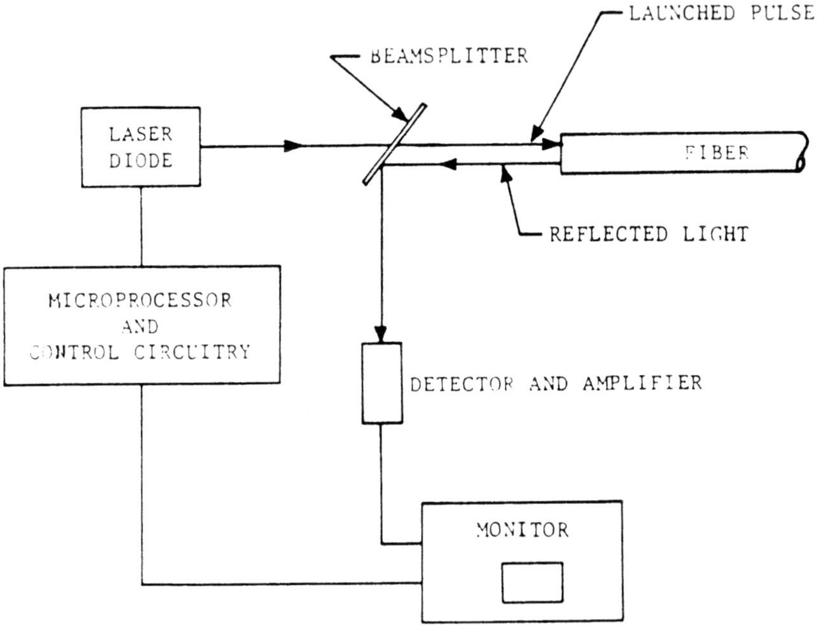

Figure 12-1. Basic OTDR functional configuration.

Fresnel reflection occurs at an interface when light is transmitted through materials with different indices of reflection. Some examples include the glass-air interface at breaks in a fiber, at the end of the fiber or at the glass-index matching gel interface in mechanical splices and connectors. Index matching gels are used to reduce reflections at interfaces by eliminating the air gap and replacing it with a gel having an index which more closely matches that of the transmitting medium. Although these gels come close, they do not precisely match the transmission medium index. The amount of Fresnel reflection can be determined by calculating the Fresnel reflection coefficient, R, using equation 12-1.

$$R = \left[\frac{N_t - N_i}{N_t + N_i}\right]^2 \qquad \text{Equation 12-1}$$

Where, R is the Fresnel reflection coefficient
N_t is the index of refraction of the transmitting medium
N_i is the index of refraction of the receiving medium

This equation is valid for light impinging normally to the fiber surface.

As an example, for N_t = 1.458 (fused silica) and N_i = 1.0003 (air),

$$R = \left[\frac{N_t - N_i}{N_t + N_i}\right]^2 = \left[\frac{1.458 - 1.0003}{1.458 + 1.0003}\right]^2 = 0.035$$

This example shows that a silica/air interface will reflect approximately 3.5% of the incident light. In a transmission system this constitutes a loss which can be reduced by using an index matching gel. The index of refraction for typical optical fibers ranges from 1.400 to 1.599. This corresponds to Fresnel reflections ranging from 2.7% to 5.3%, respectively.

The reflected and backscattered light is received by the OTDR and reflected by the beamsplitter onto the photodetector as shown in Fig. 12-1. The electrical output of the photodetector is processed and used to control the vertical displacement of the power versus distance trace on the monitor. This displacement is proportional to the optical intensity of the returned light.

Simultaneously, the elapsed time required for the reflected light to return to the OTDR is continuously measured. From this data, length calculations are performed using the formula given in equation 12-2.

$$L = \frac{ct}{2n} \qquad \text{Equation 12-2}$$

where:
- L = calculated distance
- c = speed of light in a vacuum (3×10^8 m/s)
- t = round trip travel time
- n = index of refraction of the core. (This value is entered into the OTDR before the measurements are taken or a default value stored in the instrument is used).

This data is processed and used to control the horizontal displacement of the power versus distance trace on the monitor. This corresponds to the distance light travels down the fiber.

The end result of monitoring the reflected and backscattered light and calculating the distance propagated is a trace showing the magnitude of the detected optical power as a function of the distance the light traveled. A typical trace of power versus length is shown in Fig. 12-2. Only these two parameters are plotted against each other on an OTDR trace.

From this trace, the fiber attenuation, the location of splices, connectors or breaks, and the losses incurred at splices, connectors and faults can be determined.

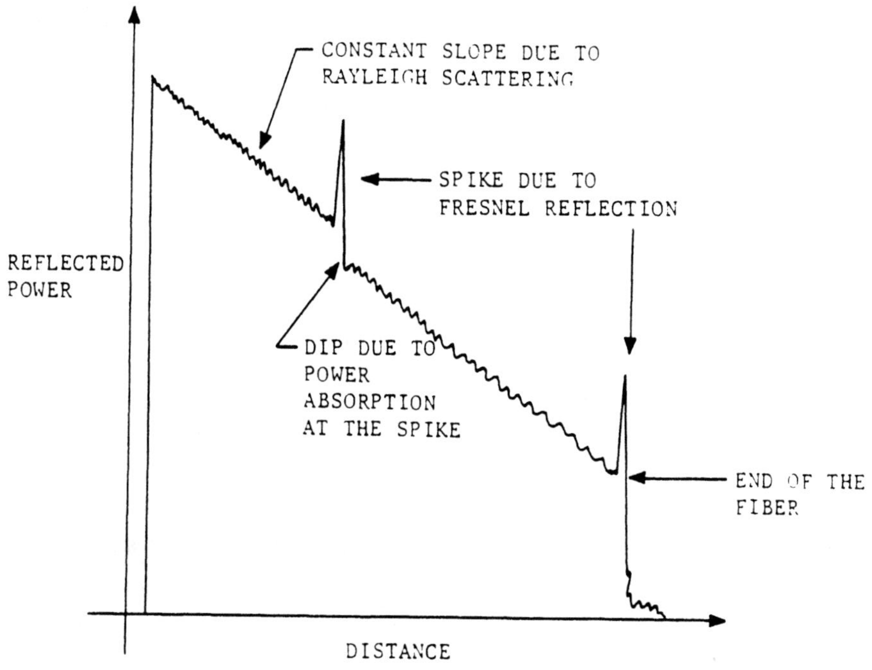

Figure 12-2. Reflected power versus distance trace.

II. IMPACT ON FIBER OPTIC SYSTEM TESTING

An OTDR is primarily used by fiber manufacturers, installation teams, and system troubleshooting teams. Fiber manufacturers use an OTDR to quantify the characteristics of fiber prior to shipment. This information is provided to cable manufacturers who compare these traces to traces of the same characteristics taken after the fiber has been cabled. The cable manufacturers can then determine the impact that the cabling process has on the optical properties of the fiber.

Installation teams use either OTDRs or Optical Power Meters (OPMs) to measure overall link attenuation, splice loss and connector loss to ensure the losses are within the specified limits. One advantage the OTDR has over the

OPM is the ability to monitor splice loss during alignment of the fibers. This results in an optimum splice loss. Traces of faults and fiber loss can then be taken and kept as permanent records characterizing the system and used for analysis at a later date should a failure occur or system testing be done.

Troubleshooting teams use the OTDR in locating faults. An OTDR gives a very accurate (within millimeters for the latest OTDRs) definition of a fault's location.

Whether or not an organization should buy an OTDR for their use is dependent upon many things. The primary consideration is what the organization is responsible for. Fiber manufacturers, installation teams and trouble-shooting teams use an OTDR regularly. These groups may purchase or lease the OTDR. OTDRs are fairly expensive ($25,000 and up) and thus, become a major capital expenditure for most organizations. Making such an expenditure is not unusual for most companies but it is unusual if the OTDR will need replacing the next year with an updated version. That is the dilemma most OTDR users are facing today, the fact that the technology is improving rapidly in the OTDR arena and every few months an improved model is put on the market.

End users of the installed fiber optic system do not need to use an OTDR for routine scheduled maintenance in most cases. The time when an end user would need an OTDR is when a system failure occurred in order to determine the location of the fault. Some users keep an OTDR in the emergency repair kit while others don't. It largely depends on the criticality of the users system, the location of the system, budget considerations and the availability of other equipment dealers as to whether or not a user needs to keep an OTDR on hand.

If the system is extremely critical and downtime must be kept to an absolute minimum, then an OTDR must be available. An OTDR alone is not enough, there must be trained personnel to operate the equipment and perform the repairs. Another consideration is the location of the system. If the system is located near a large metropolitan area where equipment rental firms carry OTDRs for rental, then it may not be necessary to keep an OTDR available. However, if the system is in a remote location such as some military applications or remote facilities, then an OTDR should be kept on hand provided the agency can afford it.

Each user must carefully weigh the trade-offs to determine whether or not to keep an OTDR on hand for emergency repairs. Once the installation has been performed, the OTDR generally will not be used until the system fails. Thus, the OTDR could be sitting in a store room for many years before it would be used again.

In the long term, an OTDR saves system users money and down time by allowing cables to be repaired rather than replaced. This is a result of the OTDR's ability to identify fault locations which can then be repaired. No other instrument provides this capability. The impact this has on maintenance and logistics is that less spare cable needs to be considered for inventory because only short segments of cable will be needed to make repairs rather than replacing cable runs. Trained personnel must be available to perform repair procedures using the OTDR. As will be detailed later in this chapter, a great deal of skill is needed to interpret trace anomalies when there are problems with the system.

III. OPERATION

Operation of the OTDR requires the ability to interpret the characteristic signatures which are indicated on the trace. The identification and explanation of these signatures is discussed below.

The easiest parameter to measure is fiber attenuation which is the slope of the trace which corresponds to the segment of fiber for which this measurement is wanted. Fibers which have no splices, in-line connectors or faults, will have a continuous slope of constant value, baring any significant variations in the backscatter coefficient as discussed in Section IV.

In order to perform measurements with an OTDR, one end of the fiber to be tested must be available. The fiber end is connected to the OTDR output port via a connector. The output port can be fitted with many connector adapters which makes the OTDR compatible with the different hardware used by most systems. This includes bare fiber adapters, that is, unterminated fiber adapters. The pulse width is then selected. A long pulse width is used for long distance measurements and a short pulse width is used for short distance measurements. The information is displayed on the monitor or it may be plotted on a strip chart recorder or it can be stored on computer diskettes which can be used at a later time for analysis. The OTDR can have one or a variety of these capabilities. That is dependent upon the model. Many OTDRs have a General Purpose Interface Bus (GPIB) port which allows the data to be inputted directly to a computer.

Operating an OTDR is fairly simple, it is the interpretation of the data that requires considerable knowledge, especially if anomalies in the trace appear. Some of the more common anomalies will be discussed in Section IV.

The signature for the discontinuity caused by a mechanical splice and connector is similar as shown in Fig. 12-3. The large spike is caused by Fresnel reflection at the interface followed by a downward displacement of the trace. This displacement is due to power absorption at the discontinuity. The attenuation of the discontinuity is the difference in vertical displacement measured at the steady state regions immediately before and after the splice as indicated in Fig. 12-3.

A fusion splice doesn't exhibit a Fresnel reflection spike since the fibers are actually melted together which eliminates the interface. It does, however, exhibit a dip in the trace due to attenuation at the splice. The attenuation of the splice is measured in the same manner as the mechanical splice.

Distance measurements can be read directly off of the OTDR horizontal scale. Most OTDRs have a feature in which a cursor is placed on a particular point in the trace and the distance from the end of the fiber closest to the OTDR to that point is indicated in a numeric readout.

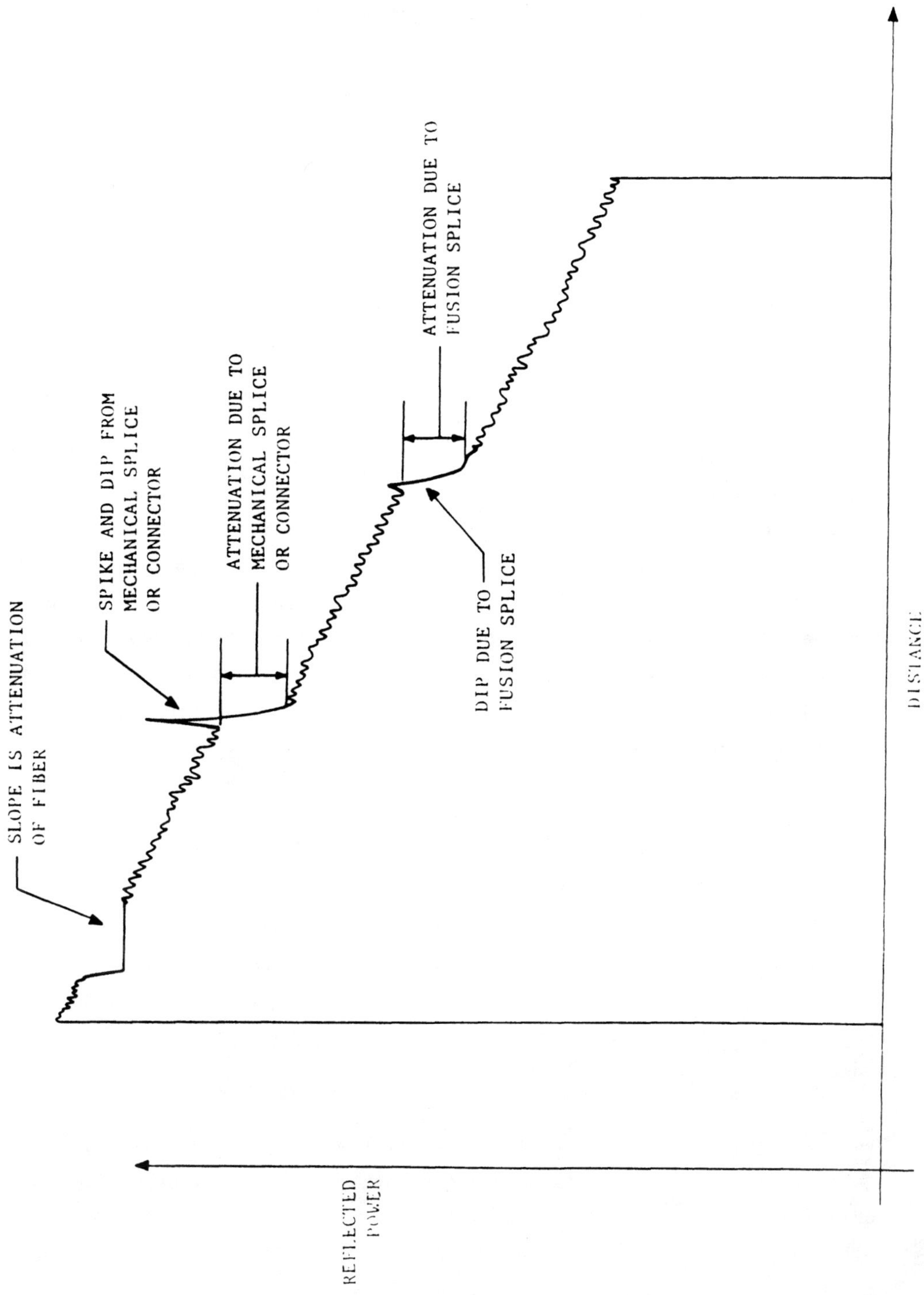

Figure 12-3. Typical OTDR trace.

IV. INACCURACIES AND ANOMALIES

A number of mechanisms contribute to inaccuracies in OTDR measurements. The following paragraphs will discuss some of the major mechanisms and methods to reduce their effect.

Fiber Rayleigh scattering is characterized by a measure called the backscatter coefficient. This parameter is an indicator of how much backscatter a fiber will exhibit. This coefficient must be known when testing concatenated fiber since it can impact how the OTDR is used.

An inaccurate loss value will result if measurements are made on a fiber with a low backscatter coefficient which has been spliced to a fiber with a high backscatter coefficient. If light is transmitted from the fiber having the low backscatter coefficient to the fiber having the high backscatter coefficient, the resulting trace of this splice will show an apparent "gain" as depicted in Fig. 12-4. The gain is seen as an upward displacement of the trace at the splice. This result is due to the fact that the OTDR only measures the magnitude of the backscattered light. Since the second fiber has a higher backscatter coefficient, it will reflect more light than the first fiber. This will appear as an increase in the trace optical power level. The result is an inaccurate loss measurement since the true loss is "buried" in the apparent gain displayed by the trace. There can not be a gain, only a loss.

Conversely, if light is transmitted from the high backscatter fiber to the low backscatter fiber, an exaggerated loss will be displayed since the second fiber will reflect less light compared to the first. Likewise, the measurement will be inaccurate since the true loss is "buried" in the exaggerated trace.

These are testability issues since inaccuracies result. However, this should not impact system performance since signal propagation does not depend upon backscatter.

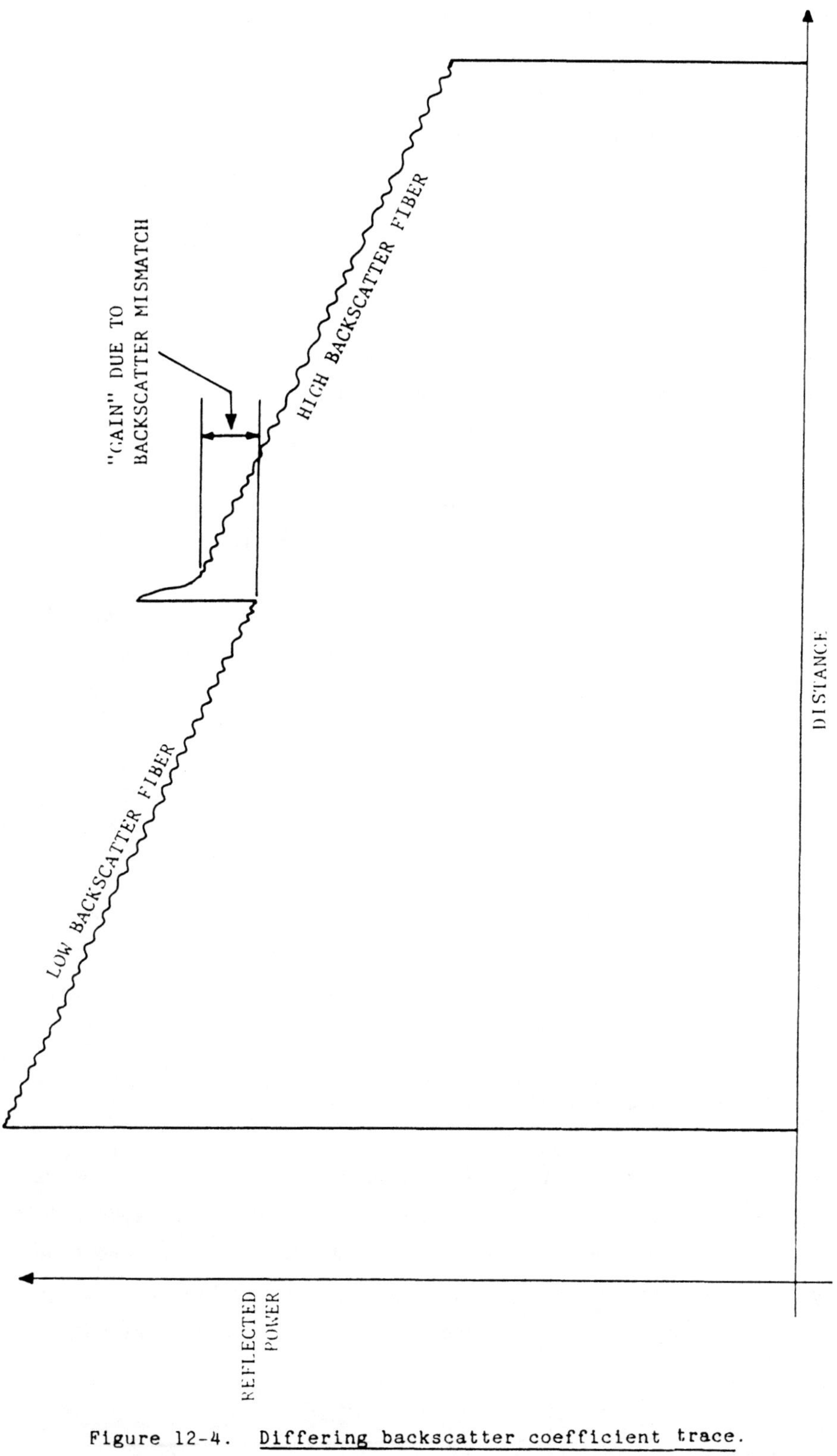

Figure 12-4. Differing backscatter coefficient trace.

If fibers with different backscatter coefficients have been concatenated, the splice loss can be determined by taking measurements from both ends of the fiber, that is, in both directions, and averaging the loss across the splice. Of course, this eliminates the single-ended testing convenience of the OTDR which is one of its major advantages. To avoid this problem altogether, only fibers with equal backscatter coefficients should be concatenated. Consult the manufacturer before purchasing the fiber in order to ensure the backscatter coefficients are equal.

Concatenated fibers with numerical aperture (NA) mismatch will result in inaccurate OTDR readings. (For detailed discussion on NA mismatch see Chapter 1.) This error is not directional and therefore can not be eliminated by changing the optical pulse launch direction. The inability to eliminate the error stems from the fact that the OTDR estimates round-trip loss. This means the loss due to NA mismatch will contribute to the overall loss regardless of the launch direction. Therefore, if the launch pulse is in the direction of system transmission (small NA to large NA), the NA mismatch will show up as a higher loss than what the system actually sees. For example, a 2.5% change in NA will introduce a 0.1 dB error in the OTDR reading.[1] This error is not a problem when an optical source/optical power meter test set is used. The test set will give an accurate reading in both directions. One direction will have a loss due to the NA mismatch and the opposite direction will not.

The need to match the backscatter coefficient and NA places additional constraints on spare fiber kept in inventory. Not only must the fiber have the same physical characteristics such as core and cladding diameters, but these optical parameters must also match. Otherwise the testability of the system will be degraded as a result of the increased maintenance time to perform two-ended measurements and the inaccuracies incurred.

The variation of index of refraction which exists in all fibers will also affect OTDR accuracy. This variation causes a typical inaccuracy of 0.1% (± 1 meter per kilometer measured) and it exists regardless of the type of OTDR used. For example, a fault located at 20 km would have a distance inaccuracy of ±20 meters.[2] The accuracy of the distance measurement is degraded by these variations since this parameter is used to calculate the distance as indicated in equation 12-2. The greater the variation, the more error there will be in identifying the location of a fault which is another one of the OTDR's primary advantages.

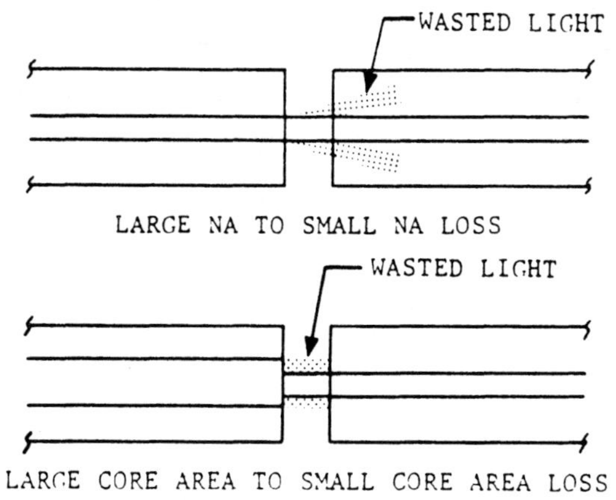

Figure 12-5. Numerical aperture and core size losses.

A spool of fiber wound with uneven tension will exhibit unequal attenuation per unit length due to microbend losses. These are losses caused by light leaking into the cladding due to microscopic sized bends in the fiber. This type of loss is only a problem when testing a fiber which is still on the spool. The problem associated with this is that these are temporary losses which can be removed by rewinding the fiber under a constant tension.

A major source of inaccuracy which affects measurements made by the OTDR is the existence of the "dead zone". The dead zone occurs in the segment of fiber immediately following the output port. The OTDR is unable to process the light reflected from this region because the optical power level is sufficient to saturate the detector. The length of the dead zone is dependent upon the launched pulse width, the receiver amplifier bandwidth, and the time required for the saturated detector to settle out.

The primary problem caused by the dead zone is the inability to perform near-end measurements. This problem occurs when the light reflected at a nearby discontinuity is sufficient to saturate the detector or arrive at the detector before it has settled out. This effect makes locating and measuring faults in the fiber near the OTDR impossible as illustrated in Fig. 12-6. This problem may be minimized by using a short pulse width.

Another solution for the problem of near-end resolution is to add a 1 km or shorter reel of launch fiber between the OTDR and the fiber to be tested. This effectively moves the beginning of the fiber under test out of the region blanked by the dead zone. However, this reduces the range of the OTDR proportionately. An example of what this arrangement would look like on the OTDR trace is shown in Fig. 12-7. Of course, if less fiber can be used and the dead zone eliminated, this length should be used.

Another problem experienced by the OTDR is poor fault resolution which occurs when the Fresnel reflection from one fault obscures the Fresnel reflection from an adjacent fault. This is caused by the faults being too close together to be individually resolved with the given repetition rate and pulse width of the OTDR. Fig. 12-8 shows a trace illustrating this problem. This is especially troublesome where splices, connectors, and couplers are closely spaced such as in aircraft, ships and where network users are in mutual close proximity. This OTDR problem can also be minimized by using a short pulse width.

The best means to prevent these two problems from occurring is to use an OTDR which is suited to the application. OTDRs can be purchased in a wide range of dead zones, distance ranges, and bandwidths which allows selection tailored to a particular application. For example, a communications link typically involves long fiber spans with devices that are spaced far apart.

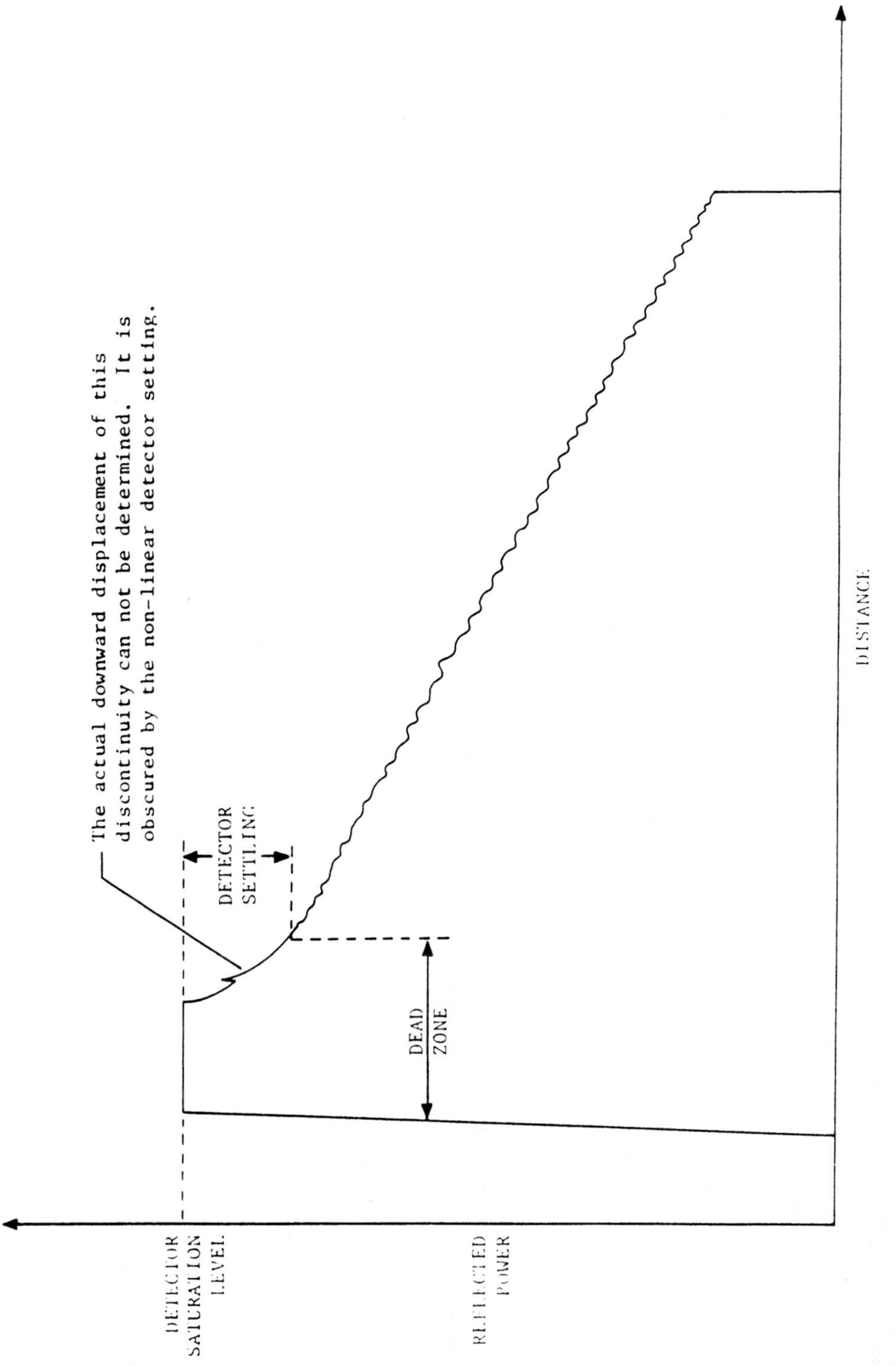

Figure 12-6. Detector saturation.

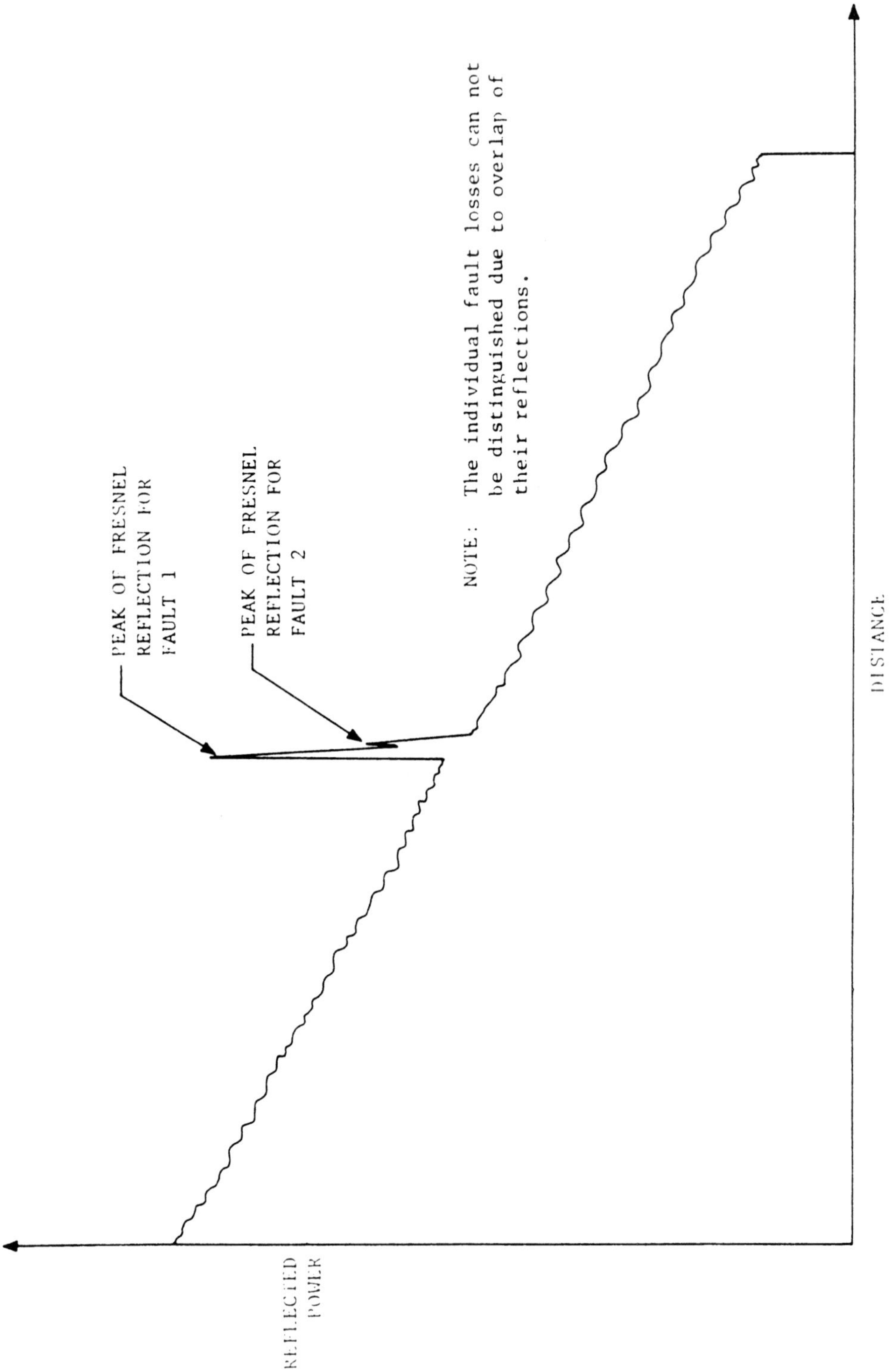

Figure 12-7. Dead zone obscuring Fresnel reflection of a fault.

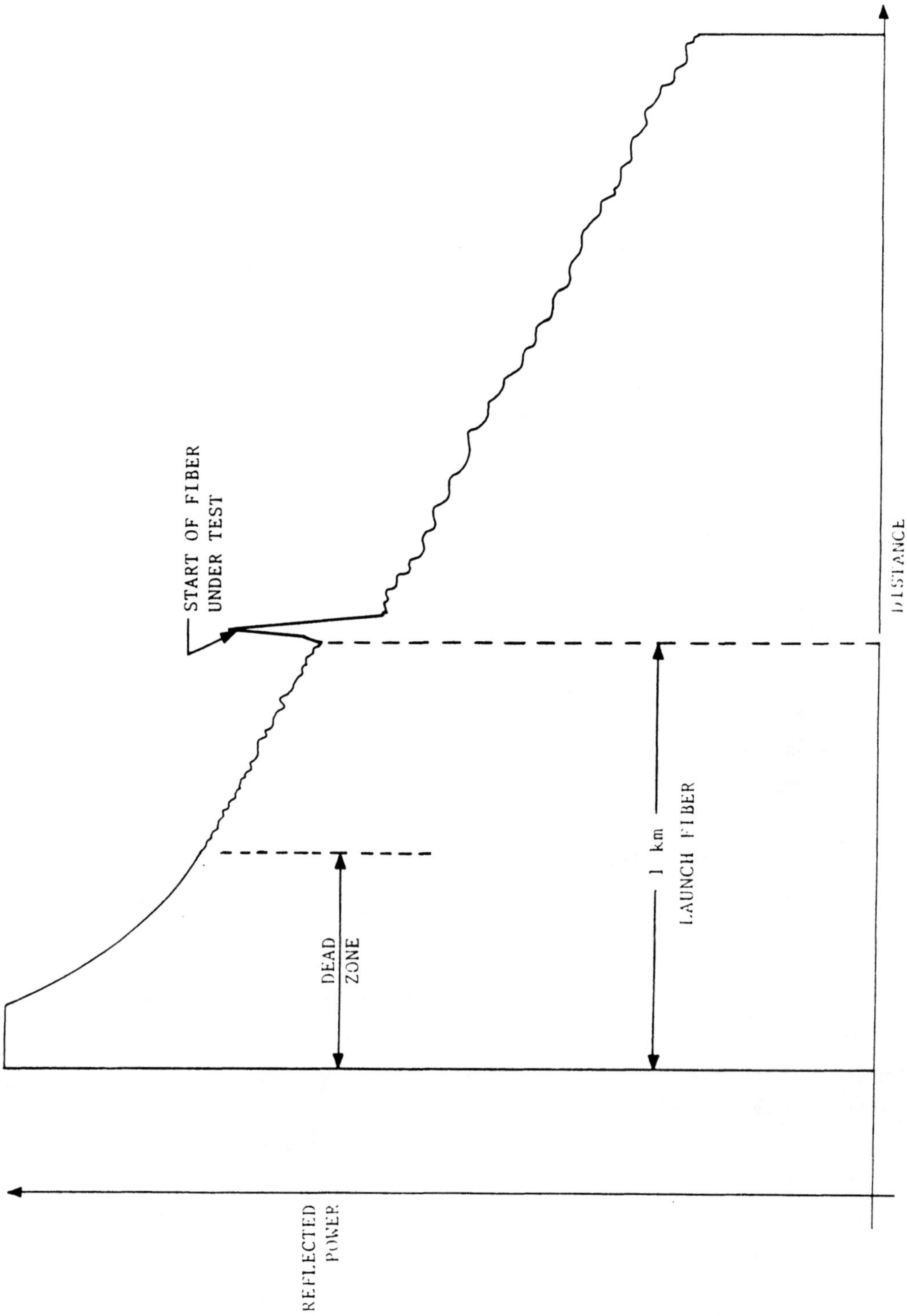

Figure 12-8. Use of 1km fiber reel prior to test fiber.

Therefore, an OTDR with a longer dead zone and a greater range would be appropriate for this system. However, a long dead zone is not a desirable feature here, only tolerable. On the other hand, a shipboard or aircraft system would involve shorter spans with connections closely spaced. An OTDR with a shorter distance range and a shorter dead zone would be better suited for these applications. Additional OTDR features such as selectable pulse widths will add flexibility to the instrument's use. The importance of this feature is related to the area under the pulse being proportional to the energy in the pulse. The wider the pulse, the greater the range it will have. However, fault resolution is sacrificed. Conversely, the narrower the pulse, the shorter the range and the better the fault resolution will be. There have been major advances in 1987 in the resolution of OTDRs. There are now OTDRs available with resolutions in the millimeter range. This type of resolution provides millimeter loss measurement accuracy and allows the determination of the exact location of features in any fiber optic network, whether it be simple or complex. OTDRs are being manufactured that are so precise that during a cable pull test, the fiber can be monitored and an elongation of only 1mm can be detected.

Another consideration for short haul systems is that the system's source is usually an LED. OTDRs, however, use laser diodes as a source which may not give an accurate attenuation measurement for the system due to the differences in mode distribution. In many instances, the difference in measured attenuation is not significant. If the OTDR will only be used to measure attenuation and not fault location, then a transmission-type power meter equipped with an LED should be used rather than an OTDR.

A final consideration in the use of OTDRs is the fact that the wavelength specified on the source of the OTDR may not be the actual wavelength transmitted. Many sources can vary by as much as 30nm to 40nm which means a 1300nm source can actually range from 1260nm to 1340nm. Since the attenuation of a fiber varies with the wavelength of the light transmitted as shown in Fig. 12-9, this effect can cause significant errors when measuring the attenuation of long lengths of fiber. To reduce this problem, the OTDR source should have a wavelength tolerance which is close to that of the system's source. Also, the same OTDR source should be used to test the entire system instead of performing measurements with different OTDRs unless the actual operating wavelength of the OTDRs have been determined to be the same. Observance of this precaution will avoid confusion in subsequent measurements which may be taken to determine the rate of degradation.

V. PREDICTF) PRODUCTION LOT FAILURE RATE AND MAINTENANCE REQUIREMENTS

Survey questionnaires were sent to all known manufacturers of OTDRs with responses received on six models.[3,4,5] Although actual field failure rate data was not available, the predicted production lot failure rates ranged from 10 to 15%[3,5]. A total of 123 OTDRs were used to calculate this data.[3,5] The age of the data varied as follows: one model used 1 month old data, 18 models used 3 year old data, 49 models used 2.5 year old data and 55 models used 1 year old data.[3,5] The components which were expected to fail the most frequently were given as follows:[3,5]

- laser
- CRT
- coupler
- detector

If any of these components fail, with the exception of the CRT, the OTDR would be inoperative. If only the CRT failed, the OTDR could probably be used as long as either a GPIB, computer diskette or plotter was a feature of the instrument. However, it would be operating in a degraded mode.

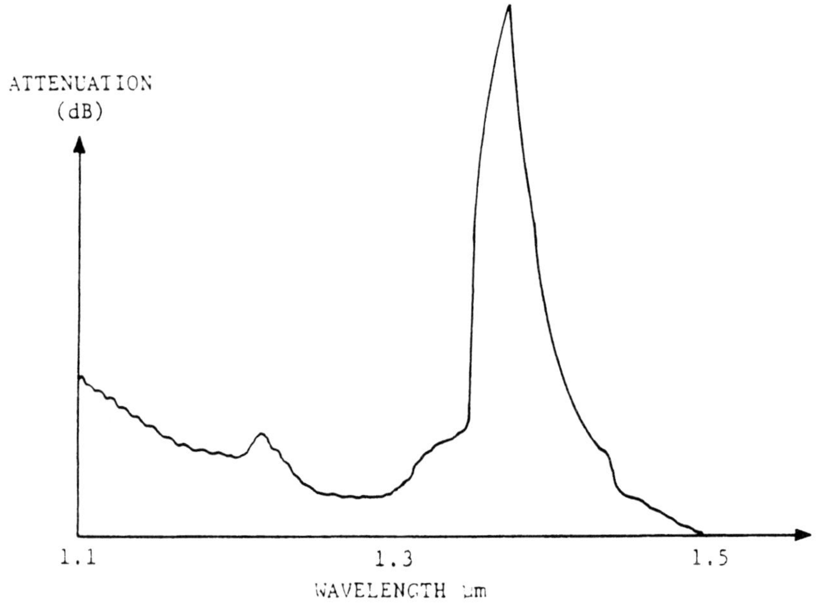

Figure 12-9. Attenuation vs. wavelength plot.

VI. MAINTENANCE AND OPERATIONAL CONSIDERATIONS

The ODTR is a delicate instrument and should not be dropped or severely jolted. This may misalign the optical elements. If the OTDR still functions after being dropped, the distance measurement would probably still be accurate since it is a digital counting function. As long as a detectable light level is received, the solid state electronics performing the calculations should function. The attenuation measurements, however, may be inaccurate as a result of optical misal.gnment and should be checked using a transmission-type power meter or another OTDR of known integrity.

Use the OTDR only within its specified temperature range since the laser diode source is very sensitive to temperature. As the temperature of the laser changes, the operating wavelength will shift. The average operating temperature range specified for six OTDRs is -8.3°C to 71°C.[3,4,5] The specified storage temperature range for five OTDRs is -35°C to 71°C.[3,4,5]

Personnel operating the OTDR must always avo d eye exposure to the infrared or visible light emitted from the OTDR. A safety interlock is usually an integral part of the output port which automatically extinguishes the laser output if an adapter is not connected to the OTDR.

All manufacturers which completed survey questionnaires recommended that preventive maintenance be performed on a yearly basis. The maintenance actions required calibration on an annual basis. One manufacturer recommended that the connector of the OTDR be cleaned twice daily during use. The mean time to repair was reported by the responding manufacturers to be 5.6 hours for one model[4] and 3 days for 3 models.[5] A loaner OTDR may be supplied to the user during repairs based on the manufacturer's policy.

VII. LOGISTICS CONSIDERATIONS

The skill level required to operate and interpret an OTDR is significantly higher than that required to operate an optical power meter/ optical source test set. However, only one operator is needed and that person need only measure from one end of the fiber. OTDRs are designed in one of two ways: they operate at only one wavelength, or they operate at switchable wavelengths using modular inserts for each wavelength. If the modular inserts cannot be stored in the case with the OTDR then they must be stored in a convenient, easily accessible location which impacts logistics. Most OTDRs have strip chart recorders and/or cameras to photograph the trace to provide a hard copy of any measurement made. This feature will also impact stocked items due to the need for replenishing paper and/or film.

VIII. SUMMARY

The OTDR provides a method of performing distance and attenuation measurements on optical fibers. The major advantages of an OTDR is the ability to measure distance to and between splices, connectors, faults and other points of interest. Splice alignments can be monitored for optimum loss values. Also, it allows measurements to be made from one end of the fiber. This reduces maintenance time and/or maintenance personnel.

It is extremely important that the application in which it will be used is well defined before purchasing an OTDR. This sophisticated instrument has its limitations just like any other piece of test equipment. Also, upon close evaluation, it may be determined that an optical power meter/optical source test set will satisfy the anticipated maintenance requirements.

Section VI

Conclusions

CONCLUSION

This report provides a tremendous amount of information in one compendium on the fundamentals of fiber optic technology and detailed information on reliability, maintainability and supportability of primary fiber optic components. Also, the performance of fielded systems which use fiber optics has been addressed. This information can be used by system planners and designers in determining which technology, electrical or optical, will have the least impact on the reliability, maintainability and logistics of future systems.

The laboratory data collected in this study has its usefulness in analyzing the physics of failure, modeling, life testing and device design, and its impact on performance. However, once the device is incorporated into a system and deployed, the environment changes and failure becomes defined by the system and its required level of performance. Only a statistically significant number of deployed sytems operating for a sufficient period of time will allow an accurate comparison to be made between laboratory life test predictions and actual operation. There is a wide variation in the range of predictions for some components. However, despite these variations, almost all of the predicted lifetime test results indicate lifetimes which equal or exceed the typical 25-30 year life cycle of military systems.

The field data used in this study was received in a form which must be used extremely carefully. In all but one case, this data was given in bulk numbers versus discrete data points and it represents information on more than one type of component. For example, couplers of all different sizes were treated as one in the source data. Likewise, cable with a variety of channels and fiber types (single mode and multimode) were treated as one in the source data. The one source from which discrete field data was obtained did not address the specific application, cause of failure, type of failure and criteria defining the failure. This significantly limits the confidence with which the data can be used. Generally, the field data represents rough order of magnitude figures and should be used cautiously in the allocation of reliability parameters.

However, it is very encouraging that these figures equal or exceed typical military equipment lifetimes. This is especially true since the majority of the lifetime data was conservatively calculated or, when predicted, was considered to be conservative. This leads to the conclusion that the primary fiber optic system components, when considered individually, should last at least as long as the intended life of the equipment. However, more failure information on operating systems must be made available to substantiate this study's findings. Presently, the supporting data is too sparse to draw detailed conclusions with a high degree of confidence.

A very strong research effort is on-going in the fiber optics industry to understand the physics behind the failures in optical components. This is especially true with the active components. This study has identified and described in detail many of these failure mechanisms for the primary fiber optic components used in systems. Many failures are a result of improper handling, while others are inherent to the particular fabrication technique. Some of these failure mechanisms can be prevented by close monitoring of the manufacturing process while others can be prevented by proper screening techniques. These techniques are discussed in detail within the individual chapters and they are correlated with the associated failure mechanism. This information can be extremely useful when selecting a component for a particular application. It provides the necessary knowledge that can be used as the basis for detailed inquiry about a component. In particular, this information can be used to better understand the specifications or lack thereof, for a given component as they relate to reliability. Also, questions can be asked of component manufacturers which will allow a determination of whether or not certain failure types are prone to occur. This will serve to minimize the occurrence of failures by providing a means of more closely tailoring the selection of a component to a given application.

In the course of pursuing information on fielded systems, several sources stated that the directive on maintenance philosophy states that no preventive maintenance is to be performed on fiber optic systems. This philosophy is partially based on experience with electrical systems. A high percentage of failures in electrical systems have been attributed to preventive maintenance activities. In fiber optic systems, failure or degraded performance can be caused by actions that may appear to be quite insignificant. Actions such as touching the end of a polished fiber in a connector and then mating the connector without cleaning, touching the lens of an active device and then reassembling without cleaning, bending the cable at the back of a connector when disconnecting without regard for its minimum bend radius, and measuring the attenuation of a system link in the direction opposite to that which the system normally transmits can all create problems. The first two actions can degrade signal integrity, the third action can break a fiber within the cable and the last can give an erroneous or meaningless measurement. If any of these events took place during scheduled preventive maintenance, their deleterious impact would outweigh the intended benefits. In each case, the impact would lead to additional downtime to rectify the "problem" whether it be real as in the case of a broken fiber, or anticipated due to erroneous test results. In some cases, component damage can occur resulting in the need for replacement. If the parts are not available in an emergency repair kit kept on site, then requisitioning the part will add to the system downtime.

This philosophy of not performing preventative maintenance may seem unwise for a new technology. However, as much of the field source data in this study indicates, the majority of failures are not inherent to the fiber optic component itself. Instead, the components have performed properly until external forces such as backhoe dig ups, improper installation practices, or improper applications have caused damage. For the components for which there is sufficient field data, the calculated Mean Time Between Failures (MTBFs) which may have resulted from failures intrinsic to the optical component have been found to be much greater than those failures due to external forces which are avoidable. For these components, a comparison of the MTBFs for the different causes of failure is listed below:

COMPONENT	FAILURE TYPE	MTBF (10^6 hrs) (90% Confidence)
Cable	Non-relevant	between 1.34 and 3.61
	Relevant Chargeable	between 5.85 and 541
Splices*	Non-relevant	between 13.5 and 54
	Relevant Chargeable	no failures reported
Couplers**	Non-relevant	between 2.64 and 4.26
	Undefined (possibly Relevant Chargeable)	between 4.67 and 9.52

* Both mechanical and fusion splices combined.
** Various types of couplers combined.

The devices are all components that, by their construction and function, do not lend themselves to preventive maintenance. Once these components are installed, generally they are not bothered unless they fail. Therefore, these components have a "hands off" maintenance philosophy by default. The data given above indicates that this treatment has not resulted in components which frequently fail. In fact, the lower confidence limits in the above data for those components where intrinsic failures were reported exceeds 500 years. These large values for MTBF lower limits support the maintenance philosophy which is becoming prevalent within the fiber optics community. This philosophy embraces the minimizing of preventive maintenance on fiber optic components in an effort to reduce failures caused by the good intentions of these "preventive" actions.

Of the fifteen individual items covered in this study, only five of them are field repairable. These five items can be repaired by replacement of piece parts, e.g., replacing connector pins, or by mending, e.g., splicing fiber. These five items are fiber, cable, connectors, some closures and some mechanical splices. Connectors are repaired by replacing piece parts of individual channels and preparing the associated fiber(s) for termination. Connectors and splices are often considered to be the thorn in fiber optic technology. This is because they are often time intensive to assemble and alignment is extremely critical. However, once connectors and splices are properly assembled and installed, they have not been reported to present a problem. Field repair is performed on fiber and cable by inserting connectors or splices.

Closures, in particular junction boxes, can be repaired by replacement of the seal, which is the most common element that fails.

The ten remaining items are emitters, transmitters, photodetectors, receivers, couplers, multiplexers, demultiplexers, switches, and some splice trays and organizers. When these items fail in the field, or are suspected of having failed, they are replaced. When these items are part of a module, the failed module would be sent to a maintenance facility for repair. However, the failed optical item would probably be discarded and replaced, for these ten items can rarely be repaired as discrete units.

For those items which can be field repaired, the greatest logistics impact they have is on cost. The assembly and installation of connectors and splices is time consuming even when done properly. Considering the critical alignment tolerances which must be met, repeated attempts may be necessary. The cost of repair time is high, both in terms of actual maintenance and system down time.

In addition to cost, another factor that must be considered when deciding to implement fiber optics is the test equipment peculiar to this technology. This equipment is needed to check a system out after installation to ensure that it meets specifications, to troubleshoot a system, and to characterize components. The test equipment required depends upon the application, the parameters to be measured, and the desired accuracy. The primary equipment used (excluding hand tools) is the optical power meter, optical source and Optical Time Domain Reflectometer (OTDR). These instruments require special training for operation and interpretation of the results. Fortunately, the instruments which are probably the most commonly used, are also the least expensive and the easiest to use. These are the optical power meter and the optical source. The OTDR is quite expensive and requires the most training to operate and interpret the results.

The results of this report indicate that, in addition to the benefits mentioned above, fiber optics is a reliable technology. The major elements have been shown to have individual lifetimes equal to, or greater than the typical 25-30 year life cycle requirement of a system. This is predicted information and its accuracy will only be known when systems have been in operation for this period of time. However, the fact that there is consistency in these predictions among different researchers gives credence to these values. Along with these predictions for high reliability, the predominant maintenance philosophy is to perform no maintenance unless repairs are necessary. This means minimizing preventive maintenance. The data collected in this study shows that this philosophy does not impair the individual component performance or reduce their lifetime. However, when repairs are needed, only five of the fifteen items studied here can be repaired in the field. The remaining items must be replaced, typically at a maintenance facility. Many of these items are very expensive due to the precise alignment between components that is necessary in their manufacture. This is especially true of devices that use bulk optics. However, this condition is expected to change as more integrated optic components are made available.

Fiber optic technology offers many well known benefits over conventional electrical technology. Among these benefits are:

1) Extremely high bandwidth
2) Reduced cable size and weight relative to signal carrying capacity
3) Dielectric transmission medium
4) Reduced electrical hazard
5) Immunity to electromagnetic interference
6) Reduced sensitivity to electromagnetic pulse damage
7) Provides more secure transmission medium
8) Low attenuation to signals, and
9) Non-interfering, non-radiative transmission medium.

Despite these many benefits, there has been a reluctance on the part of the military to use fiber optics in their systems. Quite opposite to this trend is the tremendous number of commercial systems which abound worldwide. One reason for the military's reluctance has been the unavailability of a thorough investigation into the reliability, maintainability and supportability of this technology. This report addresses this issue and provides a document that will alert system designers and planners to the areas that must be taken into consideration when developing a fiber optic system.

Section VII

Recommendations

RECOMMENDATIONS

There is no doubt that fiber optic technology will be used extensively in military systems by the year 2000. Numerous market forecasters have made this prediction and the many benefits of this technology make its integration into military systems inevitable. This emphasizes the need for a rigorous analysis of fiber optic R,M&L considerations. Although this report is a very good start, the collection of numerical data must continue in order to be prepared for the time when the use of fiber optic military systems are widespread.

The numerical data collected was used to assign reliability parameters to individual fiber optic components much like is done for electrical components in MIL-HBK-217. Also, this information can be used by engineers to improve new system designs and upgrade deployed systems.

This report already contains a substantial amount of lifetime and Mean Time Between Failure data from which decisions can be made. However, this data needs refinement. Also, in conjunction with this numerical data, there is a wealth of information on many other subjects very relevant to system design. The major areas addressed are:

1) Reliability and maintenance parameters
2) Proper installation practices
3) Proper manufacturing techniques
4) Design weaknesses
5) Environmental and mechanical impacts
6) Failure mechanisms
7) Failure modes
8) Failure prevention
9) Proper screening procedures
10) Quality assurance tests
11) Maintenance and logistics considerations
12) Status of operating systems, and
13) Fundamental fiber optic component operating principles and design.

This information will be extremely valuable to the system designer and planner. However, to bring the reliability, maintainability and supportability aspects of the fiber optics discipline up to the level that the electronics industry currently enjoys, much work remains to be done. This study is only the beginning of that effort.

Before work in this area is seriously continued, the lessons which were learned during the term of this contract should be understood. These are briefly listed below.

1) Survey questionnaires require a significant amount of effort to prepare and distribute. Our experience, along with other's, has shown that this effort far exceeds the benefits derived from the small percentage returned.

2) The identification of data sources is imperative in a study of this nature. Identifying useful sources was the most difficult aspect of this study. Many potential sources, once contacted, were unable to provide data due to its proprietary nature, while many other sources could only provide data considered incomplete for the purposes of this study.

3) The broad scope of this effort made it difficult to spend a significant amount of time on each aspect to be addressed.

4) Our efforts have shown that a very meager amount of useful fiber optic R,M&L data is available from military sources. This is primarily due to: a) the lack of a centralized data collection facility; b) the type of data available is not complete enough for analysis; and c) there are very few fiber optic military systems which have been operational long enough for there to be a lot of data available.

5) The vast majority of information on fiber optic component and system reliability; maintainability; and supportability addresses commercial quality products and systems, especially telecommunications grade.

6) Few manufacturers closely track their deployed products with regard to failures. If this is done, the information was not made available to the authors for inclusion in this study. However, the results of in-house reliability testing was made available by those manufacturers who were able to provide the data.

7) Gathering data of this nature is an extremely time consuming task. A large quantity of literature must be read and closely scrutinized before the usefulness of its contents can be determined. We have found the amount of useful failure data in the literature to be quite small compared to the amount of data in circulation.

In a nutshell, the scope of the effort should not be so large for it to become intractable, a certain amount of confidence should exist about the quality of data to be provided by potential sources, the avenue to be taken in retrieving the data should be a proven technique, and sufficient time must be allowed to extract and analyze the data. These lessons all point out that the collection of numerical R,M&L data is difficult. The most difficult type of military data to find deals with logistics. This is mainly due to the small number of operational fiber optic systems.

The cavernous gap that exists between available numerical reliability field data and information on the physics of failure/degradation must be bridged. Before this can be done, the obstacles to obtaining numerical data must be solved. This is a longstanding problem which seems to defy correction. However, it must be corrected so that important design decisions based on sound engineering analysis can be made.

The United States Air Force has initiated a plan which addresses the issue of improving the R,M&L of military equipment. The program was approved on 1 February 1985 and is called R&M 2000. It focuses on the changes which need to take place in management to expedite improvements in equipment R&M. This is an extremely important area which can significantly impact the effectiveness of any R&M effort. Once program managers are educated in the importance of R&M, beyond cost, more attention will be paid to the subject. This means more managers will understand why the proper data must be recorded and archived. As a result of this knowledge, the adherence to policy in recording failures will be more strongly enforced. This will result in improved R&M tracking which will provide the feedback necessary to design better equipment. The benefits of better equipment will come in the form of lower costs as a result of fewer repairs, increased capability and increased mission effectiveness.

There are two ways of approaching the problem of scarcity of fiber optic data, one is by implementing long term solutions and the other is by implementing short term solutions.

In the long term, two conditions must prevail before the R&M archiving situation will improve. The first condition is the development of R&M files for each fiber optic system. The second condition is the training of maintenance personnel in which parameters and circumstances must be recorded and/or measured in a fiber optic system.

The first condition requires the development of an information gathering system which can be standardized either for use by all fiber optic systems or by individual classes of systems, e.g. local area networks, shipboard applications, airborne applications, intra-office installations, etc. The accumulated data must be maintained at a central facility where it can be used as feedback for improving deployed systems and new designs. To be most effective, the R&M data collection must be carried out for the complete life cycle of each piece of equipment. In addition, the R&M parameters for each new system should be supplied to the central archiving facility. This will provide the necessary baseline from which to judge the performance of the equipment. If it performs very reliably from the onset, then the design and reliability allocations can be used to improve other systems. Likewise, if the design proved to have low reliability, this information would also be available to prevent a recurrence of this design. This type of activity will only be successful with the full support of program managers.

The second condition of training maintenance personnel of the proper parameters to record in the event of failure is essential for the data files to be meaningful. Unless complete information is recorded, an accurate assessment of the cause of failure cannot be made with a high degree of confidence. It is this information which will allow a refinement of component selection. The more that is known about the conditions which lead to failure, the more accurately such parameters as maximum optical and electrical values, derating, and cooling can be defined for optimum equipment performance. Also, equipment design techniques can be more accurately assessed. This not only requires a knowledge of fiber optic R&M priciples, but a commitment to continue tracking this data for the life cycle of the equipment. This commitment must be encouraged by program managers.

These two conditions, the development of R&M files for each fiber optic system and the training of maintenance personnel, are goals which must be established but will take time to become an integral part of the military system.

To make short term strides in building on the fiber optic R&M work contained in this report, the primary change that must occur is a narrowing of the scope of future efforts in a given contract. This can be done in several ways:

1) Concentrate on only one or two related fiber optic components, e.g., emitters and detectors, or connectors and splices.

2) Concentrate on related fiber optic disciplines; e.g., component fabrication, materials, quality assurance, and screening techniques; or installation procedures, test techniques, and test equipment.

3) Concentrate on specific types of equipment of which an adequate number are deployed and have been operational for a relatively long time. These systems must have R&M data on file.

Each of the three approaches listed can use the contents of this report as a very good point from which to start. This report can be used in outlining the approaches and in providing much of the information for filling in the outline. In essence, much of the background information has been documented and data source investigation performed.

There are good and bad points in performing any of the long term or short term approaches. In general, they either address a small portion of a broad field, e.g., several components out of many, or they encompass a wide range, e.g., developing a technician's handbook. The latter approaches would provide the greatest benefit to the widest audience in the shortest period of time. In addition, they would be building the sorely needed foundation for an effective data collection system. Also, this study has shown that the type of data needed on individual components which will benefit designers in performing R&M analysis is not currently available in a form suitable for rigorous analysis. If manufacturers do not make this information available or if manufacturers and DoD fail to begin properly tracking this data, the short term solutions are not viable approaches. Also, if manufacturers continue their reluctance to supply this information, DoD will have to take on the responsibility of tracking their own data for subsequent analysis. Therefore, this leads to the conclusion that the best approaches to follow in continuing this effort are the establishing of a filing system and the development of a handbook for maintenance personnel. It should be pointed out that gathering data on individual components should not be ignored. That is, talks should begin in earnest between DoD and manufacturers in making the appropriate data available. However, efforts should concentrate on long term solutions.

These endeavors need to be performed in parallel since they compliment each other. That is, one can not be effectively used without the other. If only the filing system were established but personnel did not know what information to put in it, the system would be used improperly. The entry of incorrect information could then become the start of a habit which may carry over when personnel are trained properly. Likewise, if the personnel are trained and there is no systematic filing procedure, the data would be difficult to locate and therefore use, which is the current situation.

Based on the above analysis, the authors recommend that follow-on activity focus on two areas:

1) The first area is the development of a filing system which is tailored to archiving fiber optic R,M&L data. This system has two major elements. One is the data sheets on which the appropriate information will be logged. These log sheets will be in a format which allows easy entry into an archiving network. Much like the goals of the Computer Aided Logistics Support (CALS) program which is now under development by DoD, all system designers, both electrical and optical, will have access to the archiving network. The second element is the central archiving network which can follow the design of current database systems like the Defense Logistics Agency (DLA) and Ships Parts Control Center (SPCC). Another alternative would be to expand or convert an existing database for the purpose of archiving fiber optic R,M&L data.

2) The second area on which to focus follow-on activity is the development of a handbook which addresses the parameters and conditions associated with fiber optic system R,M&L. This handbook will be used to train personnel in collecting the proper information for an assessment to be made on the possible cause(s) of system failure. This is the information which will be entered into the archiving network. This handbook could be used in advanced training sessions following fundamental fiber optic training of military personnel. In this case, the R,M&L maintenance handbook would become an integral part of the fiber optic training program the military will need to establish. In this way, the very important issue of R,M&L will be stressed from the beginning.

These two recommendations are consistent with the goals of R&M 2000. Therefore, this will provide a smooth transition into current Air Force programs. If these recommendations are started in the near future, by the time fiber optic systems are common place in the military, the Air Force will be in a position to easily retrieve R,M&L data which can be used to upgrade existing systems and enhance new designs.

References

Chapter 1. FIBER AND CABLE

1. Designers Guide to Fiber Optics; AMP, Inc., 1982.

2. Kreidl, J., "New Fiber Offers Improved Performance--At a Price", Lightwave, pp. 20-23, June 1985.

3. "Single-Mode Progress: From the Lab to the Field", Photonics Spectra, Staff Report, pp. 71-80, April 1986.

4. Kreidl, J., "Light Polarizing Fibers to Grab Military Sensor and Gyro Projects", Lightwave, pp. 47-48, Nov 86.

5. Nishimura, M., "The Two Modes of Single Mode Fiber", Photonics Spectra, pp. 109-116, June 1986.

6. Hatano, S., "Multi-Hundred-Fiber Cable Composed of Optical Fiber Ribbons Inserted Tightly Into Slots", International Wire & Cable Symposium Proceedings, pp. 17-23, 1986.

7. Kao, C., "Fiber Cable Technology", Journal of Lightwave Technology, Vol LT-2, No. 4, pp. 479-487, August 1984.

8. Angeles, P,, "Fiber Optic Plenum Cable", International Wire & Cable Symposium Proceedings, pp. 575-582, 1986.

9. Chamberlain, J.C., "Zero Halogen, Fire Retardant Fiber Optic Shipboard Cable", International Wire & Cable Symposium Proceedings, pp. 538-544, 1986.

10. Lakas, W., "New Fiber Optic Ribbon Cable Design", International Wire & Cable Symposium Proceedings, pp. 4-10, 1986.

11. Matsuo, T., "Composite Submarine Cable Containing Optical Fibers and Pilot Pairs", International Wire & Cable Symposium, pp. 123-130, 1986.

12. Adams, R., "How Hydrogen Attacks Marine Fibers", Lightwave, pp. 38-39, May 86.

13. Beck, W.B., et.al, "Radiation-Hardened Fibers: Asking The Right Questions", Photonics Spectra, pp. 65-72, May 1986.

14. Greenwell, R.A., "Fibers That Stand Up To Radiation Hazards, Photonics Spectra, pp. 129-140, April 1987.

REFERENCES Con't.

Chapter 1. FIBER AND CABLE

15. Kar, G., "Optical Waveguides: Fabrication and Drawing Standards", Photonics Spectra, December 1985.

16. Miller, S.E., et al, Optical Fiber Telecommunications, Academic Press Inc., New York, New York, 1979.

17. Suggs, J.W., Telephone communication concerning lifetime of optical fibers, May 1987.

18. Celwave, Response to Vitro Corporation conducted Cable Manufacturer Reliability Survey 4, 8, April 1987.

19. Corning, Response to Vitro Corporation conducted Fiber Manufacturer Reliability Study 1-10, April 1987.

20. "Fiber Optic Cable Applications and Testing", Innovations, Belden, pp 41-53, Sept 1979.

21. Nevins, R.C. and Taylor, D.H., "Testing For a Long Service Life", Telephony, February 11, 1985.

22. Electronic Industries Association Standard Test Procedures for Fiber Optic Fibers, Cables, Transducers, Connecting and Terminating Devices, EIA-RS-455, March, 1980.

23. Gulati, S.T., "Strength and Static Fatigue of Optical Fibers", Research and Development Laboratories, Corning Glass Works, Corning, NY 14830.

24. Carr, J., "Changing Attitudes on the Fragility of Optical Fibers", Outside Plant Magazine, 1986.

25. Ayre, R.W., "Measurement of Longitudinal Strain in Optical Fiber Cables During Installation by Cable Ploughing", Journal of Lightwave Technology, Vol. LT-4, No. 1, pp. 15-21, Jan 86.

26. Bark, P.R., "Stress-Strain Behavior of Optical Fiber Cables", International Wire & Cable Symposium, pp 385-390, 1979.

27. Szentesi, O.I., "Reliability of Optical Fibers, Cables, and Splices," IEEE Journal on Selected Areas In Communications, Vol. SAC-4, No. 9, pp 1502-1508, December 1986.

28. Szentesi, O.I., "Reliability and Restoration of Fiber Optic Cables", IEEE Global Telecommunications Conference, GLOBECOM '82.

29. Szentesi, O.I., "Field Experience With Fiber Optic Cable Installation, Splicing, Reliability, and Maintenance", IEEE Journal On Selected Areas In Communications, Vol. SAC-1, No. 3, pp 541-546, April 1983.

REFERENCES Con't.

Chapter 2. CONNECTORS AND SPLICES

1. Designers Guide to Fiber Optics, AMP Incorporated, 1982.

2. Chin, H., Fiber Optic Technology - 1984, #ESD-TR-85-109, Electronic Systems Division, United States Air Force, pp. 78.

3. Anderson, J. M., et al, "Lightwave Splicing and Connector Technology", AT&T Technical Journal, Vol. 66, Issue 1, Jan/Feb 87.

4. Warner, Charles, "Profile Alignment Simplifies Fusion Splicing," Lightwave, pp. 46-48, December 1986.

5. "Fiber Optic Connector Wall Chart", Fiberoptic Product News, Vol. 2, Number 7, pp. 2-10, July 1987.

6. Wagner, P., "Fiber Fusion for Terminating Optical Connectors", Connection Technology, Vol. 3, Issue 1, pp. 26-29, June 1987.

7. The Photonics Design and Applications Handbook, Book 2, Lauren Publishing Company, Inc., Pittsfield, MA, pp. H-93 thru H-103, 1987.

8. Military Specification, Splice, Fiber Optic Cable, General Specification For, DoD-S-24623, 25 March 1985.

9. Military Specification, Connector, Fiber Optic, Single Terminus, General Specification For, MIL-C-83522, 15 August 1986.

10. Norland Products, Inc., Response to Vitro Corporation conducted Splice Manufacturer Reliability Survey 01, February 1987.

11. ERICSSON, Response to Vitro Corporation conducted Splice Manufacturer Reliability Survey 02, February 1987.

12. Adams, R., "Fiber to Home Expected to Spur Mechanical Splicing", Lightwave, pp. 33-37, August 1987.

13. Szentesi, O.I., "Reliability of Optical Fibers, Cables, and Splices", IEEE Journal on Selected Areas in Communications, Vol. SAC-4, No. 9, pp. 1502-1508, December 1986.

14. Szentesi, O.I., "Reliability and Restoration of Fiber Optic Cables", IEEE Global Telecommunications Conference, GLOBECOM '82, November 1982.

15. ITT Cannon, Response to Vitro Corporation conducted Connector Manufacturer Reliability Survey 01, March 1987.

REFERENCES Con't.

Chapter 2. CONNECTORS AND SPLICES

16. ITT Cannon, Response to Vitro Corporation conducted Connector Manufacturer Reliability Survey 02, March 1987.

17. Alcoa Fujikura, LTD, Response to Vitro Corporation conducted Connector Manufacturer Reliability Survey 03, August 1987.

18. Military Specification, Connectors, Fiber Optic, Circular, Plug and Receptacle Styles, Multiple Removable Termini, General Specification for, MIL-C-28876, 3 February 1987.

19. Military Specification, Termini, Fiber Optic Connector, General Specification for, MIL-T-29504, 25 August 1987.

20. Military Specification, Connector, Electrical Circular, Miniature, High Density, Quick Disconnect, Environment Resistant, Removable Crimp and Hermetic Solder Contacts, General Specification for, MIL-C-38999, 21 September 1981.

21. Military Specification, Connectors, Fiber Optic, Circular, Environment Resistant, Hermaphroditic, General Specification for, MIL-C-83526, 19 December 1985.

22. Miller, S.E., et al, Optical Fiber Telecommunications, Academic Press Inc., New York, New York, pp. 491, 1979.

Chapter 3. EMITTERS AND TRANSMITTERS

1. Botez, D., "Single Mode Lasers for Optical Communications." Future Trends in Fiber Optic Communications, SPIE Vol. 340, pp. 32-49, 1982.

2. Hunsperger, R.G., Integrated Optics: Theory and Technology, Springer-Verlag, Berlin, Germany, 1984.

3. Suematsu, Y., Iga, K., Introduction to Optical Fiber Communications, John Wiley and Sons, New York, 1982.

4. Lebduska, R.L., Engineering Technical Order on Fiber Optic Communication Systems (draft), MIPR 1842-EEG-81-007, Naval Ocean Systems Center, San Diego, CA., 11 March 1983.

5. Lee, T.P., "Recent Development in Light-Emitting Diodes (LEDs) for Optical Fiber Communications Systems." Future Trends in Fiber Optic Communications, SPIE Vol. 340, pp. 22-31, 1982.

6. Miller, S.E. and Chynoweth, A.G., Optical Fiber Telecommunications, Academic Press, New York, New York, 1979.

REFERENCES Con't.

Chapter 3. EMITTERS AND TRANSMITTERS

7. Chin, H., Lesnick, M., Quadri, M. and Wallace, H., Fiber Optic Technology - 1984, ESD-TR-85-109, Hanscom Air Force Base, MA, July 1985.

8. Keiser, G., Optical Fiber Communications, McGraw Hill, USA, 1983.

9. Nash, F.R., et al, "Implementation of the Proposed Reliability Assurance Strategy for an InGaAsP/InP, Planar Mesa, Buried Heterostructure Laser Operating at 1.3 μm for Use in a Submarine Cable." AT&T Technical Journal, Vol. 64, No. 3, pp. 809-60, March 1985.

10. Hakki, B.W., Fraley, P.E. and Eltringham, T.F., "1.3 m Laser Reliability Determination for Submarine Cable Systems." AT&T Technical Journal, Vol. 64, No. 3, pp. 771-807, March 1985.

11. Fujita, O., Nakano, Y. and Iwane, G., "Reliability of Semiconductor Lasers for Undersea Optical Transmission Systems." IEEE Transactions on Electron Devices, Vol. ED-32, No. 12, pp. 2603-8, December 1985.

12. Hirao, M., et al, "High-Reliability Semiconductor Lasers for Optical Communications." IEEE Journal on Selected Areas in Communications, Vol. SAC-4, No. 9, pp. 1494-501, December 1986.

13. Ettenberg, M. and Kressel, H., "The Reliability of (AlGa)As CW Laser Diodes." IEEE Journal of Quantum Electronics, Vol. QE-16, No. 2, pp. 186-96, February 1980.

14. Goodwin, A.R. and Plumb, R.G., "Life Testing of Diode Laser Sources: A Case Study." Fiber Optic Technology, pp. 111-14, March 1983.

15. Mizuishi, K., et al, "Reliability of InGaAsP/InP Buried Heterostructure 1.3 μm Lasers." IEEE Journal of Quantum Electronics, Vol. QE-19, No. 8, pp. 1294-301, August 1983.

16. Enochs, S., "Packaging Laser Diodes: An Update." Photonics Spectra, Vol. 21, Issue 9, pp. 119-24, September 1987.

17. Eales, B.A. and Bricheno, T., A High Reliability 1.3 μm Single Mode Laser Package, Standard Telecommunication Laboratories Limited, Harlow, Essex, U.K., ca. 1984.

18. Hannon, P., "Thermoelectric Devices: How Telecommunications Lasers Keep Their Cool." Lightwave The Journal of Fiber Optics, Vol. 3, No. 8, pp. 19-20, August 1986.

REFERENCES (Con't)

Chapter 3 EMITTERS AND TRANSMITTERS

19. Hawkins, B.M., Phenomenological Model For Degradation of GaAlAs LEDs. Honeywell Optoelectronics Division, Richardson, TX., ca. 1984.

20. Yamakoshi, S., et. al., "Reliability of High Radiance InGaAsP/InP LED's Operating in the 1.2-1.3 m Wavelength." IEEE Journal of Quantum Electronics, Vol. QE-17, No. 2, pp. 167-73, February 1981.

21. NEC Fiber Optic Devices Hand Book Second Edition, NEC Corporation, Mountain View, CA., pp. II.1-52, July 1983.

22. Kaufman, H., "Lasers and LEDs - Lasers Gain Appeal for the Loop." Lightwave - Journal of Fiber Optic Technology, Vol. 3, No. 11, pp. 22-30, November 1986.

23. Boetz, D. and Ladany, I., AlGaAs Single-Mode Stability, NASA Contractor Report No. 3719, National Aeronautics and Space Administration, August 1983.

24. MTBF Laser Diode Optical Module at 1.3 μm, Document QF 237, Thomson Semiconducteurs, Rutherford, NJ., 15 September 1986.

25. Paul, D.K. and Greene, K.H., "Reliability of 1.3 μm InGaAsP/InP Laser Diodes." Fiber and Integrated Optics, Vol. 5, No. 2, pp. 23-63, 1984.

26. Matsushita, S., et. al., "Fiber-Optic Devices for Local Area Network Application." Journal of Lightwave Technology, Vol. LT-3, No. 3, pp. 544-55, June 1985.

Chapter 4. DETECTORS AND RECEIVERS

1. Hunsperger, R.G., Integrated Optics: Theory and Technology, Springer-Verlag, Berlin, Germany, 1984.

2. Keiser, G., Optical Fiber Communications, McGraw-Hill, USA, 1983.

3. Saul, R.H., Chen, F.S. and Shumate, Jr., P.W., "Reliability of InGaAs Photodiodes For SL Applications". AT&T Technical Journal, Vol. 64, No. 3, pp. 861-82, March 1985.

4. Lebduska, R.L., Engineering Technical Order On Fiber Optic Communication Systems, MIPR 1842-EEG-81-007, Naval Ocean Systems Center, San Diego, CA, 11 March 1983.

REFERENCES (Con't)

Chapter 4. DETECTORS AND RECEIVERS

5. Tashiro, Y., et al, "Degradation Modes in Planar Structure $In_{0.53}Ga_{0.47}As$ Photodetectors", Journal of Lightwave Technology, Vol. LT-1, No. 1, pp. 269-72, March 1983.

6. Streetman, B., Solid State Electronic Devices, Prentice-Hall, Englewood Cliffs, NJ, 1972.

7. Kuhara, Y., Terauchi, H. and Nishizawa, H., "Reliability of InGaAs/InP Long-Wavelength p-i-n Photodiodes Passivated With Polyimide Thin Film", Journal of Lightwave Technology, Vol. LT-4, No. 7, pp. 933-7, July 1986.

8. Scholl, F.W., et al, "Reliability of Components For Use in Fiber Optic LANs", Reliability Considerations in Fiber Optic Applications, SPIE Vol. 717, pp. 108-117, 25-26 Sep 1986.

9. Brian, M. and Lee, T., "Optical Receivers For Lightwave Communication Systems", Journal of Lightwave Technology, Vol. LT-3, No. 6, pp. 1281-1300.

10. Epitaxy, "Preliminary Results Show Long Life of InGaAs Planar PIN Photodiodes", Lightwave, Vol. 4, No. 4, pp. 13, April 1987.

11. Forrest, S.R., "Optical Detectors: Three Contenders", IEEE Spectrum, Vol. 23, No. 5, pp. 76-84, May 1986.

12. Williams, G.F. and LeBlanc, H.P., "Active Feedback Lightwave Receiver", Journal of Lightwave Technology, Vol. LT-4, No. 10, pp. 1502-08, October 1986.

13. O'Mahony, M.J., Justice, D.J. and Holmes, P., "A p-i-n Bipolar Optical Receiver For Submarine System Application", Journal of Lightwave Technology, Vol. LT-3, No. 3, pp. 608-11, June 1985.

14. Nakano, Y., et. al, "Reliability of Semiconductor Lasers and Detectors for Undersea Transmission Systems." Journal of Lightwave Technology, Vol. LT-2, No. 6, pp. 945-51, December 1984.

15. Snodgrass, M.L. and Klinman, R., "A High Reliability High Sensitivity Lightwave Receiver for the SL Undersea Lightwave System." Journal of Lightwave Technology, Vol. LT-2, No. 6, pp. 968-74, December 1984.

REFERENCES (Con't.)

Chapter 4. DETECTORS AND RECEIVERS

16. NEC Fiber Optic Devices Hand Book Second Edition, NEC Corporation, Mountain View, CA., pp. II.1-52, July 1983.

17. EPITAXX Advertised Data, Lightwave - The Journal of Fiber Optics, Vol. 4, No. 8, pp. 6, August 1987.

18. Recchio, A.J., Personal correspondence regarding receipt of field data on optoelectronic devices, Research Assistant, IIT Research Institute, Rome, NY, 7 August 1987.

Chapter 5. COUPLERS, MULTIPLEXERS AND DEMULTIPLEXERS

1. Ishio, H., Minowa, J. and Nosu, K., "Review and Status of Wavelength - Division - Multiplexing Technology and Its Application", Journal of Lightwave Technology, Vol. LT-2, No. 4, pp. 448-463, August 1984.

2. Erdmann, R., Perry, C. and Parmenter, C., "Fiber Optic Multiplexing and Modulation", SPIE Proceedings Reprint, Vol. 417, 7 April 1983.

3. Ealing Electro-Optics Product Guide 87-88, Ealing Electro-Optics, South Natick, MA, 1987.

4. Chin, H., Lesnick, M., Quadri, M. and Wallace, H., Fiber Optic Technology- 1984, ESD-TR-85-109, Electronic Systems Division, Hanscom Air Force Base, MA, July 1985.

5. Lawson, C.M., Kopera, P.M., Hsu, T.Y. and Tekippe, V.J., "In-Line Single-Mode Wavelength Division Multiplexer/Demultiplexer", Electronic Letters, Vol. 20, No. 23, pp. 963-964, 8 November 1984.

6. Williams, J.C. and McDuffee, F.T., An Engineering Guide to Couplers, ITT Electro-Optical Products Division, Roanoke, VA, ca. 1980.

7. Optical Networks Using Biconical Taper Couplers - The Transmissive Star, DS-012-8201, Canstar Communications, Ontario, Canada, ca. 1980.

8. Schmidt, K.M., et al, "Reliability of Fused Tapered Couplers", Reliability Considerations in Fiber Optic Applications, SPIE Vol. 717, pp. 94-100, September 1986.

9. Bond, K., Telephone communications concerning the reliability and testing of passive couplers, Canstar, Toronto, Canada, 15 September 1987.

REFERENCES (Con't.)

Chapter 5. COUPLERS, MULTIPLEXERS AND DEMULTIPLEXERS

10. Loescher, S., Telephone communications concerning the reliability of passive multiplexers, demultiplexers and couplers, Gould, Glen Burnie, MD, 15 September 1987.

Chapter 6. SWITCHES

1. Smith, P., "On the Role of Photonic Switching in Future Communications Systems." IEEE Circuits and Devices Magazine, Vol 24, No. 5, pp. 9-14, May 1987.

2. Hinton, H.S., "Photonic Switching Using Directional Couplers." IEEE Communications Magazine, Vol. 25, No. 5, pp. 16-26, May 1987.

3. Siecor Electro-Optic Products, Optical Bypass Switch environmental and mechanical test results, ca. 1986.

4. NEC Corporation, NEC Fiber Optic Devices Handbook, Optical Devices Development Department, Second Edition, pp. 121-36, July 1987.

5. Panalarm Division, Survey Questionnaire Information, 20 March 1987.

6. Harvey, G.T., et al, "The Photorefractive Effect in Titanium Indiffused Lithium Niobate Optical Directional Couplers at 1.3 μm." IEEE Journal of Quantum Electronics, Vol. QE-22, No. 6, pp. 939-46, June 1986.

7. McCaughan, L. and Choquette, K.D., "Crosstalk in Ti:$LiNbO_3$ Directional Coupler Switches Caused by Ti Concentration Fluctuations." IEEE Journal of Quantum Electronics, Vol. QE-22, No. 6, pp. 947-51, 6 June 1986.

Chapter 7. ENCLOSURES/SPLICE TRAYS AND ORGANIZERS

1. Tomita, Nabuo., "A New High Density Fiber Organizer for Optical Fiber Cable Joints." Journal of Lightwave Technology, Vol. LT-4, No. 8, pp 1223-1227, August 1986.

2. Preformed Line Products Company, Response to Vitro Corporation conducted Splice Enclosure Manufacturer Reliability Survey 01, March, 1987.

3. 3M, Response to Vitro Corporation conducted Splice Enclosure Manufacturer Reliability Survey 02, May, 1987.

4. 3M, Response to Vitro Corporation conducted Splice Organizer and Splice Tray Manufacturer Reliability Survey 01, April, 1987.

REFERENCES (Con't.)

Chapter 8. **AN/FAC-2A, 2B and 3 COMMUNICATIONS SET**

1. Fiber Optic Communications Set, AN/FAC-3, Operations and Maintenance Manual, TEMO-OA0183-018, December 1981.

2. Fiber Optic Communications Sets, AN/FAC-2A(V) and AN/FAC-2B(V), Operations and Maintenance Instructions, AF TO31W1-2FAC2-1, August 1983.

3. Fiber Optic Relocation, DEBIIB, AN/FAC-2A(V), MEA-85-24 Ramstein AB, Germany, 11-27 June 1985.

4. Russo, T., Telephone Communication Concerning The Reliability of AN/FAC-3 Communication Sets, National Security Agency, 27 March 1987.

5. Optical Wavelength Multiplexer Reliability Prediction, Final Report, MDA904-80-C-0795, (NSA-R-5130), ITT Electro-Optical Products Division, March 1982.

Chapter 9. **AV-8B CNI SYSTEM**

1. Higbee, V., Telephone communication concerning the operation of the AV-8B CNI system, McDonnell Douglas, St. Louis, MO., 21 April 1987.

2. Poppitz, R., Telephone communication concerning use of fiber optics in the AV-8B, McDonnell Douglas, St. Louis, MO., 12 March 1987.

3. Romine, R., Telephone communication concerning the fiber optic hardware used on the AV-8B, McDonnell Douglas, St. Louis, MO., 26 March 1987.

4. Romine, R., Telephone communication concerning the AV-8B total number of flight hours, McDonnell Douglas, St. Louis, MO., 14 August 1987.

5. Inman, M. and Archer, E., Telephone communication concerning AV-B CNI system hardware, operation and failures, McDonnell Douglas, St. Louis, MO., 2 April and 27 April 1987.

6. Honeywell Optoelectronics Data Book, 110-0698-001, Honeywell, Richardson, TX., ca. 1986.

7. Romine, R., Telephone communication concerning the architecture and failure history of the AV-8B CNI system, McDonnell Douglas, St. Louis, MO., 6 May 1987.

REFERENCES (Con't.)

Chapter 9. AV-8B CNI SYSTEM

8. Brand-Rex Company, Flight-Light™ Aerospace Optical Cable, Drawing No.'s C643 and C644, dated 30 September 1983 and 12 September 1985, respectively Willimatic, CT.

9. Epo-Tek 300 Series Epoxies, 20M/281, Epoxy Technology, Inc., Billerica, MA., ca. 1986.

10. MSGT. Steward, Telephone communication concerning total number of AV-8B aircraft using fiber optics, Resident ILS Station, St. Louis, MO., 6 October 1987.

10. Henson, J., Telephone communication concerning documentation on the AV-8B CNI system, McDonnell Douglas, St. Louis, MO., 14 April 1987.

11. Daum, D., Telephone communication concerning AV-8B CNI system architecture, McDonnell Douglas, St. Louis, MO., 2 April 1987.

12. Winestock, D., Telephone communication concerning maintenance and testing of AV-8B CNI system, McDonnell Douglas, St. Louis, MO., 27 March 1987.

13. Wiring Repair With Parts Data General Wiring Repair Procedures, Organizational Maintenance, Technical Manual, Navy Model AV-8B, 151573 and Up, A1-AV8BB-WRM-000, Commander, Naval Air Systems Command, 1 December 1985.

Chapter 10. FIBER OPTIC TRANSMISSION SYSTEM - LONG HAUL

1. "Army Tactical Communications Systems Fiber Optic Transmission System Long Haul (FOTS(LH))." October 1983.

2. Mondrick, Alexander G., et. al., "FOTS(LH): The Army's Long Haul Fiber Optic Transmission System." Signal, November 1983.

3. Wichansky, Dr. Howard, "Army Applicatons of Fibre Optics." Signal, August 1984.

4. Magnus, Arthur, et al, ITT, "Fiber Optic Transmission System (Long Haul)." CECOM-TR-82-J155F, October 1986.

Chapter 11. OPTICAL SOURCES AND POWER METERS

1. Testing Fiber Optic Cables Using Sources and Power Meters Seminar Workbook, FOTEC WBZ-286-5K, 1986.

REFERENCES (Con't.)

Chapter 11. OPTICAL SOURCES AND POWER METERS

2. Wendland, Paul H., The Photonics Design and Applications Handbook Book Z, "Fiber Optic Test and Measurement", pp. H-83-H-87, 1987.

3. Fiber Optics Instrumentation Product Catalog, Photodyne, 1986-1987.

4. Practical Testing of Fiber Optic Systems and Components Seminar Workbook, FOTEC WB1-885-5k, 1985.

5. Photodyne, Inc., Response to Vitro Corporation Conducted Power Meter Manufacturer Reliability Surveys 01-06, March 1987.

Chapter 12. OPTICAL TIME DOMAIN REFLECTOMETERS

1. Danielson, B. L., et al., "Measurement Procedures for Optical Fiber and Related Components." Document # RADC-TR-86-81, August 1986.

2. Rickenback, Robert, and Wechsler, Edwin R.; "Verifying OTDR Calibration Quickly." Lasers and Applications, Vol 4, No. 2 (February 1985), pp. 77-32.

3. Photodyne, Inc. Response to Optical Time Domain Reflectometer Manufacturer Reliability Survey 01-02., March, 1987.

4. Photon Kinetics, Response to Optical Time Domain Reflectometer Manufacturer Reliability Survey 03., March, 1987.

5. Siecor Corporation, Response to Optical Time Domain Reflectometer Manufacturer Reliability Survey 04-06., February, 1987.

Bibliography

Chapter 3. EMITTERS AND TRANSMITTERS

1. Suematsu, Y., "Advances in Semiconductor Lasers." ANRITSU News, Vol. 6, No. 25, pp. 2-7, April 1986.

2. Hwang, C.J., Panel Discussion - "Performance, Life and Reliability Requirements for Semiconductor Laser Applications." Laser and Laser Systems Reliability, SPIE Vol. 328, pp. 44-53, 28-29 January 1982.

3. Carni, P.L., et al, CSELT/Optical Fibre Communication, McGraw-Hill Book Co., USA, 1980.

4. Yamakoski, S., et al, "Reliability of High Radiance InGaAsP/InP LED's Operating in the 1.2-1.3 µm Wavelength." IEEE Journal of Quantum Electronics, Vol. QE-17, No. 2, pp. 167-73, February 1981.

5. Hawkins, B.M., Phenomenological Model for Degradation of GaAlAs LEDs, Honeywell Optoelectronics Division, Richardson, TX, circa 1983.

6. Speer, R.S. and Hawkins, B.M., "Planar Double-Heterostructure GaAlAs LED's Packaged for Fiber Optics." IEEE Transactions on Components, Hybrids and Manufacturing Technology, Vol. CHMT-3, No. 4, pp. 480-4, December 1980.

Chapter 4. DETECTORS AND RECEIVERS

1. Silicon Photocells Including PIN Silicon Photocells and GaAsP Photocells, Hammamatsu Corporation, USA, Jan 1985.

2. Wall, B.E., A Review of the Present Status and General Trend of Fiber Optic Technology and Its Application to Future, Large-Area Tracking Ranges, Technical Report 6426, Naval Underwater Systems Center, Newport, RI, 1 Oct 1985.

3. Glista, Jr., A.S. and Katz, R.S., Fiber Optics Technology Working Group Report (IDA/OSD R&M Study) Part II, Task Order T-2-126, Office of the Assistant Secretary of Defense (MRA &L), Washington, D.C., Aug 1983.

4. Windhorn, T.H. and Metze, G.M., "AlGaAs Optoelectronic Devices on Monolithic GaAs/Si Substrates", Optical Fiber Sources and Detectors, SPIE Vol. 587, 1985.

BIBLIOGRAPHY Con't.

Chapter 4. DETECTORS AND RECEIVERS

5. Casey, Jr., H.C., Semiconductor Laser Sources and Detectors At Wavelengths of 0.67, 1.44, 1.93 and 2.50 μm, HDL-CR-86-100-1, U.S. Army Laboratory Command, Harry Diamond Laboratories, Adelphi, Md., Sep 1986.

6. Weigand, R.M., A Wiring Strategy For Base Communications Using Fiber Optics, ESD-TR-85-146, Electronic Systems Division, Hanscom Air Force Base, MA, June 1985.

Chapter 12. OPTICAL TIME DOMAIN REFLECTOMETERS

1. Conner, Margery S., "Fiber Optic Testers Determine Faults In Short-haul Networks." EDN, Vol 31, No. 26 (December 25, 1986), pp. 98-104.

2. Dupay, Richard Ellis. "Choices and Compromises: Selecting an OTDR." Lightwave, Vol 2, No. 11 (November 1985), pp. 39-40.

3. "OTDR Distance Accuracy Vs. Index of Refraction", Anritsu Application Note: 8605-00 AEN

4. "Use of 1Km Reel of Cable For accurate Loss Measurement", Anritsu Application Note: 8603-00 AEN

5. "Using a 1Km Reel of Fiber Optic Cable to Make Near End Measurements With an OTDR", Anritsu Application Note: 8602-01 AEN.

6. "Using OTDRs for Fiber Alignment", Anritsu Application Note: 8601-01-AEN.

7. "Using OTDRs to Locate Faults and Measure Power Loss", Reprint from Electronics Test, August 1984, Morgan-Grampian Publishing Company.

Appendix

TABLE I-1. Commonly Used Elements and Their Abbreviations.

ABBREVIATION	ELEMENT
Ag	Silver
Al	Aluminum
As	Arsenic
Au	Gold
Ga	Gallium
Ge	Germanium
In	Indium
P	Phosphorous
S	Sulfur
Se	Selenium
Si	Silicon
Sn	Tin
Te	Tellurium
Zn	Zinc
AlGaAs	Aluminum Gallium Arsenide
AlGaAsP	Aluminum Gallium Arsenide Phosphide
Al_2O_3	Aluminum Oxide
GaAs	Gallium Arsenide
InGaAs	Indium Gallium Arsenide
InGaAsP	Indium Gallium Arsenide Phosphide
InP	Indium Phosphide
Si_3N_4	Silicon Nitride
SiO_2	Silicon Dioxide
TiO_2	Titanium Dioxide
ZnS	Zinc Sulfide

TABLE I-1. Conversion From Lifetime Hours to Years.

HOURS	YEARS
8760	1
10^4	1.14
10^5	11.4
1.75×10^5	20
2.19×10^5	25
2.628×10^5	30
3.066×10^5	35
3.5×10^5	40
3.942×10^5	45
4.38×10^5	50
5×10^5	57
6×10^5	68.5
7×10^5	79.9
8×10^5	91.3
9×10^5	102.7
10^6	114
10^7	1,141
10^8	11,415
10^9	114,155

TABLE I-3. Physical Constants.

Constant	Symbol	Value	Units
Speed of light in vacuum	c	299,792,458	m/sec
Planck's constant	h	6.626×10^{-34}	J-sec
		4.136×10^{-15}	eV-sec
Boltzmann's constant	k	1.381×10^{-23}	J/K
		8.617×10^{-5}	eV/K
Elementary charge	e	1.602×10^{-19}	C
Acceleration of gravity	g	9.806	m/sec^2